Beyond the Black Box

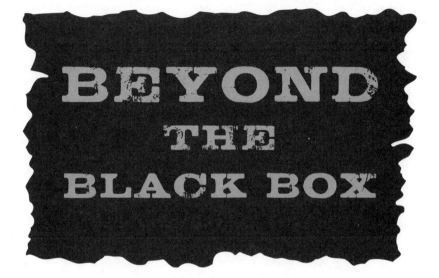

BEYOND THE BLACK BOX

THE FORENSICS OF AIRPLANE CRASHES

George Bibel

THE JOHNS HOPKINS UNIVERSITY PRESS
Baltimore

© 2008 The Johns Hopkins University Press
All rights reserved. Published 2008
Printed in the United States of America on acid-free paper
2 4 6 8 9 7 5 3

The Johns Hopkins University Press
2715 North Charles Street
Baltimore, Maryland 21218-4363
www.press.jhu.edu

Library of Congress Cataloging-in-Publication Data
Bibel, G. D. (George D.)
Beyond the black box : the forensics of airplane crashes / George Bibel.
p. cm.
Includes bibliographical references and index.
ISBN-13: 978-0-8018-8631-7 (hardcover : alk. paper)
ISBN-10: 0-8018-8631-7 (hardcover : alk. paper)
1. Aircraft accidents—Investigation. 2. Airframes. 3. Mechanics, Analytic. 4. Flight
recorders. 5. Physics. I. Title.
TL553.5.B52 2007
363.12′410153—dc22   2006101451

A catalog record for this book is available from the British Library.

To my lovely bride,
Mary Pat, who still agrees to fly away with me

Contents

Preface

Terrible news flashes across the airwaves: an airplane disaster has occurred. As the story unfolds, we hear about the events surrounding the crash, the grim statistics, the recovery operation, and the search for the black box.

Although there are many examples of "no survivor" accidents, the majority of crashes severe enough to wreck a plane, even break the fuselage into multiple sections, can be survivable. The flying public would be well advised to listen carefully to the flight attendant's preflight safety instructions.

We trust that an investigation will pinpoint the cause(s) of the accident and deliver lessons that will protect us in the future. Most aviation accidents, though, are caused by pilot error. This book is not about pilot error but about aviation accidents that illustrate concepts of physical science. Occasionally, human error will be discussed, but only when it helps illuminate important concepts.

As an educator, I have researched accident investigations to supplement my mechanical engineering classes and motivate my students. *Beyond the Black Box* arose from my desire to reveal these intriguing accident stories to a general audience and to illustrate and teach the critically relevant science and engineering principles that guide aircraft design.

Although this book is not written as a textbook, I believe it would make an excellent supplemental text for many subjects, particularly high school science classes and science courses for non-technical college students.

Most textbooks are written because the author has a vision about how to better explain basic concepts. Few are written to connect to real-life situations. For this reason, I believe that most science and engineering texts are boring. This book grew out of my attempts to make my lectures more interesting and was written to fill the void between scientifically superficial newspaper articles and government crash reports filled with excess technical jargon.

Each chapter covers a different aspect of aircraft accidents and presents many stories interspersed with scientific explanations of varying lengths and depths. Mathematical sophistication is limited to occasional high school algebra and physics.

This book is by no means an indictment of the aviation industry. On the contrary, I have flown hundreds of times and hope to fly hundreds of times more. If anything, this book should lead to an appreciation, even re-assurance, of all that is done to keep the skies safe.

I WOULD LIKE TO ACKNOWLEDGE HELP in preparing this book from the following people: Katherine Vermeersch, Jay Howard, Steve Kukla, Lars Helgeson, and Matt Cavalli for reading the manuscript; Steve, Lars, and Gary Niemeier for providing encouragement and important feedback; Gary Niemeier for his many consultations on graphics; Rowena Rae for editing; Trevor Lipscombe for supporting the project; and finally my lovely bride, Mary Pat, for her patience, understanding, and encouragement.

Beyond the Black Box

1...

The Crash Investigation Process

I t all started with Knute Rockne, the legendary football coach of Notre Dame. Rockne died in an airplane crash on March 31, 1931. The subsequent outpouring of national attention changed the crash investigation process forever and established important precedents.

Knute Rockne, age 43, was at the peak of his coaching career. It is difficult to capture his celebrated fame several generations removed. Losing only 12 games in 13 seasons (including 5 undefeated seasons and 3 national championships) was the bedrock of his popularity—everyone loves a winner. Rockne greatly enhanced his reputation and influence with a dynamic, dominant personality and coaching innovations. Rockne did not invent the forward pass and backfield shift, but he used them with innovative skill to create a livelier, crowd-pleasing game.

Rockne left Chicago by train to Kansas City. From Kansas City, Rockne was to fly in a Fokker F-10A more than 1,300 miles to Los Angeles for a two-day speaking tour. The Fokker had three 425-horsepower Pratt & Whitney piston engines, wooden wings, and carried 12 passengers at 154 miles per hour.

Shortly after leaving Kansas City, the plane was forced low by clouds. The pilot considered turning back, but a radio report from Wichita (clear skies and unlimited ceiling) encouraged him to continue. Barely 100 miles from Kansas City, eyewitnesses heard the plane in the clouds and looked up just as a section of wing fell off. The plane broke up, spilling five passengers, and dove 1,500 feet to the ground.

Historical Background

The Air Commerce Act of 1926 initiated regulation of the airline industry in the United States, specifically certification of pilots and planes. The act established the Aeronautics Branch of the Department of Commerce as an advocate and promoter of commercial aviation, similar to the role played by the Department of Commerce for business. This created a conflict between regulatory and promotional powers within the same agency. The Federal Aviation Administration has a similar conflict today.

Although safety was always viewed as an integral part of promotion, crash investigation had a different priority. Participation was considered voluntary for pilots, airlines, and aircraft manufacturers. This approach is not entirely without merit when viewed from a 1920s perspective. At that time, government regulation was not as commonplace as it is today. There was also a need to "encourage" the infant aviation industry and not hinder its growth with overregulation and potential lawsuits.

The Commerce Department viewed voluntary and secret testimony as necessary to its investigative process. It believed that honest testimony would not occur if, for example, pilots were forced to speak publicly against their employers. With fewer people involved and less corroborating information available, false testimony was easier in the 1920s than it is today.[1] Additionally, the Department of Commerce freely admitted that its role as crash investigator was to promote safety, not encourage lawsuits.

The Air Commerce Act of 1926 required the Aeronautics Branch of the Department of Commerce "to investigate, record and make public the causes of accidents in civil air navigation in the U.S."[2] No other guidance or legal powers were given concerning accident investigation.

Although there had been a few publicly released crash reports assigning "probable cause" before the Rockne crash, by 1931 (under pressure from the airlines and manufacturers) the process had become closed and secretive. The Department of Commerce chose to meet the crash investigation requirements of the Air Commerce Act by releasing generic statistics on all crashes. This approach was not without controversy. At the time of the Rockne accident, there were individual congressmen pressing for public release of crash reports for selected accidents of interest.

The national grief, presidential announcements, and front page headlines about the Rockne crash created a situation that the Aeronautics Branch could not ignore. Knute Rockne was not about to become just another generic statistic. For the first time, it was recognized, and became clearly established, that secrecy was unacceptable and full disclosure would become a way of life. The public demand for information resulted in the first release of the field inspectors' preliminary crash report.

In a rush to give the public an explanation, the crash investigators released three different theories within a week: the pilot lost control in turbulence; ice broke loose, snapping a propeller blade; and ice build-up rendered instruments inoperative, causing a steep dive that broke the wing when the pilot tried to pull up.

It turns out that the Navy had performed flight tests on the Fokker F-10A before the Rockne crash and rejected it as unstable, reporting their findings to the Department of Commerce. The concern was wing "flutter," a structural vibration that can cause the wings to fall off. Even after modification (after the crash), the Navy still considered the Fokker too unstable for use. An Aeronautics Branch inspector also uncovered pilot rumors of turbulence-induced wing vibrations and a fear of exceeding the plane's cruise speed, even briefly. These rumors appeared to be widely known among pilots, but they were never discussed publicly or even with airline officials for fear of termination and possible blacklisting within the industry.

Nick Komons writes in *Bonfires to Beacons (a History of the Civil Aeronautics Administration)* that the Aeronautics Branch had enough evidence to ground the Fokker even before the Rockne crash. Dominick Pisano in "The Crash That Killed Knute Rockne" goes so far as to speculate that the three crash theories were put forth to cover up the department's knowledge of the plane's stability problems.

Many people believe that if Rockne had not been a casualty, the Aeronautics Branch would have placed a lid of secrecy on the crash and all previous knowledge about the Fokker's problem with wing flutter.

After the crash, all F-10As were eventually grounded for safety inspections. This was the first time that regulators grounded an entire fleet of aircraft for safety. Most of the planes returned to service within a month. Anthony Fokker, the planes' designer (and famous designer of World War I fighters) was retired as director of engineering at the Fokker Aircraft Corporation, then a division of General Motors. The Fokker plane, once considered the safest in the world, and wooden planes in general were now tainted. The Rockne crash and ensuing publicity helped the growing interest in all-metal planes.

The Investigation Process Today

Through a long series of legislative actions, the National Transportation Safety Board (NTSB) (and the Federal Aviation Administration, described later) eventually emerged. The NTSB is an independent federal agency charged by Congress with investigating all transportation accidents including highway, rail, marine, oil and gas pipeline, and aviation.

The NTSB investigators are given more legal power than many government agencies. Although they rarely use this power, the NTSB can obtain subpoenas and court orders to interrogate witnesses, inspect files, enter facilities and aircraft, examine the computer data of any party involved in an air crash, and obtain any other relevant information.

The result of an NTSB investigation is a final report establishing the "probable cause" of an accident. Probable cause is not a legal standard. It is not the function of the board to assign fault or determine civil or criminal liability. The NTSB final report is not even admissible in court.

Legal liability is based on "evidence." The rules of evidence are complex and defined by many court rulings, and other legal wrangling. Lawyers would more charitably suggest that hundreds of years of jurisprudence have established rules for identifying the reliability of information. The NTSB investigation seeks to make meaningful safety recommendations as soon as possible and occasionally makes such recommendations even before the final report is issued. Public safety does not allow time for legal arguments. Engineers and lawyers simply play by different rules.

NTSB safety recommendations have no legal standing and are just that—recommendations. Despite its investigative powers, the NTSB is not actually in charge of anything except the crash investigation. Criminal liability must meet a higher legal standard. The FBI will show up at any hint of criminal activity, conduct a parallel investigation with the NTSB, and take over if any criminal evidence is found.

What Is Investigated

The NTSB investigates "accidents" and "incidents." An aircraft accident has a lengthy and formal definition. Briefly, an accident involves a "serious" or fatal injury, or substantial damage. Immediate NTSB notification is also required if any of the following incidents occur which could affect the safe operation of the plane.

- Malfunction of the flight control system
- Uncontained engine failure[3]
- In-flight fire
- In-flight failure of the electrical or hydraulic systems

- Sustained loss of thrust in two or more engines
- An emergency evacuation for any reason

What Does the FAA Do?

The Federal Aviation Administration (FAA), under the Department of Transportation, regulates all aspects of commercial aviation. This regulation includes enforcement of federal rules associated with mechanics, pilots, engines, aircraft, airports, and air traffic control. In this capacity, the FAA "certifies" all of the above. The FAA also runs the national airspace system (i.e., the air traffic controllers). The FAA has no investigative authority in its own right but typically participates in all NTSB investigations.

An agreement between the NTSB and the FAA permits the NTSB to delegate certain mishaps to the FAA for investigation. The NTSB occasionally does so after it asseses the merit of using its expertise and resources on a case in which there is minimal risk to flyers, passengers, and commercial aviation. The board usually delegates to the FAA the investigation of non-fatal, general aviation accidents involving fixed-wing aircraft of less than 12,500 pounds.

The FAA is also charged with promoting and encouraging civil aviation. Just like critics of the Aeronautics Branch in the 1920s, many critics today consider this responsibility an inherent conflict of interest with regulating safety.

Conflict between NTSB and FAA

Conflict arises between the NTSB and FAA in two ways.

First, since the FAA runs the air traffic control system and certifies pilots, planes, maintenance procedures, and airports, the NTSB crash investigation often involves areas within the FAA's responsibility. For example, the NTSB may conclude that the safety rules are inadequate or that the FAA did not provide sufficient oversight.

Second, "encouraging civil aeronautics" is understood by most people to mean that the FAA advocates for the aviation industry, but many people view this position as conflicting with safety regulation. When the NTSB makes a safety recommendation, the FAA evaluates it with a cost-benefit analysis. The cost-benefit analysis attempts to estimate the number of lives saved based on an "estimate" of prevented disasters and the associated costs (i.e., the plane and litigated payouts).[4]

Approximately 80% of NTSB recommendations are accepted by the FAA, and the rest are rejected for cost considerations. At other times, the NTSB safety recommendations may result in protracted negotiations and, of particular interest here, significant national research programs.

The NTSB "Go Team"

An NTSB "go team" is dispatched from Washington headquarters to major airplane crashes. Go teams consist of NTSB investigators who are experts in appropriate technical specialties. Under direction of the NTSB "investigator-in-charge", each NTSB investigator heads a working group for one area of expertise. The groups, described in the NTSB Investigation Manual, are staffed by "parties" to the investigation: the FAA, the airline, pilots' and flight attendants' unions, and airframe and engine manufacturers. A quick review of the various groups involved gives some insight into the complexity and thoroughness of the investigation.

Some groups are self-explanatory such as the weather group and the witness group. Other groups such as the cockpit voice recorder group and the flight data recorder group require a brief explanation.

Large commercial planes are required to have two "black boxes": the cockpit voice recorder and the flight data recorder. Because planes often crash nose first, the black boxes are located in the rear of the plane.

The cockpit voice recorder, with microphones in each pilot's headset and another overhead between the pilots, records all flight crew conversation and other ambient cockpit noises. Potential sounds of interest include engine noise, warning alarms,[5] cabin decompression noise, landing gear extension and retraction, and other clicks and pops. Sometimes recorded sounds undergo significant analysis as a clue to mechanical failure.

The amount of data recorded in the flight data recorder varies greatly depending on the age of the recorder. Older models might only record 11 basic flight parameters such as time, altitude, speed, heading (orientation with respect to the three-space axis), and engine settings. Newer recorders can record the position of all control surfaces (rudder, ailerons, flaps, and trim tabs), accelerations, engine sensor readings (temperatures, pressures, vibrations, and rpm), autopilot settings, smoke alarms, and over 1,000 other data sources. By FAA regulations, new planes must monitor at least 88 specified parameters.

Older, magnetic tape recorders record the last 30 minutes of sound in the cockpit in a continuous loop. Newer, solid state technology records

2 hours. Since a recent crash could involve a plane with 30-year-old technology, the data available for analysis varies considerably.

The operations group reviews the detailed history of the accident flight and crewmembers' duties. Also reviewed are the plane's weights and balances. Control settings (flap and trim settings, engine power, and airspeeds) are compared with flight manuals published by the plane's manufacturer, the more specific operating manuals written by each airline, and the pilot's training manuals. One might think that procedures for flying the plane are well established; however, it is difficult to verbalize all possible situations. Accidents by their nature often represent unusual circumstances.

The air traffic control group reconstructs all communications with the plane and all radar data.

The human performance group studies crew performance and all factors that might result in human error, including fatigue, medication, alcohol, drugs, medical histories, training, workload, equipment design, and illumination.

The maintenance group reviews all maintenance records and FAA oversight. The investigation may extend into design, certification, manufacturing, and maintenance management. All Airworthiness Directives (FAA-mandated requirements for supplemental maintenance) and all historical problems with similar planes are reviewed for relevance.

The aircraft performance group defines the motion of the airplane, which may involve the use of simulations and animations. The initial task is to define the impact motion of the airplane using all available data including cockpit voice recordings, flight data recordings, radar data, witnesses, ground scars, aircraft damage data, aerodynamic data, aircraft control data, and engine data. Next, the performance group determines what events produced the defined motion, such as weather disturbances, engine and flight control anomalies, and pilot actions. The performance group may perform preliminary calculations related to impact velocities, acceleration and stopping distances, and trajectories of separate parts.

The systems group documents all control lever positions, switch settings, and instrument readings in the cockpit. This group also determines the integrity of the major systems (hydraulic, electrical, and pneumatic) and associated subsystems (autopilot, communications, fire protection, instrumentation, landing gear, ventilation and oxygen, de-icing, and navigation).

The powerplants group documents the condition of the engine, propellers, and engine-related components. Initial efforts focus on documenting the normal operations of the engine on impact. If any evidence suggests the engine may be a factor, the following are considered:

pre-impact fire
uncontained failure
separation of an engine or engine components

The engines are also inspected for internal damage from foreign object ingestion (ice, birds, tools, stones, etc.), internal blade breakage, rub marks, and molten metal impingement. If conclusive evidence of normal engine operation before impact cannot be established, additional testing, disassembly, and laboratory examination may be required.

The survival factors group documents impact forces and injuries, the performance of seats and restraint systems, emergency evacuation, airport emergency planning, and crash fire-rescue efforts. This group also evaluates the cabin and cockpit deformations and all medical reports, including postmortem examinations. The study of injuries and patterns of injuries often relate to the crashworthiness of the cabin interior and fuselage.

The postmortem examination may give many clues about the physics of the crash. For example, if a post-crash fire results in soot deposits in airways, breathing occurred after impact and the crash forces were survivable. Excess carbon monoxide in the blood indicates failure of pressurization systems. If a bomb is suspected, the bodies are x-rayed for fragments.

The structures group is in charge of securing the crash site and documenting the wreckage. This includes the calculation of impact angles to help determine the plane's pre-impact course and attitude (the plane's orientation). The structures group must account for the total aircraft structure, document the aircraft damage and wreckage, and determine the pre-accident integrity of the aircraft (did any mechanical defect cause the crash?). This group will also document the wreckage distribution and condition; impact marks; ground scars; soot, heat, and fire patterns; crush lines; cabin deformation; floor disruptions; and locations of separated seats.

The structures group coordinates with the systems group to document the condition of all flight surfaces, examine all movable mechanisms for condition prior to impact, and trace control systems from cockpit to control surface for integrity.

The Party System

The NTSB designates other organizations and corporations as parties to the investigation. Except for the FAA, which is automatically a party to the investigation, the NTSB has complete discretion over who it designates. Only those organizations and corporations that can provide expertise to the investigation are granted party status, and only those persons who can provide the board with needed technical or specialized expertise are permitted to serve on the investigation. All party members report to the NTSB.

Outside experts are assigned to the various NTSB groups. For example, a member of the pilot's union is assigned to the human performance group; airline maintenance personnel are assigned to the maintenance group. The structures group has representatives from Boeing or Airbus (or a number of smaller commuter plane companies), and the powerplants group has members from General Electric, Pratt & Whitney, or Rolls Royce. A major disaster might have 90 investigators assigned to the various NTSB groups.

The party system, in spite of the obvious conflicts of interest, works surprisingly well. Presumably this occurs because everyone keeps an eye on everyone else and the natural rivalries cancel. For example, the equipment manufacturers instinctively blame the pilots, and the pilots return the favor. Hopefully, the NTSB successfully arbitrates any disputes. Therefore, according to the theory, the expertise of the engine and airframe manufacturer is required, in spite of any obvious conflicts, since nobody knows these complex systems as well as the original designers. Also, everyone is uncomfortable with a failed investigation. In many ways, an understood (and fixable) problem is more reassuring than an unexplained crash. The party system is not without critics, mainly plaintiff lawyers and some consumer groups.

The individual working groups remain at the accident scene as long as necessary, which varies from a few days to several weeks. Some groups move on—for example, the powerplants group moves to an engine teardown at a manufacturer or overhaul facility; the systems group to an instrument manufacturer's plant; the operations group to the airline's training base. Work continues at Washington headquarters and forms the basis for later analysis and drafting of a report that goes to the NTSB, typically 12 to 18 months from the date of the accident. Safety recommendations may be issued at any time during the course of an investigation.

Investigations that are particularly complex have been known to last years. A few are described in this book.

After the on-scene examination of the wreckage concludes, each group writes a factual report. For a major crash, there is usually a public hearing. On the first day of the public hearing, all factual reports and the cockpit voice recorder transcript are released to the public for the first time and are entered into the "public docket." The public docket is the formal collection of all documents related to the crash investigation. Witnesses are also questioned under oath during the public hearings. Witnesses might include FAA regulators, air traffic controllers, design engineers, and any survivors.

After the investigation team completes a formal technical review the factual reports become the basis of the final report, which contains the following sections: factual information, analysis, probable cause, conclusions, and recommendations.

Crash Investigation Leads to Continual Safety Improvements

Modern airplane disasters usually result from a series of improbable, almost random events and pilot-system interactions that are difficult to predict. Mechanical failures are usually a series of events that never happened before. There are few clear-cut, general statements about how planes crash because all the obvious problems have long since been addressed.

It is somewhat reassuring to realize how many things have to go wrong for the typical crash to occur. It is even more reassuring to know that multiple improvements usually result from the investigation, any one of which would prevent similar future accidents.

Fundamentally, large commercial aircraft obtain increased safety from having redundant systems and procedures; there are at least two of everything, including pilots and engines. One event, error, or failure will not crash the plane.

As an interesting aside, the concept of redundancy is often credited to Howard Hughes and his "Spruce Goose." The plane, flown only once, was designed to ferry troops across the Atlantic during World War II. The concept was based on German submarines making passage by ship impractical, a situation that nearly (but never quite) developed. With a gross weight of 400,000 pounds, this 218-foot-long monster had, for the first time, the capacity for redundant electrical and hydraulic systems. The idea caught on and greatly contributes to modern safety.

Safety continues to improve in large part because of the findings and recommendations of crash investigators. One of the many purposes of this book is to tell the story of the investigation, demonstrate the extensive forensic and scientific analysis that takes place, and illustrate many basic principles of physics.

2...

How Planes (Often) Crash

British Midlands Flight 092

O n the evening of January 8, 1989, just 19 days after the terrorist bombing of a Boeing 747 over Lockerbie, Scotland,[1] British Midland Airways Flight 092 departed from London bound for Belfast, Northern Ireland. Because of IRA (Irish Republican Army) terrorism,[2] all flights to Belfast routinely flew under heightened security rules. The wary residents of Kegworth, England, immediately suspected the IRA when a wounded Boeing 737 screamed just 50 feet over their village before crashing nearby. Adding to speculation, it was later learned that 25 British soldiers bound for Northern Ireland were on the plane. However, the IRA, famous for bombings in London and Ireland, had never attacked an airplane.

Airline and government officials quickly and repeatedly reassured the jittery public that terrorism had been ruled out. Early clues can hint at a bombing, but in this case there was direct evidence of engine trouble. It was immediately announced that the pilot reported severe vibrations in one of the engines only eight minutes after takeoff from London's Heathrow Airport. Later, the pilot reported an engine fire but shortly after said it was under control. The 737, which has two engines, is safe to fly with only one engine. Shutting down a troublesome engine is always an option available to the pilot. With sufficient altitude, planes can even glide many miles and land safely without any engines operating. However, the wounded plane is at least partway to a disaster, because redundant design features have been compromised the plane now has diminished tolerance for additional problems and errors.

With one engine out, the standard procedure calls for an emergency landing at the nearest airport. The plane requested and received permission to land at East Midlands Airport, just 100 miles north of London.

A surviving passenger told his story. They had just started to serve the meals when the passenger heard a loud crash and bang and saw sparks on the left side of the plane near the no. 1 engine. The pilot announced he'd lost an engine and was requesting an emergency landing at East Midlands. The passenger recalls the second engine going out during the descent. There was a shudder and a bang. Then there was a loud crunch.

Just 2.7 miles from the airport and at a height of 900 feet, the second engine abruptly lost all power. The flight crew immediately attempted to relight the engine that had previously been shut down, but it was too late. Seventeen seconds later, the fire warning alarm for engine no. 1 sounded, followed by the ground proximity alarm. Ten seconds before impact, the captain announced on the PA system, "prepare for a crash landing." The plane crashed about 1,100 yards from the runway.

Witnesses on the ground said the left engine was on fire. A ham operator reported the pilots frantically saying, "we've got problems with the other engine."[3]

A gouge in the field suggests that the plane bounced and then lifted and clipped a row of trees before plowing into an embankment and breaking into three sections. The nose cone, containing the cockpit, snapped off, as did the tail section.

Because the plane crashed near an airport, firefighters and rescue personnel were quickly on the scene, pumping foam into the jet to prevent the leaking fuel from catching fire. It took over seven hours to remove all the survivors from the wreckage. Surgeons had to perform amputations inside the wreckage.

Said a rescuer, "The whole of the inside of the aircraft was just absolutely wrecked. The bottom of the aircraft and the hold area must have collapsed. At the front, all the seats had shot forward because of the force of the impact. It was a mass of metal and seats. There was no sign of fire."[4] Another rescuer stated that "people were strapped into their seats, dead in the tangled body of the plane."[5]

Of the 126 people on board, 39 died on impact, 8 died later in the hospital, and 73 were seriously injured. At the hospital, most of the wounded suffered from head injuries and broken bones. Forty-three of the patients at the hospital had suffered "an episode of impairment of consciousness."[6] In some cases, survivors were unaware that they had undergone surgery.

Because of the crash patterns associated with impact, some sections of the plane were relatively intact and passengers in those sections had a substantially higher survival rate. After establishing a few facts about the physics of airplane crashes, I will use the Kegworth incident to illustrate many basic principles about how planes crash. First, though, it is interesting to compare conclusions stated in the final report with the unfolding investigation reported in the press.

Flight 092: Initial News Reports

Reported January 9. The pilot, badly injured with a broken back and legs but in stable condition, was praised as the hero who steered a crippled aircraft over a nearby village.

It was quickly established that the crash appeared to have followed a rare double-engine failure. The retired head of Britain's Air Accidents Investigation Branch said the odds against both engines failing were "10 million to 1. I would look for some inadvertent technical mistake such as something incorrect being done to the engines during turnaround, either inadvertently or deliberately."[7]

The Boeing 737-400, introduced only four months earlier, was the newest Boeing plane. At the time of the accident, only 17 were in service worldwide. The British Midland plane had been delivered just 12 weeks prior and had flown just 520 hours.

British Midland Airways grounded all of their 737-400 planes for immediate inspections.

Reported January 10. Consistent with eyewitness accounts, investigators said the left engine showed fire damage. The flight data recorder indicated that the right engine was shut down before impact. There was no sign of fire or mechanical failure in the right engine. Investigators then tried to determine why the plane's apparently functional right engine was turned off during the flight. Initial speculation was focused on crew error and faulty wiring.

Ed Trimble of the Air Accidents Investigation Branch said the flight recorders had been examined and "we now have a fairly clear idea of what happened." He refused to elaborate. Trimble said the right engine appeared to have been turned off "at a previous point in the flight,"[8] rather than while the pilot tried to make an emergency landing in the moments before the plane crashed.

A Transport Department spokesperson, requesting anonymity, confirmed a newspaper report that debris, "probably engine fragments,"[9] was

Fig. 2.1. Boeing 737 engine. *Photo*: Gary Niemeier

found some distance from the crash. The *Evening Standard* newspaper said that engine blades were found six miles from the crash site.

Meanwhile, the National Transportation Safety Board listed only three cases of double-engine failure on commercial jets flying in the U.S. since 1962. Two incidents were caused by the engines ingesting excess water in a severe storm; the third incident involved a flock of birds.[10]

Reported January 11. British authorities asked for immediate inspections of engine-monitoring circuitry in jets using similar engines: Boeing 737-300, 737-400, and Airbus A-320. The inspections included the instrumentation for fire monitoring, overheating, and vibrations in each engine. Figure 2.1 shows the Boeing 737 engine.

Investigators determined that the pilot apparently shut down the right engine even though the plane was having trouble with its left engine. Experts said it was highly unlikely the pilot could have confused the two engines, given the system of double checks between pilot and copilot and the cockpit layout.

Meanwhile, in London, the pilot of Flight 092 was questioned again by investigators. The captain was interviewed for 45 minutes at the hospital after answering preliminary questions the previous day.

The British authority said the FAA was being kept advised of the situation.

Reported January 13. Investigators completed a preliminary examination of the engines on Flight 092 and said they found no evidence of faults in the engine warning systems. The British Civil Aviation Authority ordered British airlines using similar engines to inspect the fire and vibration warning systems for "left-right sense."

The U.S. FAA also ordered checks of the warning systems of 300 Boeing 737s operated by U.S. carriers. An FAA administrator said the inspections were being ordered "even though we have no evidence whatsoever that this is a problem in the airplanes, including the one in the accident."[11] The FAA further reported that there had been "very, very isolated incidents"[12] of cross wiring of warning systems in U.S. aircraft, but none had led to an accident. A Boeing spokesperson called the FAA's directive "a precautionary measure in the tradition that has made the system so safe."[13]

The next day, in an apparent unrelated incident, the FAA directive was expanded to 757s when cross wiring was found in two cargo compartment fire extinguisher systems.

Reported January 17. In expanded inspections for faulty wiring, five Boeing 757 jetliners were found with crossed wires leading to cargo hold fire extinguisher bottles. However, this circuit is not related to the Kegworth Flight 092 crash of a 737.

The Final Crash Report

Simply put, after the left no. 1 engine failed, the pilots shut off the wrong engine. However, the details are a little more charitable to the flight crew.

Because of metal fatigue, the no. 1 engine lost a fan blade, sparked, smoked, caused the airframe to shudder, and later caught fire and failed completely. The initial failure led to a series of compressor stalls, a jet engine's version of an engine misfire (compressor stalls are explained in Chapter 5). The engine coughed repeatedly causing rattling or vibrations that could be felt throughout the plane and in the cockpit. Once the fan blade broke free, the rotating, unbalanced weight vibrated continuously. These vibrations were recorded by the flight data recorder.

The flight crew also could smell smoke. The pilot stated that he looked at the engine instruments, but he got no clear indication of which engine was the problem. (The vibration indicators on each engine should have answered this question.) He also stated that the smoke and fumes were coming from the passenger cabin. Based on his belief that ventilation air came from the no. 2 engine compressors, he suspected a fire in that engine. This would have been correct for earlier versions of the 737, but it was not on this latest model, the 737-400.

The first officer said he also monitored the engine instruments, and when asked by the pilot which engine was causing the fire, he said "it's the le . . . it's the right one," to which the pilot responded "okay throttle it back."[14] The right, no. 2 engine was incorrectly throttled back. The first officer later stated he could not remember why he decided the problem was with no. 2 engine.

When the no. 2 engine was throttled back to idle, the auto-throttle was simultaneously disengaged, which interrupted the aircraft's climb. Less power was required from both engines. The damaged engine stabilized at the lower power settings and the plane stopped shaking, convincing the flight crew that the correct engine had been shut off. However, excess vibrations were still being recorded by the flight data recorder and should have been evident to the flight crew. The flight data recorder also confirmed that the pressure surges (which shook the plane) stopped in the damaged no. 1 engine after auto-throttle disengagement.

Why did the flight crew miss the excess vibration readings in the damaged engine on the instrument panel? This brand new model 737-400 had just been introduced by Boeing the previous October. The updated 737 had a new style of solid-state cockpit display instrument panel. The engine vibration electromechanical pointer had been replaced by a less conspicuous light-emitting diode (LED).

The captain stated he rarely included vibration monitors in his instrument scan because he believed they were unreliable. The investigators confirmed that earlier vibration meters did have a history of unreliability, but that was not the case for this 737 with newer instrumentation.

Also, the captain and first officer of Flight 092 had only 23 and 53 hours of experience respectively with the updated instrument panel. Neither pilot had had a chance to practice interpretation of engine problems on a simulator. The first time they saw abnormal engine indications was in flight. It is believed the instrument panel correctly warned of excess vibration problems with engine no. 1, but the flight crew missed this information.

Although three flight attendants and many passengers saw "flames" or "sparks" coming from the left engine no. 1, this information was not communicated to the flight crew, perhaps because flight attendants know that intrusions during busy phases of flight are unwelcome, especially during an emergency landing. Nonetheless, a flight attendant told the pilot "the passengers are very, very panicky,"[15] so the pilot announced that there was trouble with the right engine and that the engine had been shut down. They could expect to land in 10 minutes. Many passengers who saw the fire in the left no. 1 engine were puzzled about the pilot's reference to the right engine. The flight attendants stated that they did not recall any announcement about the right engine being shut down.

The flight engineer started the "engine failure and shutdown" checklist. Completion of the list might have correctly diagnosed shutting down the wrong engine, but this activity was repeatedly interrupted by landing the plane. Several activities served as distractions, such as communicating with air traffic control, checking the weather, changing radio frequency, lowering the landing gear, adjusting the flaps, and attempting unsuccessfully to reprogram the flight management system to display the required landing pattern at the new airport.

When the aircraft was 13 miles from the airport, air traffic control advised a right turn to bring the aircraft back to the runway center line. When an airplane banks its wings to turn, lift is reduced on the tilted wings. To compensate and fly level, engine thrust is routinely increased. The increased thrust probably initiated the ingestion of a blade broken during the initial fatigue failure.

The broken fan blade had temporarily lodged within the engine's acoustic lining panels. The period of high vibration, caused by the unbalance of a missing blade, shook the broken blade free. The broken blade passed through the engine, doing significant damage and causing the fire seen by witnesses on the ground. Sections of fan blade were found on the ground below this point of the final approach. At about 900 feet, the no. 1 engine lost almost all thrust. The plane was too low to glide very far, and with the no. 2 engine already shut down, the plane crashed just short of the runway.

This crash is of particular interest because it illustrates many of the principles addressed in this book including engine failure (Chapter 5), metal fatigue (Chapter 6), and seat failure (Chapter 8). The Flight 092 crash sequence, ground impacts, and damage inside the plane will be explained later in this chapter after a general discussion of how planes crash.

How Do Planes Crash?

This chapter discusses how, not why planes crash. The issue here is the consequences of unplanned contact with the ground. Why a plane crashes usually involves the details of flying the plane (often pilot error) and is not the focus of this chapter.

A review of crash data and basic physics will reveal some common patterns of airplane crashes. The crash data also demonstrate that an airplane crash, contrary to popular opinion, is a survivable event.

Most people assume an airplane crash automatically involves 100% fatalities. This certainly can happen. The phrase "severe impact, no survivors" is an accurate description of some, but not most, crashes. The total loss of life scenario gets ten times the publicity and has ten times the impact on our psyche, but it is not the most common crash scenario that threatens passenger safety.

At this point, it is useful to examine a few statistics.

Survivability Data:
The Statistics Are Much Better Than You Think

In addition to investigating aircraft accidents, the National Transportation Safety Board also maintains the government's database of civil aviation accidents. The NTSB defines an accident as an occurrence in which any person suffers death or serious injury, or in which the aircraft receives substantial damage. Based on this broad definition, aircraft accidents are surprisingly survivable; over 95% of all passengers on commercial aircraft survive investigated accidents. Between 1983 and 2000, there were 568 accidents involving 53,487 passengers and crew and resulting in 2,280 (4.3%) fatalities. (The statistics do not include small general aviation planes, which crash more frequently.) These numbers are somewhat optimistic, however, because an injury sustained during severe turbulence can be an example of an NTSB-investigated accident. Most people would not consider severe turbulence resulting in a broken arm as a life-threatening event. (While on a plane, though, one could easily consider the terrifying, jolting, jarring turbulent motion to be life threatening.)

Over that same period (1983–2000), the NTSB identified 26 serious accidents in the U.S. involving 2,739 occupants. A serious accident was defined as one involving fire, serious injury, and either substantial aircraft damage or complete destruction. Slightly more than 55% of passengers and crew survived. Seven of the 26 serious accidents were judged nonsurvivable. Of the remaining 19 accidents, the survival rate was over 76%. The 19 survivable accidents involved 465 fatalities, which were categorized as 66% impact/trauma, 28% fire/smoke, and 6% other (drowning, trauma from engine parts, etc.). These numbers seem too pessimistic, however, because accidents serious enough to destroy the plane often result in zero or low fatalities. This fact raises two questions: (1) how should the statistics be interpreted, and (2) what is a serious accident?

Another approach to evaluating crash data is to look at "total hull loss" accidents, or accidents in which the aircraft is damaged beyond economic repair. As an example, consider the hull loss data for McDonnell Douglas DC-10s over a 34-year-period from 1972 to 2007.

The DC-10 was in production from 1972 to 1989. During that time, 446 planes were delivered. As of 2007, 29 planes had been written off as a total hull loss due to accident. Of those 29, two were destroyed in hangar fires with zero loss of life, 15 involved fatalities, and four involved 100% loss of life. All DC-10 hull loss accidents and a list of fatalities are given in Table 2.1.

A survey of serious accidents with DC-10s illustrates three important points about airplane crashes:

1. Airplane crashes serious enough to destroy the plane are highly survivable.
2. Even when the fuselage breaks during impact, the probability of survival is good (perhaps not excellent, but good).
3. Most crashes occur at takeoff, landing, or approach.

During takeoff and landing, there are more human, instrument, and equipment inputs and activities, all of which are subject to failure or error. However, the speeds are lower, increasing the chances of survival.

Excluding the two hangar fires, 27 DC-10 accidents put passengers at risk. Because of the risks of impact and crushing injury and of post-crash fire, total hull loss is a serious situation with great potential for disaster. Yet, there are many cases of survivors walking away from impacts great enough to break the plane into two or more sections. Two such incidents are on the DC-10 hull loss list. On January 23, 1982, a plane skidded off an icy runway and broke off the nose section, resulting in only two fatalities. On July 19, 1989, a burst engine destroyed all hydraulic systems resulting in total loss of all control surfaces. The pilot steered the plane by altering engine thrust and barely managed to land in a cartwheel. The fuselage broke in four places. Amazingly, most passengers survived. This famous Sioux City crash is discussed in detail in Chapter 5.

Table 2.1. Hull Loss Accidents for All DC-10s

Date	Fatalities of Passengers + Crew	Reason
12/17/73	0/157	Undershot landing
03/03/74	345/345	Explosive decompression
11/12/75	0/139	Bird strike
01/02/76	0/363	Landed short
03/01/78	2/184	Tire burst
05/25/79	271/271	Engine separation at takeoff
10/31/79	72/88	Landed on closed runway
11/28/79	257/257	Hit Mt. Erebus in white-out
02/03/81	0/0	Hangar fire
01/23/82	2/198	Skidded off icy runway
09/13/82	48/405	Tire burst
12/23/83	0/3	Hit small aircraft
08/10/86	0/0	Hangar fire
09/17/87	1/3	On-ground explosion
01/10/88	0/9	Caught fire
05/21/88	0/254	Aborted takeoff
07/19/89	111/296	Uncontained engine failure
07/27/89	72/199	Crashed on approach in fog
09/19/89	171/171	Terrorist bomb
12/21/92	58/340	Heavy landing/wind shear
04/14/93	0/202	Heavy landing
06/13/96	3/275	Engine failure, aborted takeoff
09/05/96	0/5	In-flight cargo fire
12/21/99	17/314	Overran runway on landing
04/30/00	0/7	Hydroplaned on landing
12/18/03	0/7	Skidded on landing, fire
04/28/04	0/3	Overran runway
07/01/05	0/15	Ran off runway
06/04/06	0/3	Overran runway

In 24 of the 27 accidents (89%), almost 90% of the passengers and crew survived. If all accidents are counted, 69% survived—not a great percentage, but it is considerably better than common public misconceptions about airplane accidents, in which most people consider total loss of life to be the norm. If one pays attention to the national news, minor crashes will occasionally be reported. However, human nature being what it is, a disaster with 100% fatalities gets significantly more attention.

Of the four DC-10 accidents that occurred with total loss of life, one resulted from a terrorist's bomb, and another occurred during controlled flight into a mountain. This flight crew lost track of their location with respect to the mountain—obviously a serious situation, but also a preventable accident. The possibility of this type of accident has been greatly reduced with improved instrumentation. In the remaining two DC-10 crashes with no survivors, described below, it is somewhat reassuring to know how many improbable events had to occur before the plane crashed. Even more reassuring are the numerous design changes made to prevent similar accidents.

DC-10 Paris Crash, March 3, 1974

In this accident, an improperly latched door explosively blew out. Without pressure in the cargo hold, the unbalanced pressure in the passenger cabin ruptured the cabin floor and damaged control cables running to the tail. There were 345 fatalities. Explosive decompression and this accident are discussed in Chapter 4.

DC-10 Chicago Crash, May 25, 1979

This crash is a classic example of how a series of improbable events occurred to create a situation that had never happened before. Remove any single event, and the plane would have landed safely.

During takeoff, the engine on the left wing fell off, tearing several feet of leading edge slats. The slats are essentially movable wing extensions used during takeoff and landing to create more lift at a lower speed (Figure 2.2). Also destroyed were sections of hydraulic lines and electric cables. In spite of this damage, the pilot was able to take off safely. This illustrates how planes can take off with only one engine once the critical takeoff speed has been reached.

Normally during takeoff, the slats remain extended with hydraulic pressure. In the Chicago crash, however, the damage resulted in total loss of hydraulic fluid (and hydraulic pressure) in that section of the wing, and unknown to the flight crew, the slats retracted. Normally, sensors would indicate slat retraction. Additional sensors should have warned the pilot that the slats on the right wing "disagreed" with the slats on the left wing, but because of damage to the electrical lines, these sensors did not work.

None of this damage prevented the plane from flying safely. The flight crew, following correct procedure, flew at the prescribed speed for a plane with an engine out during takeoff. Unfortunately, this speed was 7 miles per hour below the stall speed for a retracted leading edge slat. With the

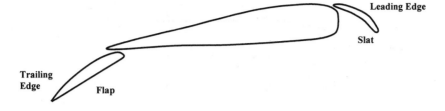

Fig. 2.2. Schematic of wing cross section with leading edge slat shown extended to increase lift at lower speeds

leading edge slat retracted, the wing did not produce enough lift to maintain safe flight at the speed at which the plane was being flown. After reaching an altitude of only 300 feet and just short of one mile from departure, the plane stalled, dived, and crashed with 271 fatalities.

The DC-10 should have been able to fly safely out of any potential stall condition, but the plane had multiple component damage and failure of the stall warning system. Remove any one of these unfortunate failures (slat retraction, slat sensors, and stall warning systems not working), and the flight crew would have been able to make the correct adjustments to safely fly the plane.

The stall warning system also lacked sufficient redundancy. When a stall condition is imminent, the pilot's stick shakes. However, much of the pilot's instrumentation was damaged. There was, in fact, a second stall warning system, but it measured conditions only on the right side of the plane (the wing with properly working slats). Later mandated improvements required each of the two stall warning systems to take readings on both wings. In addition, a shaker stick was subsequently required on the copilot's stick.

System redundancy is a major design feature in modern transport aircraft. In the Chicago crash, however, multiple component failures overwhelmed system safety. Initial design analysis did indicate a problem if loss of engine thrust occurred simultaneously with an erroneous slat retraction, but the likelihood of this happening was considered remote. Also, the analysis assumed a functioning stall warning system. Engine pylon failure was considered unlikely based on historical precedent at the time. The vulnerability of the hydraulic lines and slat positioning sensors might have been recognized if pylon failure had been considered in the hazard analysis. Ultimately, the engine pylon fell off because of a series of improbable maintenance events.

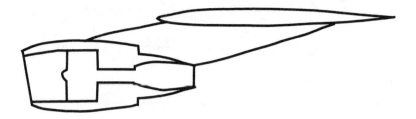

Fig. 2.3. Engine-pylon assembly for wing mounted DC-10 engine

Many design changes were required to prevent similar accidents.

- The stall warning computers were required to process sensors from both left and right wings.
- To give a redundant stall warning, a second stick shaker was added to the copilot's controls.
- Mechanical locks were added so that the slats would fail in the extended position, even if all hydraulic fluid were lost.
- Numerous changes were made to engine pylon maintenance and inspection procedures to prevent the engine from falling off.

It is startling to see just how much the engine of a DC-10 (or any jet engine) hangs off the wing (Figure 2.3). This configuration was first developed by Boeing at the request of the United States Air Force. The Air Force wanted the engines, prone to exploding, as far as possible from the bombs and the flight crew.

McDonnell Douglas recommended a two-step procedure for removing the engine pylon assembly from the wing for normal maintenance. The procedure was to remove the engine first and then the pylon. However, most airlines removed the entire engine-pylon assembly in one step using a fork lift. Misalignment of the forklift resulted in overload and damage to the pylon attachment fitting.

Because the FAA was uncertain of how widespread the maintenance procedure and related damage were, they took the unprecedented step of grounding the entire fleet of DC-10s for 37 days for immediate inspections. During this time, nine additional cracked pylon fittings were found. The grounding resulted in severe economic pressure on McDonnell Douglas and the airlines flying DC-10s. Some airlines even sued the FAA to get the grounding released. The Chicago crash and the subsequent grounding of

DC-10s directly relate to the cancellation of new orders for the DC-10, the end of its production in 1989, and the eventual takeover of McDonnell Douglas by Boeing in 1997.

A review of DC-10 crash data demonstrates that airplane crashes are in fact highly survivable. This raises the following questions: What is required for survivability, and how do survivable crashes (accidents with at least one survivor) occur?

Requirements for Survivability

The Naval Flight Surgeon's Pocket Reference to Aircraft Mishap Investigation states that survivability requires the following four conditions:

1. Continued existence of a safe volume (a person is not crushed)
2. Tolerable deceleration forces or G forces (see Chapter 9)
3. Adequate occupant restraints (the seat belt or seat does not break)
4. No post-crash fire (see Chapter 7)

In a search for safety improvements, crash investigators evaluate the above conditions, but usually survivable accidents are identified by virtue of someone surviving.

The connections securing the passenger to the aircraft structure must be maintained. These connections include the integrity of the fuselage and the floor, the attachment of the seat to the floor track, and the restraint of the occupant in the seat. If the seat-to-fuselage connection is broken, the passenger becomes a dangerous projectile similar to the victim of a car crash when not using a seat belt. The Kegworth Flight 092 accident is an example of serious injuries and fatalities attributed to a breakdown of the seat-to-floor connections.

In one of the more remarkable air disasters, a Lockheed L-1011 crashed in the Everglades on December 29, 1972. The crash velocity was 240 mph forward and 25 mph down (37 feet per second). The fuselage broke into four major sections, with additional debris scattered over an area 1,600 × 300 feet. In all but the most severe impacts, an aircraft is left substantially intact. It may be broken into two or three sections, but almost always it is recognizable as a fuselage. In this crash, however, the destruction was so complete that no circular cross section remained of the fuselage. The crash was judged "non-survivable." The only problem with this declaration was that 77 of the 176 occupants survived.

In spite of there being actual survivors, the crash certainly met the definition of non-survivable. A safe volume of the fuselage was not maintained, and the occupants were not restrained. Although many of the seats remained attached to fuselage structure (often an important requirement for survival), the fuselage was not sufficiently intact to consider the occupants adequately restrained.

The NTSB was at a loss to explain how anyone survived. The board believes survival was related to seats remaining attached to large floor sections that were semi-intact and to some passengers being thrown clear and landing at a reduced velocity. The seat design, with energy absorbers in the support structure, was also considered a possible explanation.

Normally if the seat breaks off, the passenger is thrown and subject to additional crushing inside the plane or other types of serious injury if thrown outside the plane. (If the passenger is thrown outside the plane, as with not wearing a seat belt and being thrown from a car crash, a broken neck upon landing is a serious possibility.) The design goal is prevention of seat failure. Seat failure was a severe problem with the Kegworth Flight 092 crash.

It is theoretically possible for a seat and passenger to break free during a non-survivable impact, be thrown from the plane, and survive if the final landing of the passenger-seat unit somehow lessens the crash force.

Being labeled non-survivable, the Everglades accident is not included in the survivable crash databases. This crash is considered a freak occurrence unworthy of further study. It is not considered practical (or useful) to study crash forces that fragment a plane to this extent.

Takeoff, Climb, Approach, and Landing: The Major Sources of Accidents

Accidents are studied to improve safety. A Boeing study of all airplane accidents from 1953 to 2003 gives a breakdown of accident rates by phases of flight. Boeing defines an accident as an event involving hull loss or fatal injury. The percentage of accidents for the various phases of flight is given in Table 2.2.

Phases of flight are important pilot designations. For example, the distinction between descent, initial approach, and final approach is pilot jargon specific to the plane and airport. Different aircraft will have different approach requirements to safely clear obstacles unique to a given airport. The approach speed will even vary for the same aircraft type because of variations in fuel, passenger, and cargo weights. Also specified is a series of

Table 2.2. Accidents and Phases of Flight

Phase of Flight	% of Flight Time	% of Accidents	% of Fatalities
Taxi	0	5	0
Takeoff	1	12	8
Initial climb (flaps down)	1	5	14
Climb (flaps up)	14	8	25
Cruise	57	6	12
Descent	11	3	8
Initial approach	12	7	13
Final approach	3	6	16
Landing	1	45	2

predetermined maneuvers to effect an orderly transfer from instrumented flight to visual contact. The procedures define interaction with air traffic control, landing gear and flap deployment, and prearranged missed approach procedures. For our purposes, a plane's approach is an interruption of the steady-state cruise phase of flight and involves significant activity and procedures in the cockpit, all of which are subject to human error and equipment failure.

Landing mishaps are the number-one cause of accidents (45%), with almost four times as many accidents compared to takeoff (12%), the number-two cause. Almost all accidents (94%) occur during the phases of flight associated with takeoff and landing. In addition to increased cockpit activity, takeoff and landing also have less margin of error for stalling (it is more difficult to maintain the minimum required lift at lower speeds). However, the lower speeds associated with takeoff and landing make accidents during these phases more survivable.

The cruise phase of flight, a relatively steady-state activity with few human inputs, has only 6% of all accidents but a disproportionate number of fatalities. An accident from cruising altitude is more likely to involve a high-speed impact or an in-flight breakup.

Defined Crash Scenarios

In the late 1970s the FAA contracted with the major aircraft manufacturers of the time (Boeing, Lockheed, and McDonnell Douglas) to study airplane crashes for the purposes of improving crashworthiness and identifying areas for future research. Final reports were issued by all three companies in 1982, with a subsequent summary report written by NASA and the FAA in 1983.

Although these reports are relatively old, they serve our purpose by summarizing crash patterns and other important trends. Planes still crash in the same ways today, and many of the same planes are still flying. (For example, Boeing introduced the 737 in 1968 and sold over 700 of these planes in 2006.) Even older, discontinued, and lesser-used planes are structurally similar and crash in the same way as contemporary planes.

The analysis excluded severe-impact, no-survivor accidents and mid-air collisions. Nothing can be learned from these types of accidents in that design changes will not affect survivability. Also excluded were ground accidents and any fatalities due to flight turbulence. The only accidents studied were ones in which at least one occupant did not die from impact trauma.

The defined survivable accident scenarios are as follows:

- Air-to-surface hard landing
- Ground-to-ground takeoff aborts and landing overruns

Air-to-surface hard landings are typical of approach and landing accidents. An accident on approach often occurs when the plane is not where the pilot thinks it is. The plane lands too hard or even misses the runway. The Kegworth accident is an example of unusual circumstances combining to create an approach accident in which the plane crashed just short of the runway.

The forward speed of the aircraft in an air-to-surface hard landing is between the stall speed (perhaps 138–145 mph) and the flap[16] deployment speed (representative speeds are 185–200 mph). The sink rate at crash (or vertical downward motion) is typically 10 to 40 feet per second. (Sink rate crash data are shown in Figures 2.11 and 2.12.)

Ground-to-ground accidents are characteristic of landing overruns and takeoff aborts. A landing overrun occurs when the landing distance used exceeds the distance available. If an accident occurs when the aircraft speed exceeds the specified landing speed for the given wind conditions, plane weight, runway length, and surface condition, it is usually an overrun accident. A takeoff accident occurs when the pilot aborts the takeoff for whatever reason and runs out of runway to safely stop the plane.

These accidents occur at relatively low forward speeds, perhaps 60 to 150 mph. The airplane sustains damage as it traverses a ditch, road, or mound. Fuselage breaks can result from impact with sloping ground.

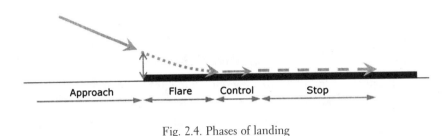

Fig. 2.4. Phases of landing

A quick review of the phases of landing (Figure 2.4) gives some insight into what can go wrong.

- The approach must be at the proper speed and angle. The plane must end up at the appropriate height to start the "flare."
- During the flare, the plane is rotated into the position for landing. Initial contact must be made within strict limits on sink rate.
- The control phase involves the safe deployment of brakes, spoilers,[17] and thrust reversers[18] to safely manage the deceleration of the plane.
- Stopping is a braking function, but it still takes many thousands of feet, depending on the conditions.

Although there is a long and complicated list of piloting "issues" that can lead to an unsafe landing, there is a much shorter list describing exactly what can go wrong during initial contact with the runway. The plane can

- hit with a vertical sink rate that is too high;
- overrun the runway, because
  a. the plane landed too far down the runway;
  b. the plane landed too fast;
  c. the plane landed with some combination of the two.

Often overrun accidents occur on wet runways. An examination of 111 landing overrun accidents (1970–1998) with Western-built jet airline aircraft showed that only 30% occurred on dry runways.

The following two accidents illustrate some of the conditions associated with a ground-to-ground takeoff abort accident and a landing overrun accident. Also illustrated are example speeds and distances involved, and the decisions made by pilots.

Takeoff Abort Accident

The takeoff speed for a DC-10, depending on conditions, might be as high as 215 mph, as compared with a cruising speed of 600-plus mph.

The critical speeds associated with takeoff are referred to as V1 and V2. V1 is the maximum speed below which a rejected takeoff can be safely aborted within the remaining length of the runway, a decision point of go/no-go on a takeoff. V2 is the actual takeoff speed. V2 is greater than V1. The pilot is not supposed to abort above V1 unless, for some reason, he believes the plane will not fly safely. V1 depends on the takeoff weight, ambient air conditions, and length of runway. The plane is expected to take off safely (even if one engine fails) once V1 speed is reached. If the pilot aborts the takeoff above V1, a crash past the end of the runway often occurs.

The following minor accident that occured during an abortive takeoff of a DC-10 (October 19, 1995, in Vancouver) illustrates some of the decisions made by the pilot during takeoff.

For the particular circumstances associated with this accident, the V1 speed was calculated to be 189 mph. This speed was reached after 6,200 feet of the 11,000-foot runway had been used up. At 6,750 feet down the runway and while traveling at 195 mph, a very loud bang was heard. At this point, the pilot had milliseconds to determine if the noise was a bomb or a burst engine, how much additional damage had occurred to the plane, and whether it was safer to abort the takeoff above V1 or to risk taking off. The pilot chose to abort the takeoff at 197 mph with 3,250 feet of runway left. This decision resulted in the plane unsafely stopping past the end of the runway. The nose-wheel gear collapsed as the aircraft rolled through the soft ground beyond the end of the runway, and six passengers sustained minor injuries.

The loud bang was a compressor stall (explained in Chapter 5), so the plane could easily have taken off safely. Modern jet engines are so reliable that the experienced flight crew had no experience with this frightfully loud, yet benign, noise.

Landing Overrun Accident

An accident that occurred on January 23, 1982, at Boston's Logan Airport illustrates a range of landing parameters.

Because of various piloting issues, Flight 30H (a 365,000-lb DC-10-30) landed too fast and too far down the icy runway. The plane skidded off the runway, eventually crashed into Boston Harbor, and broke off its nose section. Two passengers sitting near the fuselage rupture were ejected and drowned. The NTSB determined that "airport management failed to exercise maximum effort to assess the condition of the runway to assure continued safety of landing operations, . . . [and] air traffic control failed to transmit the most recent pilot reports of braking action."[19] Also contributing to the crash were the plane's speed and touchdown location.

A plane's stopping distance depends on many variables including aircraft weight, landing speed, flap settings, glide slope, braking conditions, and wind speed. Manufacturers are required to demonstrate landing distances for maximum weight on dry pavement. Adjustments are made by analysis for the other variables listed above. These adjustments are presented in the flight manuals and reviewed by the FAA.

The textbook landing calls for the plane to be 50 feet above the start of the runway. With some minor variations from weight, flap settings, etc., the air portion of the landing for a DC-10-30 remains nearly constant, at about 1,100 feet, for every landing. The required runway length equals the expected stopping distance added to the air portion of the landing. Figure 2.5 shows that the minimum landing distance for a 365,000-pound DC-10-30 landing on a dry runway is 1,131' + 2,392' = 3,523'. To provide an added margin, the FAA divides this distance by 0.6. Flight 30H required, under the best of circumstances, at least 3,523'/0.6 = 5,872' of runway length per FAA regulations.

To calculate the stopping distance on wet pavement, the FAA increases the dry stopping distance by 22% for the DC-10-30. The maximum weight at which this DC-10 is certified to land is 421,000 pounds. The maximum length needed to land this plane on a wet runway per FAA regulations is shown in Figure 2.6. If these conditions are not met, the pilot can decrease the required landing distance by dumping fuel to reduce the weight of the plane, find a dry runway, or find a longer wet runway.

Compare the landing distances described above to the conditions for Flight 30H shown in Figure 2.7. Because of an autopilot setting error, Flight 30H flew over the start of the runway at 184 mph (17 mph faster than appropriate) and touched down at 169 mph (14 mph faster than optimal). On a noncontaminated runway, these excess speeds are easily adjusted for.

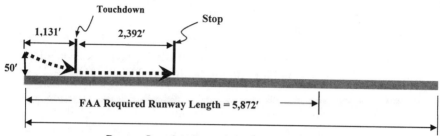

Fig. 2.5. Sample FAA-approved landing profile for a 365,000-pound DC-10 on a dry runway. Stopping distance is certified by testing on dry pavement.

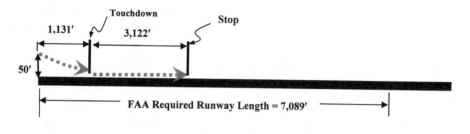

Fig. 2.6. FAA-required runway length for maximum weight (421,000 lb) DC-10 landing on a wet runway

However, the excess speed did result in the plane being farther down the runway.

Besides braking forces there are other forces acting on the plane, trying to slow it down. The aerodynamic drag and drag added by extension of the spoilers (flaps that extend from the wing to increase drag during landing) decrease significantly as speed decreases. Also, the spoilers "spoil" the lift from the wings thereby pressing the weight of the plane into the tires and allowing the brakes to become more effective. In the case of Flight 30H, however, the braking forces did not increase sufficiently because of the icy conditions. The braking forces are summarized for icy conditions in Table 2.3.

These forces could be evaluated for every increment of time to calculate the plane's acceleration with Newton's Second Law, force = mass × acceleration. The plane's initial velocity and acceleration could be converted, by calculation, into speeds and distances for each increment of time. These cal-

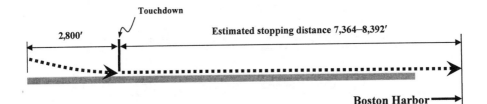

Touchdown

2,800'

Estimated stopping distance 7,364–8,392'

Boston Harbor

Fig. 2.7. Landing distances for Flight 30H

COEFFICIENT OF FRICTION

The frictional force from sliding friction is given by the force pressing the sliding object into the surface multiplied by a coefficient of friction.

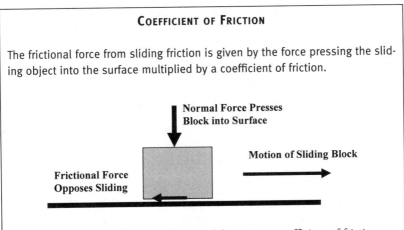

Normal Force Presses
Block into Surface

Motion of Sliding Block

Frictional Force
Opposes Sliding

Fig. 2.8. Frictional force equals normal force times coefficient of friction.

Because of mechanical brakes inside the wheels, the tires on an aircraft's landing gear do not rotate freely. This slowed motion causes "scrubbing" of the tires on the concrete runway, which creates sliding friction like the block in Figure 2.8. For a 365,000-pound DC-10 and a 0.5 coefficient of friction, the brakes result in 0.5 × 365,000 = 182,500 pounds of sliding frictional force trying to stop the plane (and rip the tires apart).

On dry concrete, the coefficient of friction varies with speed from 0.2 to 0.53. For 100 knots (115 mph), the coefficient of friction is about 0.43 on dry concrete. On wet concrete, the value degrades to 0.1 and on ice to 0.07. This means that the braking force on ice is just 16% of the value for dry concrete. On wet ice, the value is expected to be even lower. Icy pavement, called "contaminated" by the NTSB, is not certified for landing.

For a 365,000-pound plane (and coefficients of friction from 0.07 to 0.53), the braking force will vary from 25,550 to 193,450 pounds.

Table 2.3. Braking Forces for a 365,000-lb DC-10 on an Icy Runway

Speed (knots)	Airframe Drag (lb)	Spoiler Drag (lb)	Thrust Reversers (lb)	Brakes* (lb) (much higher on dry runway)	Coefficient of Friction
140	25,996	22,034	10,246	377	0.0007
110	16,180	13,699	18,364	5,967	0.0168
80	8,681	7,335	12,905	20,984	0.0607
50	3,486	2,954	6,337	23,350	0.0693

*The weight of the plane is reduced slightly by lift from the wings, making the braking force slightly less than the coefficient of friction times the weight of the plane. The lift at this point is minimal because the spoilers are raised to "spoil" the lift.

culations were made as part of the crash investigation to estimate the required stopping distance given in Figure 2.7. The difficult part of that analysis was predicting how the various forces changed at different aircraft speeds.

How Planes (Often) Crash

For a variety of flight control reasons, a plane can hit the ground with any orientation. Assuming the plane is not tumbling out of control, is flying relatively straight and level, and impacts somewhere along the length of the fuselage (a condition approximated by many takeoff and landing mishaps), there are definitive crash patterns that often occur.[20]

Consider a plane flying with a horizontal speed relative to the ground. Simultaneously, the plane approaches the ground with a vertical sink rate. Ideally, the horizontal velocity can be dissipated gradually during a crash by sliding thousands of yards. Of course, depending on the heading of the plane and the local topography, the ground can rise up and create a horizontal impact, essentially what occurred in the Kegworth Flight 092 accident.

As the impact angle changes from shallow (nearly parallel with the ground) and becomes steeper, the impact force becomes more violent. This concept can be illustrated in the "survivability envelope" for small planes. Although not directly applicable to large plane crashes, the survivability envelope for small planes does illustrate "hits and skips" versus "hits and sticks" phenomena.

At shallow impact angles, the plane hits and skips with relatively small velocity changes. At higher impact angles, the plane hits and sticks with greater changes in velocity and deceleration forces. Figure 2.9 is based on

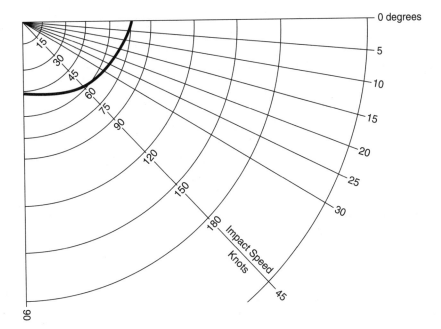

Fig. 2.9. Small plane survival is expected for combinations of impact angle and speed inside the heavy line. *Source*: General Aviation Crashworthiness Project, Report NTSB/SR-85/01

a study of 535 accidents that occurred in 1982. Inside the envelope (the heavy line on Figure 2.9) represents combinations of impact angle and impact velocity expected to result in survival.

Some improvement should be expected in planes designed since 1982. Small planes, just like automobiles, have become more crashworthy over time. For example, restraint systems and removal of sharp (and rigid) objects inside automotive and airplane cockpits have significantly improved safety in the last 20 years or so.

Hits and sticks versus hits and skips can also be illustrated with Newton's Second Law,

$$\text{force} = \text{mass} \times \text{acceleration},$$

and the definition of acceleration. Average acceleration is the average rate of change of velocity, so

$$\text{average acceleration} = (\text{final velocity} - \text{initial velocity})/(\text{time increment}).$$

If the object decelerates from 100 ft/sec to zero in two seconds, the average rate of change of velocity is

$$(0 \text{ ft/sec} - 100 \text{ ft/sec})/(2 \text{ sec}) = -50 \text{ ft/sec}^2.$$

The negative sign indicates the acceleration is in the opposite direction of the motion of the plane. If the plane is moving forward (and crashing to a stop) the acceleration is backwards, or in the opposite direction of the plane's motion.

The rate of change of velocity (or deceleration) of a plane during a crash is expected to be higher in a hit and stick crash than in a hit and skip. For example, if the plane is traveling 322 ft/sec, hits at a 60 degree angle, and crashes to a zero velocity in 1 second, the average acceleration is

$$(0 - 322 \text{ ft/sec})/(1 \text{ sec}) = -322 \text{ ft/sec}^2$$

or ten times normal gravity of 32.2 ft/sec^2 (10 G's) of impact forces.

Assume the same plane crashes at the same speed, but with a shallow impact angle of 10 degrees, and takes 2 seconds to stop because of ground slide. This crash results in acceleration five times normal gravity. Since the acceleration has been reduced by one half, the impact forces (related by $F = ma$) will also be reduced by one half.

The crash forces can also be reduced by the ability of the fuselage to crush and to absorb energy. In that case, the crushing reduces deceleration (and the crash forces) by increasing the crash pulse duration. It is theoretically possible, albeit impractical, to encase the entire plane in enough foam rubber that the crash pulse period is increased until crash forces are reduced to any "gentle" level desired.

It is difficult to actually determine the impact pulse duration without special instrumentation. The simple examples above serve to correctly illustrate the relationship between acceleration, rate of change of velocity, Newton's Law, and impact forces.

Horizontal and Vertical Motion of the Plane

Most crashes are associated with takeoff or landing when the pilot is flying at a lower speed than during the cruising phase of flight. During an emergency situation, the pilot may be reducing speed and attempting an emer-

Fig. 2.10. Horizontal and vertical components of plane's velocity immediately before crashing

gency landing (or at least a "managed" crash). At lower speeds, the wings produce less lift and the plane sinks. Even though the plane appears to be flying straight and level, it is still sinking vertically.

The relationship between the motion of the plane and lift is complicated. Lift greatly depends on the angle of attack or the orientation of the wings to the airflow. For our purposes in studying crashes, the plane has a horizontal motion and a vertical sink rate (Figure 2.10). It is convenient to describe the motion of the plane immediately before impact in terms of these horizontal and vertical velocity components. Crash forces, damage, and, most important, human tolerances differ for vertical and horizontal impacts.

Although the downward motion of the plane, called sink rate, is usually significantly less than the forward motion, a crash in the vertical direction is much less forgiving. There is nothing in a vertical impact analogous to a long, gradual horizontal slide out. Vertical impact, unless the sink rate is sufficiently low, is violent and abrupt.

A sink rate of greater than 25 ft/sec is more or less a demarcation line for significantly increased injury potential. In a set of 46 crashes with identifiable sink rates studied by McDonnell Douglas and Lockheed, 16 had sink rates of 25 ft/sec or higher. Of those 16 crashes, 75% involved fatalities. Of the remaining crashes, all with a sink rate less than 25 ft/sec, 50% had fatalities.

However, most of the fatal crashes also involved post-crash fire. If all crashes with fires are excluded, the trend becomes more definitive. Of the 10 non-fire crashes with sink rates of 25 ft/sec or higher, 6 had fatalities. For the non-fire crashes with sink rates below 25 ft/sec, only 2 of 14 crashes (14%) had fatalities.

Post-crash fires are difficult to categorize for a number of reasons. Fuel is stored in many places throughout a large plane. Fuel leaks and ignition tend to be related to circumstances that are hard to generalize and unique to each crash. Also, it is difficult to differentiate fatal injury from fire versus fatal injury from impact. Often, the crash reports do not make this distinction, because there were not enough medical examiners available, because the data were not adequately documented, or simply because the distinction is difficult to make. The impact trauma may be so severe that cause of death is difficult to assign. The crash data reflect these difficulties, with 45% of the fatalities being classified as cause unknown. Fuselage damage and passenger injury also greatly affect the ability to evacuate in the event of fire.

Figure 2.11 shows the relationship between fatalities, sink rate, and fire for the crashes studied by Lockheed. Figure 2.12 shows a similar plot prepared by McDonnell Douglas.

Recall the sink rate of the Everglades crash (described earlier in this chapter) of 37 ft/sec. The plane was seriously fragmented and the crash was judged nonsurvivable, yet 77 passengers miraculously survived.

Vertical drop tests have been done on entire planes or sections of planes to test response to excess sink rates. Since everyone worries about falling

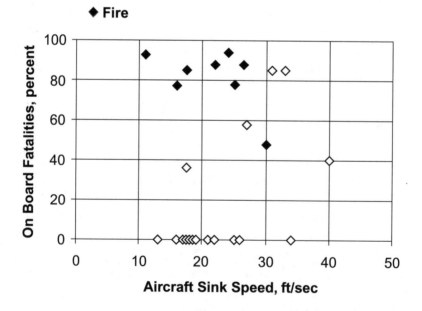

Fig. 2.11. Relationship between sink rate, fatalities, and fire. *Source*: Wittlin et al., Report DOT/FAA/CT-82/69

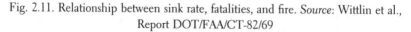

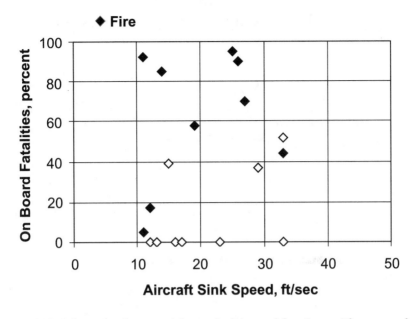

Fig. 2.12. Relationship between sink rate, fatalities, and fire. *Source*: Thomson and Caiafa, Report DOT/FAA/CT-83/42

from 40,000 feet, it is somewhat remarkable to learn that a drop from 6.2 feet (with 20 ft/sec impact) provides useful information and is considered worthy of further study. One has to consider how gently a jet aircraft is supposed to land.

The impact speed for various drop heights can be calculated from conservation of energy. The work done on the plane by gravity when it is lifted to some height (Figure 2.13) is

work = force $\times$ distance = weight $\times$ (height lifted).

If the plane is released from a given height, the work to lift it will be converted to the kinetic energy of motion. The kinetic energy of a moving object is equal to

kinetic energy = 1/2 mass $\times$ velocity^2.

Kinetic energy can be conceptualized with the impact of a baseball. A faster or heavier ball will impact with greater kinetic energy and will result in correspondingly greater pain to the person hit by the ball.

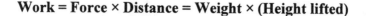

Work = Force × Distance = Weight × (Height lifted)

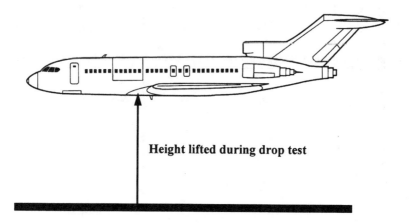

Height lifted during drop test

Fig. 2.13. Work done by gravity to lift plane = Weight × Height

Equating the work to lift the plane to the kinetic energy of the plane just before impact (the principle of conservation of energy) and making the appropriate conversion between weight and mass,[21] the impact speed for a mass dropped from various heights is

impact velocity = square root $[(64.4 \text{ ft/sec}^2) \times (\text{drop height})]$.

A plot of this equation is shown in Figure 2.14.

Using Figure 2.14, the sink rate can be interpreted as a drop from a particular height. For example, an accident with a vertical sink rate of 25 ft/sec is equivalent to dropping the plane from a height of about 10 feet.

Another useful reference for sink rate accidents is the design sink rate of the landing gear. The landing gear acts like the shock absorbers on a car and absorbs a limited amount of vertical impact energy.

Commercial airplanes are designed, by regulation, for a 10-ft/sec sink rate at the maximum design landing weight and 6 ft/sec for the much heavier takeoff weight. Typical weights for a 747-400 might be 399,000 pounds empty and 850,000 pounds at maximum takeoff weight, including 250,000 pounds of cargo and about 200,000 pounds of fuel. The 10-ft/sec and 6-ft/sec sink rates correspond to dropping the plane from

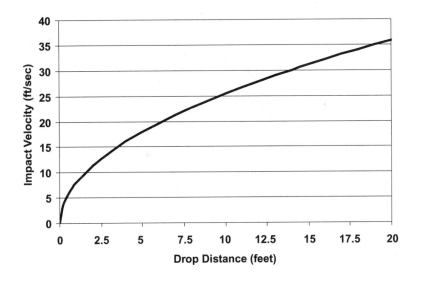

Fig. 2.14. Impact velocity for an object dropped from various heights

heights of 18.7 and 6.7 inches. These loads should be sustainable without permanent deformation to the aircraft structure. Additionally, the landing gear must demonstrate during testing that it has a reserve strength to sustain a one-time sink rate of 12 ft/sec without failing (equivalent to a drop of 27 in). Serious damage to the plane is expected at sink rates above 12 ft/sec.

A "hard landing" inspection is recommended if these values are approached or exceeded. The sink rate of the center of gravity is measured continuously on the plane. If the plane is rotating or rolling, the actual sink rate on the left and right landing gears will be higher or lower than the center of gravity sink rate depending on the direction of rotation (Figure 2.15). Boeing service experience indicates that most flight crews report a hard landing when the sink rate exceeds approximately 4 ft/sec.

An example of this type of accident occurred on December 18, 2003, during a hard landing of an MD-10.[22] Because of a crosswind, the plane landed rolling, with the left main landing gear touching down at 12.5 ft/sec and the right main landing gear touching down at 14.5 ft/sec. The maximum permitted landing weight is 375,000 pounds. The reserve (without failure) design energy dissipation of the two main landing gears combined was stated to be 843,750 foot pounds of energy. This corre-

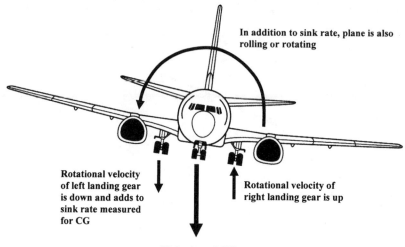

In addition to sink rate, plane is also rolling or rotating

Rotational velocity of left landing gear is down and adds to sink rate measured for CG

Rotational velocity of right landing gear is up

Velocity of CG

Fig. 2.15. Counterclockwise rotation increases sink rate on left landing gear and decreases it on right landing gear (CG = center of gravity).

sponds to lifting the maximum landing weight to the reserve drop height of 27 inches.

$$\text{Since work} = \text{force} \times \text{distance,}$$

$$843{,}750 \text{ foot pounds of energy} = 375{,}000 \text{ lb} \times 27 \text{ in} \times \frac{1 \text{ ft}}{12 \text{ in}}.$$

Boeing analysis concluded that the landing gear, at 14.5 ft/sec and 12.5 ft/sec, actually had to absorb 473,478 foot pounds (ft lb) dissipated in the left gear and 563,478 ft lb dissipated in the right gear, or 34% greater than the design reserve.

This excess energy collapsed and fractured the right main landing gear. A post-crash fire on the right side destroyed the plane. After a safe evacuation of the crew, there were no injuries on this cargo plane accident.

The Work and Energy of Crashing

A crash event can be described in terms of energy and work. Just as it is useful to speak of the horizontal and vertical components of velocity during impact, the kinetic energy of the plane can be described by its horizontal kinetic energy and its vertical kinetic energy.

$$\text{Horizontal kinetic energy} = 1/2 \text{ mass} \times (\text{horizontal velocity})^2, \text{ and}$$
$$\text{vertical kinetic energy} = 1/2 \text{ mass} \times (\text{vertical velocity})^2.$$

Of particular interest is the velocity-squared term. If the velocity is doubled, the kinetic energy is increased fourfold. Assuming identical crash scenarios at two different velocities (with one twice as fast as the other), the doubled velocity would result in crash forces four times as great. This is essentially what happens when the sink rate doubles from a modest value of 12 ft/sec, a situation within the design envelop for the landing gear, to 24 ft/sec, a situation with great potential for damage.

For the plane to stop, the kinetic energy of the plane's motion must be absorbed. In a collision with the ground, fuselage crushing, ground compression, plowing of dirt, or frictional sliding all result from the crash forces. These crash forces must do work to absorb the plane's kinetic energy.

The crash force work can be a large force applied over a small distance or a smaller force applied over a greater distance. Examples of a large crash force are when a plane crashes downward into the earth with a high sink rate, or when a plane crashes horizontally into an embankment (e.g., the Kegworth Flight 092 crash). A smaller (and more desirable) crash force applied over a greater distance happens when the horizontal motion of the plane is dissipated by the plane sliding many hundreds of feet. Larger crash forces are, of course, associated with more damage, higher accelerations, and higher G loads on the plane and passengers.

A crude estimate of crash forces is sometimes made by applying conservation of energy and equating the kinetic energy of the airplane to the work done by the crash forces:

$$1/2 \text{ mass} \times \text{velocity}^2 = \text{crash force} \times \text{distance}.$$

The stopping distance of the aircraft can be estimated by adding the ground scar to the fuselage crush zone. The crash force, the only unknown in the equation, is easily solved for. The decelerations or G forces are usually of more interest and can be found from

$$F = ma, \text{ or}$$
$$(\text{crash force}) = (\text{plane's mass}) \times (\text{acceleration during crash}).$$

G loads are an important design parameter used to define the required strength of internal aircraft components (seats, seat attachments, overhead bins, etc.), as well as the limits of human endurance.

An average crash pulse of about 22 G's was estimated for the Kegworth Flight 092 crash by using this method. The above calculation of the crash forces assumes constant crash forces. The actual crash force pulse can be quite complicated. Computer simulation attempted to identify a more complex crash pulse and estimated maximum loads of up to 28 G's. The crash acceleration (or deceleration) was estimated with computer simulation to have a peak value of 22–28 G's.

Energy Absorption

Energy absorption is provided by the landing gear, controlled structural collapse of the fuselage, and energy-absorbing seats. To improve energy absorption, the goal is to design structures, particularly the seats, which will undergo controlled deformation and thereby reduce the G loads to a level that the human body can safely tolerate. This reduction in G load is accomplished by increasing the time and distance over which impact forces are dissipated.

Rotation about the Plane's Center of Gravity

Unless the initial impact is at the center of gravity, the crash force will result in a displacement of the center of gravity and a rotation about the center of gravity. Rotation of an object about its center of gravity can be demonstrated by throwing a baseball bat. The bat's center of gravity will trace a path identical to a point mass with the same weight and thrown with the same conditions of velocity and angle. For example, a small lead ball that weighs the same as the bat will trace the same path as the center of gravity of the bat. A thrown baseball bat, assuming a two-dimensional plane motion, will also rotate about its center of mass. The motion of the bat can be described as the sum of a translation of the center of gravity and a rotation about the center of gravity. The bat shown in Figure 2.16 is seen to rotate about 180 degrees. With a faster rotational speed, the bat could have any number of rotations.

Common Crash Patterns

Again, it needs to be emphasized that a plane can impact with any orientation. However, if the pilot maintains some control of the plane, flies straight (no tumbling), and impacts along the fuselage, there are "expected" crash patterns that can be described, especially during takeoff and landing accidents.

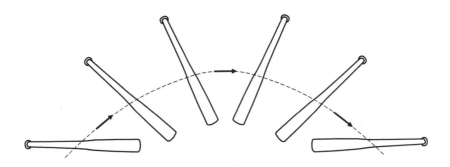

Fig. 2.16. 2D plane trajectory of baseball bat. The bat's center of gravity traces the same path as a point mass (e.g., a baseball); the bat also rotates about its center of gravity.

WHERE IS THE CENTER OF GRAVITY?

The center of gravity is the point where the plane will balance on the tip of a very strong finger without tilting. If the fuselage is approximated as a cylinder and all weight loads (fuel, passengers, cargo, etc.) are uniformly distributed along the length of the fuselage, the center of gravity or CG will coincide with the geometric center.

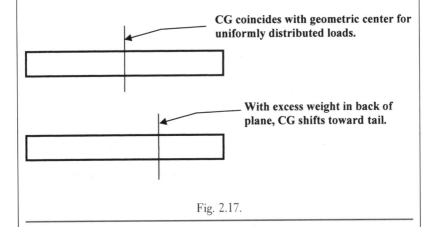

CG coincides with geometric center for uniformly distributed loads.

With excess weight in back of plane, CG shifts toward tail.

Fig. 2.17.

Determining CG for an aircraft depends on how fuel and cargo are distributed. Stating that it is somewhere near the geometric center is close enough for our purposes.

CG shifted towards the nose causes problems for controlling and raising the nose, especially during takeoff and landing. CG shifted towards the tail affects longitudinal stability and reduces the airplane's ability to recover from stalls and spins. Violating these limits—a major error (as bad as running out of fuel)—occasionally causes an airplane to crash. The location of the CG is investigated after all accidents.

If an airplane flying reasonably straight strikes the ground nose first, the crushing, friction, soil compression, and plowing forces will rotate the plane until it slides along the ground (Figure 2.18). If the impact force is large enough, the first break will be at the front of the plane.

If the tail strikes first, the rotation is counterclockwise. In this case, the first break (if any) will be at the back of the plane.

Depending on the rotational speed of the primary impact, the secondary impact on the opposite end of the plane may also cause significant crash damage. Figure 2.19 shows an exaggerated primary impact on the front of the fuselage and a secondary impact on the rear of the fuselage. The primary and secondary impact locations could just as easily be reversed if a tail strike is the first contact. In this case, there might be two fuselage breaks,

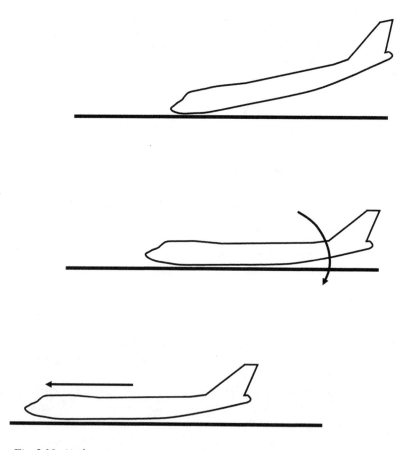

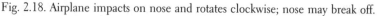

Fig. 2.18. Airplane impacts on nose and rotates clockwise; nose may break off.

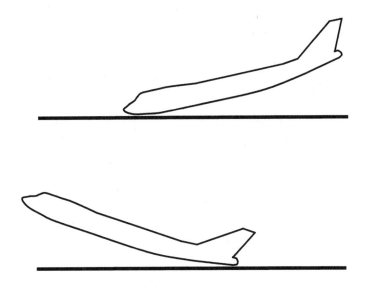

Fig. 2.19. Primary impact on nose rotates airplane, and rotation causes secondary impact. This can result in two fuselage breaks: nose followed by tail. Primary impact on tail reverses the potential breakup sequence.

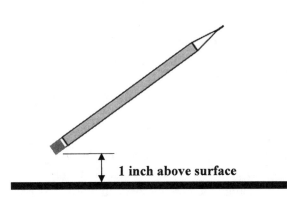

1 inch above surface

Fig. 2.20. Pencil dropped eraser tip down on hard surface results in two distinct impacts.

first in the back and then in the front. The secondary impact is less violent and less likely to result in a fuselage break.

The primary and secondary impacts can be illustrated by dropping a pencil. A pencil with an eraser tip works best to distinguish the first impact from the second impact. If the eraser tip is pointed down and dropped an inch or two from a hard surface (Figure 2.20), the impact of rubber can

Fig. 2.21. Depending on the orientation of the airplane relative to the ground, impact can occur anywhere along the length of the fuselage.

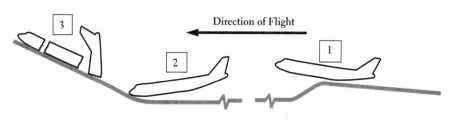

Fig. 2.22. Impact sequence for Kegworth Flight 092 crash

be heard before the impact of wood. This is approximately what happened with Kegworth Flight 092.

If the ground is not level, it can affect where the fuselage makes contact with the ground (Figure 2.21). The topography often interacts with the crash event, as seen in the Kegworth Flight 092 crash sequence shown in Figure 2.22. The sequence of events for the Kegworth Flight 092 crash was as follows:

1. The plane approached the ground, and the tail scraped and rotated the plane counter-clockwise about the center of gravity.
2. After the rotation, the nose hit first and broke. The nose strike caused clockwise rotation.
3. The tail struck the ground and broke off. All the pieces slid up the hill. The broken tail scooped up dirt and flipped upside down.

Information from the flight data recorder and the ground scars was used to estimate that the first impact occurred with the nose up 13 degrees, a forward velocity of between 119 and 128 mph, and a sink rate of 8.5 to 16 ft/sec.

Flight data recorder information was not available for evaluation of the second impact, so ground scars and a calculation of the trajectory from the first impact were used to estimate the second impact. This path was later adjusted for assumed lift and drag effects. The two estimates give the following ranges for the second impact:

nose down 9 to 14 degrees
forward speed of 84 to 109 mph
sink rate of 36.4 to 47 ft/sec
vector sum of horizontal and vertical velocities of 88 to 112 mph

The first impact in the Kegworth accident was a relatively minor "hit and skip." The second longitudinal impact was the major crash event (a "hit and stick"). The plane hit an embankment that prevented it from sliding forward, and the horizontal kinetic energy was absorbed with violent crash forces.

Based on the measured crush distance of approximately 8.5 feet along the direction of motion in the nose area, the British Air Accident Investigation Board estimated an average deceleration pulse of about 22 G lasting around 60 milliseconds.

Fuselage Crushing

Crash forces often cause local crushing of the fuselage at the point of contact with the ground. The aircraft structure needs to provide an intact shell around the passengers for a survivable impact. If an aircraft is not a good "container," it will collapse inward, losing the space required for survival. Small general aviation planes are less safe[23] in this regard because of the proximity of the passengers to the fuselage structure. Larger planes are considered safer because of the larger crushable volume protecting the passengers.

Unless a large plane flips over, the cargo hold usually provides a crush zone. Table 2.4 gives below-floor crush distances for variously sized air-

Table 2.4. Crush Distance Available Below the Floor

Number of Passengers	Plane Weight (lbs)	Examples	Fuselage Diameter (in)	Fuselage Length (in)	Crush Distance Below Floor (in)
90–115	125,000–200,000	B737	144–168	1,200–1,800	40–84
115–363	175,000–300,000	B757	168–216	1,644–2,112	40–84
250–500	420,000–785,000	DC-10 B747	224–264	2,160–2,772	108–120

Source: Wittlin, Report DOT/FAA/CT-88/15

craft. Of course, if the crash is violent enough, the entire plane will crush and eventually fragment. The September 11, 2001, terrorist attacks are examples of high-speed impacts and fragmentation.[24]

Size Effects on Fuselage Crush

Because of the larger volume available for crushing, the crash pulse or duration of the crash event is expected to be longer for a larger plane. More metal available for deformation permits larger deformations over a longer period of time. Recall that

$$\text{Average acceleration} = \frac{(\text{initial velocity} - \text{final velocity})}{(\text{time increment of velocity change})}.$$

As discussed earlier, a longer crash pulse results in lower crash forces. Therefore, a larger plane with more volume to crush will produce a longer crash pulse with correspondingly lower G loads on the passengers. The crush zone acts like a coil spring, absorbing energy and reducing the crash forces on the passengers.

The relationship between crash duration and plane size is borne out in the limited data available. In drop tests conducted by NASA on general aviation planes (i.e., small planes), the crash pulse duration was 0.05–0.15 seconds. Larger transport planes are expected to have a crash pulse greater than 0.20 seconds. For example, a 116-foot-long, 159,000-pound Lockheed L-1649 Constellation had a crash pulse of 0.26 seconds during a test crash.

To illustrate the size effect, Boeing did computer modeling of the Kegworth Flight 092 crash and of a 747 crashing under similar crash conditions. Because it is more difficult (and less accurate) to analyze breakup forces, Boeing evaluated the gentler first impact of the Kegworth crash sequence. The results of the comparison are given in Table 2.5. The 747,

Table 2.5. Kegworth Flight 092 Boeing 737 Crash versus Similar Crash in a Boeing 747 (computer simulation)

Plane	Weight (lb)	Forward Velocity (ft/sec)	Sink Rate (ft/sec)	Duration (sec)	Crush (in)	Peak G
737	120,800	250	7.2	0.13	14.6	16
747	629,400	same	same	0.80	36.4	4.7

Source: Jamshidiat et al., SAE Paper 922035

with more than five times more metal to crush, had a crash pulse over six times longer and a crash deceleration almost four times lower.

Breakup Forces at Impact

Engines and Landing Gear

The most likely components to break off a plane are the engines and landing gear. The engines and landing gear are designed to tear away cleanly to minimize any damage. They often do tear away and permit a relatively safe landing. However, the landing gear and engines themselves present hazards.

The landing gear is a structural hard spot that can penetrate into the wings and fuselage, rupturing fuel tanks or causing other serious damage. An engine being torn away with less than ideal circumstances can also rupture fuel, hydraulic, and electrical lines. This damage can result in control problems (hopefully redundant systems save the day) and fires.

Tearing off an engine or a landing gear is considered a "minor" event with respect to stopping the plane in a crash scenario. The energy associated with fracture of one of these secondary structures is insignificant compared to the kinetic energy of a 200,000-pound aircraft traveling at 150 mph. Nothing will stop the aircraft until the fuselage hits the ground.

In the Kegworth Flight 092 crash, all three landing gear mounts and both engine mounts failed without rupturing the fuel tanks. It is much easier for these components to tear off cleanly with horizontal impact. With vertical impact, the landing gear can punch up into the aircraft structure and rupture fuel lines.

Fuselage Breaks

The type of fuselage failure depends on the severity of impact. During a landing overrun or aborted takeoff, fuselage separation is mild and the airplane hangs together. During a hard landing, separation becomes more noticeable and may occur at one or two locations. For more severe impact, the breakup can occur at several locations. Whole sections may separate by hundreds of feet.

The fuselage breakup forces are inertial loads caused by acceleration from the impact. If a person standing on a scale accelerates up in an elevator at two times the acceleration of normal gravity (64.4 ft/sec^2), the scale will read twice the person's normal weight.

If the fuselage is simplified structurally as a long tube being accelerated up by a cable connected to the center of gravity, the fuselage will sag under an apparent increased weight from the inertial loading, as illustrated in Figure 2.23. If the center of gravity of the fuselage is accelerated up at 10 times the acceleration of gravity (322 ft/sec^2), the fuselage will sag as if it weighed 10 times its normal weight. Eventually, there will be a value of upwards acceleration that will break the fuselage into pieces.

Baseball also illustrates the concept of inertial loading and fracture. A 5-ounce ball can break a 36-ounce bat because the tremendous accelerations of the ball require large forces. A tennis ball, however, with less mass, cannot create enough inertial resistance to break the bat.

When a plane impacts the ground, the crash forces accelerate the plane just like the cable described above. However, unlike the acceleration from the cable, the crash impact forces are almost never through the center of gravity but offset by distance (Figure 2.24). The offset impact force accelerates the plane upward and rotates the aircraft about its center of gravity.

Because of rotation, every piece of fuselage mass is accelerating differently depending on its distance from the center of gravity of the fuselage. A mass has inertia, a property that resists linear acceleration. A mass also has rotational inertia, a property of mass that resists rotation. Rotational inertia is easy to illustrate with a spinning basketball. To spin a basketball on the tip of a finger, it takes a force applied to the outer radius of the bas-

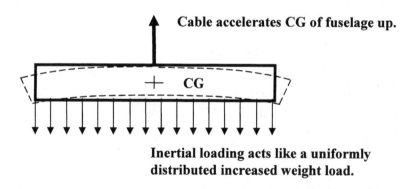

Cable accelerates CG of fuselage up.

+ CG

Inertial loading acts like a uniformly distributed increased weight load.

Fig. 2.23. A cable attached to the center of gravity of the fuselage accelerates the fuselage up. The fuselage sags from inertial loading, which acts like a virtual uniformly distributed weight load.

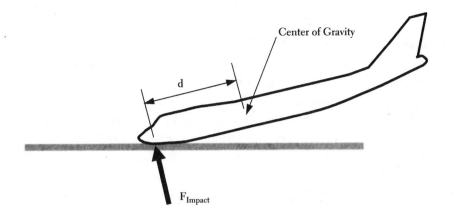

Fig. 2.24. The impact force rotates the plane about its center of gravity. The plane's rotational inertia resists this rotation, resulting in inertial loads.

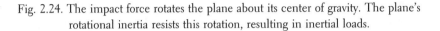

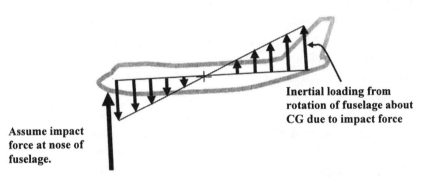

Fig. 2.25. An impact force not through the airplane's center of gravity will cause up acceleration and rotational acceleration (clockwise rotation for impact on nose as shown here).

ketball. If the basketball is made of solid lead, it will take a much larger force to spin it. The solid lead basketball has more rotational inertia. Spinning a 747 about its center of gravity will take a very large force.

Typically, a plane is moving down with some constant sink rate. The impact force violently brings the plane to a stop, creating an impact acceleration and G loads on the plane and occupants. Since rotational speed is zero at the center of gravity and increases with distance from the center of gravity, the G loads on the fuselage will also vary in a similar manner (Figure 2.25).

If a fuselage break occurs, the most common scenario is one break from the primary impact. This break can be in the front or back, depending on where first impact occurs. Just like in the pencil experiments described earlier, the plane will impact at least twice. If a second break occurs, it is usually from the secondary impact. Beyond these simple patterns, the process becomes less predictable. The plane can bounce up, becoming slightly airborne, and impact again. Tumbling can occur for a variety of reasons, including an off-center impact, sloping ground, or plowing of the nose. Also, inertial loading continues even after part of the fuselage breaks off. The end result is that the fuselage can have multiple breaks. Crashes are sometimes rated for degree of severity by the number of fuselage breaks, with more breaks indicating a more severe crash.

A few statistics on fuselage breakups from a 1982 Boeing report are given in Table 2.6. Note that because they were studying survivability, only crashes with at least one survivor are considered in these statistics. Amazingly, two crashes at high impact velocities (216 mph and 311 mph horizontal) had survivors. The results show a strong correlation between increased fuselage damage and increased probability of fire.

In partially survivable accidents, the survivors are usually sitting in the part of the fuselage that remains intact. Seat failures will occur when the fuselage breaks, and if the seat fails, a passenger can be ejected from the plane. In general, the location at which an aircraft breaks is somewhat predictable. The most likely locations are at major structural discontinuities in aircraft stiffness, such as behind the nose, immediately fore and aft of where the wings connect to the fuselage, and where the tail connects to the fuselage. At these locations, there are major structural reinforcements to connect the cockpit, wings, and tail to the fuselage.

Table 2.6. Fuselage Breaks

Fuselage Break	Number of Accidents	Number with Severe Fire	Impact Speed, mph (average for entire set)	% Fatalities
Slight*	12	2	66	6.3
Clean†	20	15	95	29.3
Considerable destruction‡	16	15	156	77.8

Source: Widmayer and Brende, Report DOT/FAA/CT-82/86
*Fuselage does not separate.
†Fuselage breaks and opens up. Break is large enough for a person to walk out or be ejected.
‡Fuselage breaks and moves off. In most cases, the sections slid many feet after separation.

United Flight 232, a DC-10, crash landed on July 19, 1989.[25] Over 60% of the passengers and crew survived, in spite of the plane breaking into three pieces.[26] The fuselage breaks more or less occurred in the locations described above. (This crash is described more in Chapter 5.)

Unless the plane inverts and crushes in through the roof, one of the safest places on a plane is over the wings. The fuselage is extensively reinforced where the wings are attached to support wing loads. Although the fuselage may break before and after the wings, the area directly over the wings typically remains intact. Because of the potential of a post-crash fire, sitting next to an exit seat is always advisable. Sometimes, when crash damage prevents exit doors from opening, breakup of the fuselage can actually be beneficial by providing the only means of escape.

Because planes often crash nose-first, the tail is considered another safe (or safer) location. For this reason, the black boxes are located in the tail section. The wings and the tail sections were the safest locations in the Kegworth Flight 092 crash.

Detailed Description of Kegworth Flight 092 Fuselage Damage

The Kegworth Flight 092 crash initially hit tail-first and then crashed nose-first into an embankment (Figure 2.22 shows the crash sequence). The second impact did most of the damage with longitudinal or forward inertial forces trying to throw the passengers forward, as shown in Figure 2.26.

If a survivable volume is maintained and impact forces do not exceed human tolerance, the next requirement for survivability is to maintain the tie-down chain (the seat anchors, seat structure, and seat belt). The seats are attached to floor tracks, and the tracks are attached to the floor structure, which typically divides the cabin from the cargo hold. The seats slide on the tracks to provide easy movement of the seats within the cabin.[27] In the Kegworth accident, the tie-down chain failed in 31 of the 52 triple seats. Seat failure occurred from disruptions in the floor structure, at a broken floor track, and from a seat ripping out of the floor structure. Of the 21 triple seats that remained attached, 14 were over the wings and the remaining 7 were in the rear of the plane.

The greatest incidence of injury and death in the Kegworth crash (and some other partially survivable crashes) was related to seat failure. When the seats broke from the floor attachment, the passengers, still strapped into their seats, crashed forward and sustained crushing injuries.

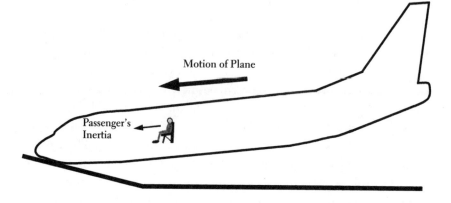

Fig. 2.26. Kegworth Flight 092 crash impact was mostly longitudinal. Passenger's inertia results in forward motion after plane stops.

Not every seat detachment resulted in a fatality and not every fatality was associated with a seat detachment. It is not possible to determine whether or not each seat failure resulted in a fatality, because many of the seats were cut out or damaged during rescue of the survivors. However, it is known where all the fatalities were seated and where the greatest floor damage occurred. Seats are most likely to be released when the floor is damaged.

The floor was most disrupted in the forward section, rows 1 through 9, when about 8.5 feet of fuselage crushed during the second impact. All floor beams were broken in seat rows 1 through 9, and all of the seats became loose and piled up on each other when forward inertia threw them to the front of the plane. The deformation[28] of seat attachments indicates impact loads in excess of 16 G's. Analysis with computer simulation estimates impact loads of up to 28 G's.

Most of the fatalities (68%) occurred in rows 1 to 9. Structural damage was greatest in rows 6 to 9 where the aircraft broke apart in front of the wing box. Fifteen of sixteen passengers sitting in rows 6 to 9 died. The floor damage was greatest in rows 6 to 9 because the floor was damaged twice, first by crushing and then by the fuselage break. The floor attachment structure in the immediate area of any fuselage break is destroyed. Also, seats located near a fuselage break may be subject to high acceleration pulses when the fuselage ruptures and "snaps" like a loaded spring. As the fuselage bends, the strain energy is stored like energy in a compressed spring. Fracture of the fuselage can release the stored energy in a violent shock load that can rupture the seat tracks and seat track attachments. The Kegworth crash had fuselage breaks in front of and behind the wings.

Except in the most severe impacts, the wing box (where the wings attach to the fuselage) usually remains intact. Only two of the forty-two seats over the wings broke loose, and only four of thirty-two passengers were fatalities. (Two of them were attributed to impact from overhead luggage and were unrelated to seat failure.) Based on examination of seat deformation, the loads on the seats in the wing area were about the same as the loads on the seats in rows 1 to 9. Clearly, the difference in injuries had to do with floor damage and the resulting seat detachment.

Twelve of thirty passengers were fatalities in rows 18 to 24, where the fuselage broke behind the wings and disrupted seat attachments. Also, the plane hit the embankment with a slight yaw or offset. This offset resulted in some crushing on the right side of the plane behind the wing box. There was a lot of floor damage associated with the crush and fuselage break. Again, floor damage predicts a higher incidence of seat detachment and fatalities.

The back of the plane, rows 25 to 27, had all seats intact and all thirteen passengers survived. The crushing of the forward section absorbed much of the impact energy, and the tail section suffered a less severe shock.

3...

In-Flight Breakup

Most airplane accidents are variations of bad landings or bad takeoffs. Pilot error is often a major factor and the focus of the investigation, so the crash investigators essentially become forensic pilots.

With in-flight breakups, pilot error is rarely an issue, and wreckage investigation takes on added significance. In-flight breakups are extremely rare, but they are also extremely fatal and usually provide dramatic lessons in physics.

Wreckage Investigation

Without minimizing the importance of data from the flight data recorder, air traffic control, radar, cockpit voice recorder, and any eyewitness accounts, much can be learned from physical examination of the wreckage. A thorough investigation requires that all data be consistent with the physical evidence. Any conflicts must be resolved by further investigation. Often the two—recorded data and physical evidence—complement each other as the puzzle is pieced together.

Typical questions addressed by wreckage investigation include the following:

- What were the flight path, impact angle, and speed?
- Did the plane break up in the air or on the ground?
- Was there an in-flight fire, or did fire occur after impact?
- Were the engines working on impact?

One of the first jobs of the crash investigator is to locate the four corners of the plane and determine if the plane broke up in the air or upon impact. Usually, a crashed aircraft remains substantially intact and is clearly recognizable as an airplane. The fuselage sometimes breaks into two or more sections. (Common crash patterns were discussed in Chapter 2.) An extremely severe impact might scatter the wreckage thousands of feet, while in-flight breakup disperses wreckage over many square miles. Wreckage investigation becomes particularly important for an in-flight breakup. The breakup sequence is traced backward to determine what broke first.

One immediate clue to in-flight breakup is the presence of naked bodies in the wreckage field. A body falling from 10,000 feet or higher is expected to reach a velocity somewhere between 120 and 600 mph; at these speeds, the windblast shreds clothing. The terminal velocity[1] of a sky diver in a spread eagle and "controlled" horizontal position is about 120 mph. Speeds up to 600 mph have been reached by divers in a more aerodynamic, straight-up-and-down diving position.

TWA *Flight 800*

The worst in-flight breakup in American history, TWA Flight 800, occurred on July 17, 1996. Taking over four years to complete, the investigation was the longest and most expensive, complex, and controversial in the history of aviation.

Shortly after 8 p.m. on July 17, 1996, TWA Flight 800 left New York's JFK Airport bound for Paris. Just 8 miles from Long Island and only 12 minutes into the flight, the Boeing 747 exploded in the evening sky at about 13,700 feet.

Because of early reports of an explosion, the FBI initiated a criminal inquiry. An inquiry becomes a formal investigation when criminal evidence is found. The distinction is important and defines who is in charge. Until a criminal investigation is triggered, the NTSB is legally in charge. However, in reality, the 26 NTSB investigators were overwhelmed by the 500 FBI agents working on the case.

For some time, this unbalance between NTSB and FBI personnel led to a tug-of-war over competing explanations for the crash. For example, the *New York Times* reported that at a White House briefing in September, the NTSB presented mechanical failure as their leading theory because there was no evidence to support terrorism. The FBI said the exact opposite. At the time, shortly after the crash, both agencies should have stopped after stating "no evidence."

There was also a bit of a culture clash. The FBI was focused on preserving evidence and catching criminals, while the NTSB was focused on identifying safety violations and methodically working through their process.

Meanwhile, Ramzi Yousef was in the eighth week of his trial in New York City, accused of plotting to blow up U.S. planes. After unsuccessfully trying to blow up a 747 over the Philippines in 1994 with a bomb placed under a seat near the center fuel tank, Yousef was convicted and sentenced to life in prison for plotting attempts on U.S. planes. Later, he was also convicted of the 1993 World Trade Center bombing.

Three leading theories were quickly identified for Flight 800: bombing, missile attack, and massive mechanical failure of the 25-year-old plane. At the time, Flight 800 was one of the oldest 747s flying, but there was no precedence for a 747 blowing up without being bombed. When enough wreckage was recovered, it was determined that the center wing fuel tank had exploded, but the cause was still an open question.

The bomb theory slowly evaporated under the weight of expanding evidence (or lack thereof). For example, in August of 1996, the last of four luggage containers was found without any blast damage. Eventually, sufficient recovery of the cockpit and passenger cabin allowed investigators to reach the same conclusion. Recovery of those parts of the plane all but eliminated the most obvious places to hide a bomb. After numerous announcements by public officials implying criminal activity, no evidence was ever found. The FBI started hinting at altering their position in May of 1997 and formally ended their investigation on November 12, 1997.

Everyone initially assumed it was a terrorist attack. Eyewitness reports added to the confusion. Some heard zero, one, or two explosions; some saw one or two fireballs. Some spoke of flaming streaks and nearby flashes of light, adding credence to the missile attack theory. Except for conflicting eyewitness reports, no evidence ever surfaced to support a missile attack. Upon further review, most witnesses were considered too far away to see anything accurately.

The Flight 800 investigation is also famous for persistent rumors of a government cover-up. Enter Pierre Salinger, former ABC correspondent and press secretary for President Kennedy. Mr. Salinger announced possession of a "secret" document that proved a government cover-up of accidental friendly fire from a Navy missile. The document, essentially a

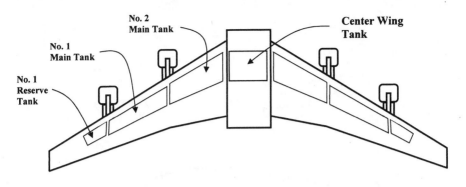

Fig. 3.1. The fuel tanks of a 747. Only the center wing tank is hot enough to support combustion.

report about rumors posted on the Internet, was widely discredited by all government officials.[2]

This crash investigation remains controversial because of eyewitness accounts, persistent rumors of cover-up (and publication of a few books), and frequent testimony by "experts." Actually, they often were bona fide experts, just not the official investigators with access to all the information. Also, the crash report reflects many different types of expertise, beyond the capacity of any single person.[3]

TWA Flight 800: The Official Conclusion

From the fracture and soot patterns, radar data, ballistic trajectory analysis (discussed later), wreckage investigation, and explosion studies, investigators concluded that fuel vapors in the center wing tank exploded (Figure 3.1). However, a definitive ignition source was never identified, thus giving a very small boost to the conspiracy theories.

Explosion of a fuel tank requires a proper fuel to air ratio and an ignition source. The requirements for an explosive mixture will be reviewed later under a general discussion of combustion in Chapter 7. Discussed here are the breakup sequence, recovery, and reconstruction.

The center wing tank of a Boeing 747 sits between the wings underneath the passenger cabin. This fuel tank is about 20 feet long, 20 feet wide, 6.5 feet high in the front, and 4 feet high in the back, and it holds over 13,000 gallons of fuel (Figure 3.2).

The partition plates of the center wing fuel tank are baffles that prevent the fuel sloshing. The plates are also major structural members that

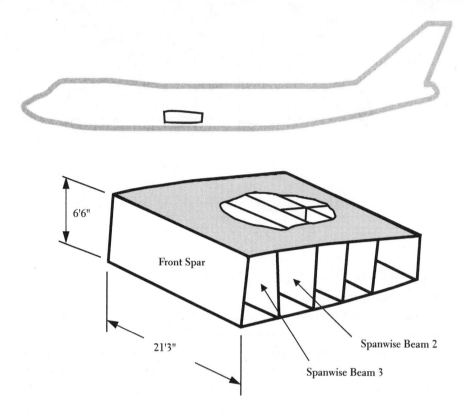

Fig. 3.2. Center wing tank and partition plates. Spanwise beam 3 blew out, destroying the front spar.

reinforce the wing-fuselage connection. Since the front bay was dry, spanwise beam 3 became the front wall of the fuel tank. Investigators believe that the explosion blew spanwise beam 3 into the front spar, destroying both. "Witness marks" (markings from one component impacting another) were found corresponding to where beam 3 blasted into the front spar.

Most of the front spar, a large portion of beam 3, and the access door to beam 2 were recovered in the most westerly recovery zone. Since the plane was traveling west to east, finding those pieces at this location strongly suggests they were among the first pieces to break off.

Boeing engineers estimated that a pressure of 20 to 30 psi would rupture the fuel tank. Subsequent analysis of the explosion (discussed in Chapter 7) concluded that pressure pulses in excess of 20 psi occurred during the explosive event. This relatively modest pressure was applied over an area approximately 20 feet wide $\times$ 6 feet high, or an area of 17,280 in.[2] The force available to rupture the tank from the pressure thrust equals

$$force = pressure \times area,$$
$$345,600 \text{ lb of force} = 20 \text{ lb/in}^2 \times 17,280 \text{ in}^2.$$

This force is almost equal to the empty weight of the plane. The explosive force was concentrated on one location designed for very little loading in that direction, easily causing breakup of the plane. The over-pressurization structurally exploded the tank and violently released the energy of compressed gas, a process described in Chapter 4.

The initial explosion blew out a 30-foot section of fuselage under the center wing tank, causing the front of the plane to fall off. The remaining part of the plane, including the wings and engines, continued to fly for about 40 seconds, actually gaining almost 3,000 feet in altitude. Shortly after, during the ensuing uncontrolled dive, the two major sections broke into additional pieces.

Because of persistent rumors of a missile strike and possible eyewitness reports supporting the rumors, the missile theory was studied extensively. There was no radar evidence of a missile strike, all Navy missiles in the area were accounted for, and the engines showed no damage consistent with heat-seeking missiles. The investigators "explained away" most eyewitness reports based on distances, line of sight, and other reasons, but still witnesses remain who insist on what they saw. The controversy caused the FBI to extend the wreckage reconstruction beyond what the NTSB investigators considered prudent. However, the reconstruction, discussed later, did not find any evidence of a missile strike.

Blast patterns from a missile strike are well understood. In an attempt to dispel any rumors, additional (and perhaps even unnecessary) testing was done for this investigation. Testing specific to this investigation indicated that a blast strong enough to penetrate the fuel tank skin would produce petaling of the surface, pitting of adjacent surfaces (from small blast fragments), and hot gas surface "washing". All of these effects are readily apparent to the experienced investigator.

In addition to a direct missile strike, penetration of the fuel tank by a missile fragment (a "near miss") was also considered. According to the Naval Air Warfare Center, a shoulder launch missile would have to detonate within 40 feet for a fragment to penetrate the aluminum skin. At that distance, numerous other distinctive high-velocity fragments would leave telltale impact marks in a starburst pattern. No such blast patterns were found.

One hundred and ninety-six relatively small holes were also examined for characteristics of high-velocity penetrations. Once again, the characteristics of a high-velocity impact hole are well understood, but nevertheless, specific testing on panels prepared by Boeing was performed. The metallurgical characteristics (from microscopic examination) of a high-velocity impact are

- splashback of material around the perimeter of the hole on the entry side, with the splashback material moving opposite to the motion of the penetrating fragment (Figure 3.3);

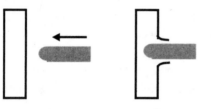

Fig. 3.3. High-velocity penetration shows splashback deformation.

- melted and resolidified material on the walls adjacent to the entry side;
- breakout deformation of material around the perimeter of the hole on the exit side (Figure 3.4).

Fig. 3.4. High-velocity penetration shows breakout deformation.

Two holes did show some evidence of high-velocity impact, but they were made by objects traveling from inside to outside.

A meteor, expected to leave a high-velocity impact as described above, has never been known to strike an airplane. However, a meaningful estimate of the probability of a meteor strike can be made. This estimate considers the known frequency of meteors striking cars and dwellings, the percentage of earth's surface covered by cars and dwellings, and the area of all airplanes in the sky at any one time. A meteor expert calculated that a meteor is expected to strike a plane every 59,000 to 77,000 years.

Terrorist Bombs

A terrorist bomb can, of course, cause an in-flight breakup and was the presumed cause of the Flight 800 explosion until proven otherwise. Surprisingly, modern aircraft structures survive internal explosions fairly well.[4] An aircraft bombing is survivable if the expanding gases are rapidly vented. The blast hole often serves this purpose. There are several cases of bombs blowing holes measuring several square feet in the fuselage and the plane still landing safely.

For example, on April 2, 1986, a bomb ripped a 9-foot × 3-foot hole in the cargo section of a Boeing 727 flying over southern Greece. The hole was in front of and above the plane's right wing. Four passengers were sucked out by the decompression as the air in the pressurized fuselage blew out at 11,000 feet. The plane managed to land safely about 10 minutes later. The bomb might have done greater damage had it gone off at higher altitude when the fuselage is more stressed with higher cabin pressure.

With even greater blast damage, Aloha Flight 243 (a Boeing 737) safely landed after a non-bomb explosive decompression blew out an 18-foot × 14-foot section of the fuselage (Chapter 4).

A detonation can result in structural failure if the bomb explodes near a structural joint and the overpressure is not rapidly vented. This approximately describes the situation of Pan Am Flight 103 over Lockerbie, Scotland.

Pan Am Flight 103

The most recent, and most studied, example of a terrorist bombing occurred on December 21, 1988. Pan Am Flight 103 left London bound for New York City and blew up over Lockerbie, Scotland. The breakup sequence and reconstruction are discussed here; the detonation[5] mechanism is described in Chapter 7.

Thirty-seven minutes after takeoff and just two minutes after last radio contact, the secondary radar response disappeared from the controllers' screen. Something had cut the power to the Boeing 747's transponder.[6] Simultaneously, the primary radar indicator showed the aircraft breaking into several sections at 31,000 feet. Wreckage came down in a path 10 miles long in and around Lockerbie, Scotland. Personal effects (papers, clothing, etc.) had drifted up to 80 miles away. Later, when the black boxes were recovered, they indicated an abrupt power cutoff, which implied that the failure was so massive and instantaneous that multiple electrical systems powering the recorders were simultaneously wiped out. A similar abrupt cutoff of all systems occurred in TWA Flight 800.

A terrorist bomb was immediately suspected in the Lockerbie crash, but officials refused to rule out structural failure and explosive decompression. However, in the previous year, this particular 18-year-old aircraft had undergone structural modification with reinforcements to carry heavy military equipment. The structural integrity of the aircraft at the time of the accident was essentially better than when it was first delivered.

Unlike for TWA Flight 800, distinctive detonation damage patterns very quickly emerged for Pan Am Flight 103. Investigators found suitcases ripped by metal fragments, heat damage on the Kevlar cargo container liner, and slivers of metal in passengers. Just seven days after the crash, it was announced that explosive residue had been found. (Trace amounts of three different explosive residues were found on Flight 800. This evidence was later discounted when it was learned the plane had been used for a dog sniffing exercise.)

Subsequent analysis for Flight 103 eventually led to a specific suitcase. Amazingly, a few fragments of the original suitcase were located with "finger tip" searches of the crash site. More traditional detective work traced all luggage and eventually determined the suitcase in question was loaded on board, suspiciously, without an accompanying passenger.[7]

The initial shock blast of extremely high-velocity gas shattered and disintegrated material immediately opposite the explosive charge. This created an approximately 20-inch × 20-inch blast hole, or "shatter zone." The blast hole did not destroy the aircraft. A plane is designed to be safe, or "damage tolerant," with even larger holes. In fact, planes have landed safely with much greater damage.

For the Lockerbie crash, experts initially estimated a 10- to 100-pound bomb. It was later concluded that the explosive was smaller than a hamburger patty and weighed less than a pound. The mechanism by which a relatively small blast hole caused structural breakup of the fuselage required further study. A 65-foot section of fuselage was reconstructed—the largest reconstruction at the time—to study the breakup sequence.

Most of the energy of the blast shock wave is transmitted out the blast hole and into the atmosphere, but some is reflected off the shatter zone and back into the fuselage. This reflective shock wave can interact with the initial pressure pulse to produce what is called a "Mach stem shock wave," or a recombination shock, which can have pressures greater than the initial pressure pulse. The recombination shock wave acts like an expanding high pressure gas bubble. Essentially, the expanding gas bubble unzipped the skin of the fuselage with a starburst fracture pattern emanating from the blast hole, and the bent and torn fuselage skin petaled outward.

As the blast pressure dissipated, long fractures or tears, driven by normal internal pressure stresses in the fuselage, continued to propagate away from the blast hole. The explosive energy of the expanding gases was also transmitted remotely through fuselage cavities (created by voids between the 12 baggage containers and the cargo hold). The expanding gases produced pressure that damaged sites remote from the blast. In comparison, no blast hole, starburst fracture pattern, petaling, shards of metal embedded in passengers (or luggage), or significant explosive residue was ever found on TWA Flight 800.

Investigators concluded that the nose of Flight 103 broke off within two or three seconds of the explosion. As a further clue to the speed of the propagating damage, the flight crew did not have time to initiate any emergency procedures. Radar indicated that the plane had broken into four pieces separated by up to a mile within only eight seconds of the blast.

The blast hole did not bring down the plane, but the pressure pulse from the explosive charge did. With a different location of the bomb and with better venting of the explosive gases, the plane could have survived.

The U.S. government has done extensive studies on the structural response of different fuselage designs to explosive charges. The results of these studies are restricted. The breakup sequence of Flight 103 described here is in the final crash report and is widely available on the Internet and elsewhere.

Wreckage Reconstruction

A reconstruction of part of the fuselage is performed after an in-flight breakup to help determine the sequence of structural failure. The first such reconstruction occurred in 1954.

First Reconstruction: de Havilland Comet

Being the first commercial jet, the de Havilland Comet had taken the aviation business by storm, displacing the American companies Boeing, Lockheed, and Douglas. After two in-flight breakups of de Havilland Comets (described in Chapter 4), the most intense, and perhaps most desperate, crash investigation to date began.

The recovered wreckage from the first Comet breakup was assembled on a wooden frame. When the tail section was recovered investigators found that it had been scored by a metal object that left traces of blue paint. The blue streak matched a similar streak on the back of the fuselage. Chemical analysis proved that the paint belonged to a cabin chair. Also embedded in the tail was a piece of carpet. The front cabin interior had blown out and across the back of the fuselage and into the tail. This information helped investigators conclude that the interior blew out in an explosive decompression caused by metal fatigue.

During the de Havilland investigation, many important precedents and standards were set in terms of deep-sea salvage and reconstruction. Almost every future in-flight breakup would advance the art and science of wreckage reconstruction.

Pan Am Flight 103 Reconstruction

Although telltale indications of a bomb blast were recognized early in the Flight 103 investigation, the reconstruction of 65 feet of fuselage helped explain how a relatively small blast hole led to breakup of the plane. Reconstruction of the cargo container helped locate the suitcase that had contained the bomb, and partial reconstruction of this suitcase helped identify the bomber.

TWA Flight 800 Reconstruction

The remains of TWA Flight 800, a 747 destroyed by an exploding fuel tank, initiated the most extensive reconstruction ever attempted (Figure 3.5).

The U.S. Navy spent over nine months recovering more than 20,000 pieces of wreckage, approximately 95% of the total plane. Ships with sonar began by mapping the concentration of metal objects on the ocean floor. Divers and ROVs (remotely operated vehicles) then systematically removed all visible debris from a water depth of 120 feet. When nothing else could

Fig. 3.5. Reconstruction of Flight 800; 94 feet of the fuselage were reconstructed from over 1,600 pieces of wreckage. *Photo:* NTSB

be found by divers, trawling began with scallop dredges. (A scallop dredge is a steel sled trailed by a net.) About 40 square miles of ocean floor were scoured. After trawling 13,000 times for almost six months, an additional 1–2% of the structure was recovered. Trawling continued, over 30 times in some locations, until no new material was found.

After wreckage recovery, the next daunting task was to identify the parts. Boeing created a computer database for part identification. The NTSB investigator would typically sit at a computer screen scrolling through computer drawings and try to identify a particular part.

Occasionally, a part is labeled with a serial number or its station number, indicating its location relative to the nose. Paint finish and sheet metal thickness can also be used to guide assembly of the jigsaw puzzle. The paint on top of the aircraft is more exposed to the weather and is noticeably duller than the paint on bottom surfaces. Thicker parts have more load bearing capacity and hint at location on the fuselage.

A contractor was hired to fabricate a steel structure to hang the fragments on (Figure 3.6). The framework resembled the ribs of a large animal and was promptly christened "jetosaurus rex." Once the wreckage was

Fig. 3.6. Inside "jetosaurus rex," the reconstruction framework of Flight 800
Photo: NTSB

recovered and identified, it took an additional three months to reconstruct 94 feet of the fuselage.

As discussed earlier, this crash investigation remains controversial because of persistent rumors and partial eyewitness accounts (discounted in the official report) of a bombing or missile attack by terrorists or an accidental military strike. The reconstruction did demonstrate a fuel tank explosion, but investigators could not identify the source of ignition. By failing to find any evidence of a bomb or missile, the reconstruction helped disprove the theory of criminal acts.

Ballistic Trajectory Analysis

Ballistic trajectory analysis has been used to analyze all in-flight breakups for a number of years. Sometimes, the analysis starts with the last known radar location of the plane and attempts to calculate a trajectory to help predict where the wreckage might be found. More commonly, trajectory analysis is used to assist in determining the breakup sequence. In this case, radar data starts the analysis with known speeds and locations, and then certain assumptions are made about air resistance[8] (and winds) to calculate trajectories. The assumptions might be adjusted to get a better match with the known wreckage distribution. This process is somewhat similar to taking all the wreckage on the ground and assigning reverse trajectories to it until the plane comes back together. The purpose is to help determine what broke off the plane first. For the case of TWA Flight 800, ballistic trajectory analysis was also used in an attempt to reconcile all witness statements.

The method is considered approximate at best, because air resistance of complex shapes is not known definitively. For example, an airplane, with a smooth and continuous shape (at least compared to jagged wreckage), requires extensive wind tunnel testing to define its air resistance. (Today, very powerful computer methods supplement wind tunnel testing.)

Results from a trajectory study are but one piece of the puzzle and are combined with other information including radar data[9] of the breakup sequence, wreckage recovery locations, and metallurgical examination.[10]

Temporarily ignoring air resistance, the basic principles of ballistic trajectories can be explained with concepts from high school physics.

Ballistics and Projectile Motion

The field of ballistics is concerned with the motion of objects. A projectile is a projected (thrown) object that continues in motion guided only by Newton's First Law (a body in motion tends to stay in motion) and gravity. By definition, the only force acting in projectile motion is gravity. Air resistance, at least in a high school physics book, is considered negligible.

A cannonball shot horizontally illustrates the interaction between the ball's horizontal inertia (inertia is discussed more in Chapter 9) and the effects of gravity. Inside the cannon, expanding gases accelerate the cannonball to its exit velocity. Once the cannonball is outside the cannon, only gravity acts on the cannonball.

Without gravity, the cannonball moves horizontally according to Newton's First Law. Gravity acts downward on the cannonball adding a vertical velocity and acceleration component. Even though the cannonball is initially projected horizontally, it falls downward with the acceleration[11] of gravity, 32.2 ft/sec^2, the same acceleration it would have if dropped. The cannonball is simultaneously moving horizontally and vertically, and the net effect is a path the shape of a half parabola (Figure 3.7).

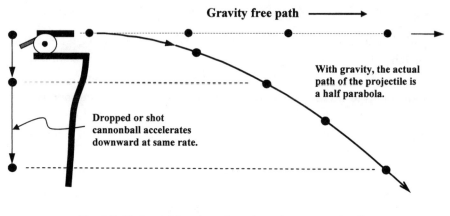

Gravity free path ──────►

With gravity, the actual
path of the projectile is
a half parabola.

Dropped or shot
cannonball accelerates
downward at same rate.

Fig. 3.7. Horizontal inertia and gravity acting on a projectile

The trajectory can be mathematically described for any initial angle and velocity—a fact understood by programmers to make computer-generated motions look more realistic, using the following procedure.

1. Determine the initial vertical and horizontal velocity components of the projectile.
2. Alter the vertical velocity by the constant (and always down) acceleration due to gravity.
3. Calculate the horizontal and vertical position (X and Y coordinates) for every increment of time until the projectile motion stops, usually by contact with the ground.

Projectile motion, or motion of objects without aerodynamic lift, is precisely defined with high school physics. The effect of air resistance is the great unknown and is difficult to predict, even for simple shapes.

Air Resistance

Air resistance (or air drag) was first studied in the nineteenth century to determine the effect on bullet speed. A clever experiment to measure the speed of a bullet was devised by suspending a gun as a pendulum. From conservation of momentum,[12] the motion of the gun pendulum could be used to calculate the muzzle velocity of the bullet. Bullets were then shot into a block of wood suspended as a pendulum and located at various distances from the gun. The farther the bullet traveled, the more air resistance decreased its velocity. Air drag was discovered to be a force slowing the bullet down and to be proportional to the velocity of the bullet squared. In other words, if the bullet's velocity doubles, the air resistance increases by a factor of four.

Drag forces also depend on air density, the cross-sectional area of the bullet, and a "drag coefficient." The drag coefficient is a parameter used to match the experimental velocity with a calculated prediction. The drag coefficient must be experimentally measured and varies greatly with surface texture and speed.

The ballistic trajectory of a component from an in-flight breakup is estimated from its mass and aerodynamic characteristics, essentially projectile motion corrected for air resistance.

Two balls the size of bowling balls will be used to illustrate the principle. One ball is made of lightweight foam and weighs 1 pound, and the other is a traditional bowling ball weighing 10 pounds. Assume that both balls are covered with the same coating so that their air drag is the same. Having the same size and surface coating, both balls will experience the same air drag forces. From Newton's Second Law,

$$\text{force} = \text{mass} \times \text{acceleration},$$

air resistance becomes just another force acting on the two balls:

$$\text{air resistance force} = \text{mass} \times \text{acceleration}.$$

The heavy bowling ball, having 10 times more mass, will have 10 times less acceleration from air resistance. Stated differently, air resistance affects heavier objects less. This is the concept behind the use of lead (the heaviest non-radioactive element) in bullets.

Even though the air drag is the same on the two balls, the ability of air resistance to slow down (or decelerate) the two balls depends not only on air drag, but also on the weight of the object. These two parameters are combined into a single expression known as the ballistic coefficient.

$$\text{Ballistic coefficient} = \frac{\text{weight}}{\text{drag coefficient} \times \text{area}}.$$

The ballistic coefficient is commonly used to describe the motion of bullets (and wreckage) and describes how far bullets will travel before being stopped by air resistance. Bullets with greater mass or with a lower drag coefficient will have a higher ballistic coefficient and will travel farther.

The physical interpretation of a foam ball being released from an in-flight breakup at 500 mph is obvious: the foam ball will rapidly decelerate and fall slowly, blowing with ambient wind conditions. A bowling ball, on the other hand, has a high weight-to-air drag ratio and a very high ballistic coefficient. Air resistance has much less effect on the bowling ball. The bowling ball will have a trajectory much closer to projectile motion than the foam ball.

Objects with a low ballistic coefficient essentially blow in the wind, whereas objects with a high ballistic coefficient continue on their ballistic trajectory (Figure 3.8). Heavy components, such as engines, will fall with little disturbance from the wind, but fuselage panels may blow considerably with the wind.

A ballistic coefficient attempts to estimate motion for objects in between the extreme cases of the foam ball and the bowling ball. The traditional non-aerodynamic projectile motion equations are modified with additional forces (and decelerations) from air resistance. The object, with a constant downward acceleration from gravity, now has an additional vertical and horizontal acceleration from air resistance (Figure 3.9). The forces from air resistance are not constant; they change as the velocity changes.

The drag coefficients are estimates based on size and shape of the object being evaluated. Again, it needs to be emphasized that the method is

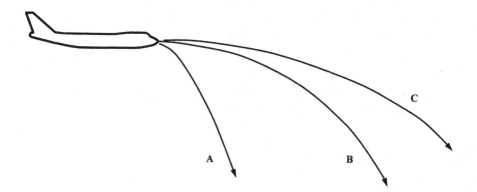

Fig. 3.8. Trajectory of wreckage coming off a plane: (A) low ballistic coefficient, (B) high ballistic coefficient, (C) same as B except higher initial horizontal speed

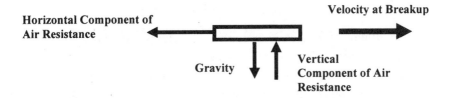

Fig. 3.9. Ballistic trajectory analysis involves analyzing the initial velocity and the forces on a wreckage component. The forces include air resistance and gravity.

an estimate limited by the accuracy in predicting air resistance for complex wreckage shapes. Predictions for wreckage with high ballistic coefficients are expected to be more accurate.

Ballistic Trajectory Analysis of TWA Flight 800

For TWA Flight 800, investigators analyzed the ballistic trajectory of some sections of the center wing tank and portions of the fuselage in front of the tank. Shortly after the explosion, the nose section broke off completely. The remainder of the airplane, including the wings and engines, actually flew for about 40 seconds after the initial explosion before exhibiting ballistic (i.e., no lift) behavior.

The separation of the forward fuselage resulted in significant changes to the weight, balance, and aerodynamic characteristics of the remaining aircraft. The NTSB conducted a series of computer simulations to examine the flight path and attempt to reconcile the radar data and eyewitness reports. In addition to aerodynamic lift on the wings, additional drag from the open fuselage was considered in the computer model.

After destruction of the fuel tank and separation of the fuselage into two main sections, additional breakup of the two sections occurred during the uncontrolled dive.

Ballistic Trajectory Analysis of Pan Am Flight 103

Pan Am Flight 103 was also analyzed with trajectory analysis to provide additional insight into the breakup sequence. Individual trajectory calculations were done on selected key items of wreckage. Analysis indicated that the nose section, weighing about 17,500 lb, had an impact speed of around 140 mph; the engines, weighing about 13,500 lb, hit at about 300 mph; and the wing sections, with 100,000 lb of structure plus about 200,000 lb of fuel, could have flown into the ground as fast as 750 mph.[13]

The location of the victims was also used to study the breakup sequence. Scottish officials sealed off Lockerbie and the immediate surroundings for many days until forensic workers had photographed all the victims. The location of each body was coordinated with the seat location to further aid identification of the breakup sequence.

Other Sources of In-Flight Breakup

Explosive Decompression

Several examples of explosive decompression of the fuselage have occurred over the years. The most recent example is China Airlines Flight 611, a 747 that broke up over the China Sea in May of 2002. Because of metal fatigue and structural failure, the cabin pressure was violently released in an explosive decompression.

Explosive decompression is analogous to popping a balloon. This situation can happen in an aircraft with an improperly latched door or a fatigue crack that reaches a critical length. With explosive decompression, the compressed air inside the pressurized fuselage is rapidly released with a loud bang or explosion. This topic is discussed at length in Chapter 4.

Aerodynamic and Inertial Forces

Bad things can happen when a plane goes into an uncontrolled dive. The most obvious is that the plane fails to pull out of the dive. However, a commercial plane is not a dive bomber. Aerodynamic and inertial forces can rip a diving plane apart before it reaches the ground. Breakup during an uncontrolled dive is usually a secondary cause of any crash. Some other error or failure caused the plane to go into the steep dive.

If a plane exceeds its design speeds in a dive, excess wind blast forces can do structural damage. At high speeds, flutter—a type of vibration—can also occur on any flight surface (wings, ailerons, rudder, stabilizer, etc.). Flutter usually results in rapid structural damage of the affected component. An uncontrolled dive is already an emergency situation; any minor mechanical damage can only make things worse.

A plane trying to fly out of a dive will also pull excessive inertial loading or G loads. A "G" is a unit of force on a body undergoing acceleration. The normal load factor on a plane in straight and level flight is 1 G. If the plane is accelerating at twice the acceleration of normal

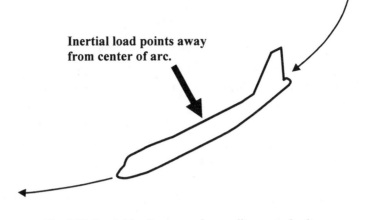

Inertial load points away from center of arc.

Fig. 3.10. Inertial loading on a plane pulling out of a dive

gravity (32.2 ft/sec^2), 2 G occurs. (Inertial loads are discussed further in Chapter 9.)

Flying in a large arc generates centrifugal loading similar to the forces that keep water in a bucket when swung overhead. When pulling out of a dive, the acceleration of the plane creates downward G loads that add to normal gravity, as shown in Figure 3.10. Additional aerodynamic loads on the wings must balance the inertial loading, effectively forcing the plane to fly in an arc just like a string forces a weight to swing in an arc. These additional loads can potentially break off the wings.

The acceleration associated with flying in an arc is described by the equation

$$\text{acceleration} = \text{velocity}^2/\text{radius}.$$

This situation is similar to the effect of riding in a roller coaster at the bottom of a hill.

If velocity^2/radius is five times the acceleration due to gravity, or 5×32.2 ft/sec^2 = 161 ft/sec^2, the plane will act as if it weighs five times its normal weight. For example, a plane flying at 500 mph in a 3,340 foot radius would have acceleration equal to

$$\left(\frac{500 \text{ miles}}{\text{hour}} \times \frac{5,280 \text{ feet}}{\text{mile}} \times \frac{\text{hour}}{3,600 \text{ seconds}} \right)^2 \times \frac{1}{3,340 \text{ feet}}$$
$$= 161 \text{ ft/sec}^2.$$

Required Strength of the Wings

Pulling out of a dive can break off components, including the wings. To guard against this possibility, wing certification requires a minimum "load factor" of 3.75 times normal gravity, but only 1.5 G in the opposite direction (i.e., a plane diving upside down).[14] In other words, the wings are not expected to break off until velocity^2/radius exceeds 3.75×32.2 ft/sec^2. This load factor is a minimum strength requirement that may be exceeded by a particular design.

Before a new plane design is certified as airworthy by the FAA, it must pass many structural tests. One such test bends the wings with hydraulic actuators to verify that the wings can in fact support the minimum required loads. For example, the 777 wing tips were bent 24 feet before snapping off.

Uncontrolled Dives

Turbulence Can Trigger a Dive

With the single exception in 1966 of a Boeing 707 over Japan, no modern, large commercial jet aircraft has been destroyed by wind gusts in the last 50 years. Modern design methods produce robust aircraft, and accurate weather forecasts make it relatively easy to avoid the most severe storms. For an aircraft to be destroyed today, it would have to fly directly into a severe thunderstorm.

Wind shear (local downdraft turbulence) has been a serious problem resulting in many crashes during takeoffs and landings. Wind shear is a flight control problem; the plane's structural design loads are not exceeded. A plane affected by wind shear during takeoff or landing is slammed into the ground when it is too low to recover. (Damage associated with vertical impact velocity is discussed in Chapters 2 and 8.) More recently, improved Doppler radar has significantly reduced this risk by "seeing" the wind shear at airports.

In the following accident, severe turbulence did not damage the plane, but it resulted in loss of control. The plane entered a steep dive that exceeded its structural limits (Figure 3.11), and it disintegrated about 45 seconds later.

On February 12, 1963, a Boeing 720 (Northwest Flight 705), just 12 minutes after takeoff from Miami, entered a turbulence-induced dive at 19,000 feet and began breaking up at around 10,000 feet. The flight data recorder logged G loads up to minus 3.7 G's (greatly exceeding the minus

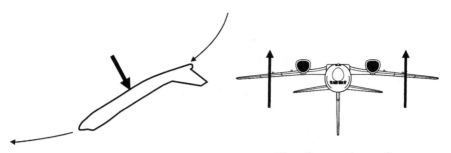

Wings flex up and snap off.

More upward lift is required to support a plane
experiencing upside down G loads.

Fig. 3.11. Upside down dive creates negative inertial loads. Wings flex up and
eventually snap off if inertial loading exceeds critical values.

1.5 G wing certification load) before recording stopped. The wings and
horizontal stabilizers broke off.

Thrust Reversers Trigger a Dive

On May 26, 1991, Lauda Air Flight 004, a Boeing 767, broke up 95 miles
from Bangkok when a single thrust reverser unexpectedly deployed during
flight at 31,000 feet. Thrust reversers, mounted near the engine exhaust,
block the jet blast and turn it around in the opposite direction. In this way,
the forward thrust of the jet engine is turned into a massive brake. Thrust
reversers are only used during landing. The unbalanced engine thrust on
Flight 004 rotated the plane, as shown in Figure 3.12. After the rotation,
the nose blocked airflow over the left wing, which resulted in more lift on
the right wing, and the plane rolled, as shown in Figure 3.13. The plane
eventually rolled into a dive. Subsequent simulator studies by Boeing indicated that the plane could have recovered if corrective action was initiated within six seconds of this unexpected mechanical failure.

Analysis of the major damage showed that failure of the aircraft structure was probably the result of turbulent buffeting and excessive accelerations. Parts of the airplane that separated from buffeting overload appeared
to be pieces of the rudder and the left elevator. These pieces were followed
by separation of most of the right horizontal stabilizer from excessive accelerations as the crew attempted to control the airplane and arrest the
high-speed descent. The loss of an airplane's tail results in a sharp nose-
over dive (similar to the upside down dive shown in Figure 3.11), which

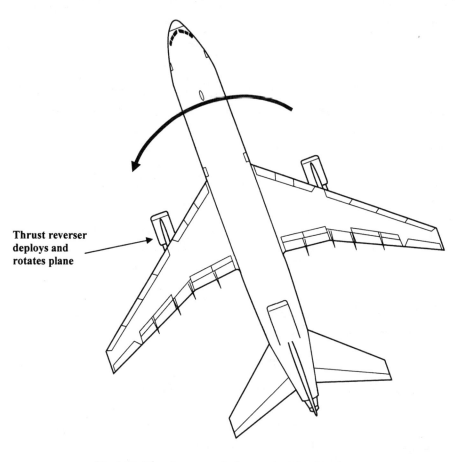

**Thrust reverser
deploys and
rotates plane**

Fig. 3.12. Thrust reverser deployment rotates the plane.

produces excessive negative loading on the wings. There was evidence for
Flight 004 of reverse bending wing failure. This sequence was probably followed by the breakup of the fuselage. The complete breakup of the tail,
wings, and fuselage occurred in a matter of seconds.

Other Uncontrolled Dives

Two other in-flight breakups from uncontrolled dives have been documented. However, many of the details of these crashes are somewhat murky.

On December 19, 1997, a Boeing 737 went into a dive from 35,000 feet
over Indonesia. The tail broke at about 12,000 feet from suspected flutter
(vibration from excessive speed). The U.S. NTSB suggested that the captain may have committed suicide by switching off both flight recorders and
intentionally crashing the Boeing 737 in a dive.

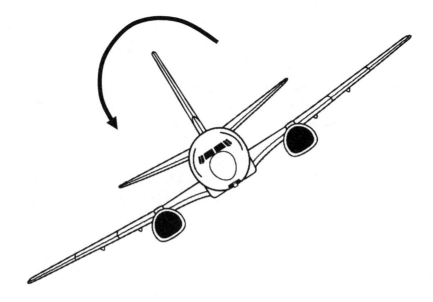

Fig. 3.13. Unbalanced lift caused plane to rotate or roll.

On the night of June 6, 1992, a Boeing 737 slowly rolled over, went into a dive, and broke up at around 13,000 feet over the Panamanian jungle. It was theorized that the artificial horizon was not working on the instrument panel. After entering a normal turn and roll (a plane rolls or tilts its wings during turning), the plane continued to roll until it flipped over and plunged into an uncontrolled dive. Without any visible references in the featureless jungle, the pilot was unable to recognize the problem.

Recovery from Uncontrolled Dives

Not all uncontrolled dives result in disaster. On February 19, 1985, China Airlines Flight 006, a Boeing 747, lost power in one engine at 41,000 feet over the Pacific. While diagnosing the engine problem, the pilot failed to detect the airplane's increasing bank angle. The plane flipped over and entered an uncontrolled dive. The plane descended from 40,000 feet to 30,000 feet in 33 seconds, and then a second dive from 30,000 feet to 9,500 feet lasted for almost two minutes. The G forces became so great that the flight crew could not lift their arms or heads. During pull-up, the flight data recorder registered 5.1 G's of vertical acceleration. Amazingly, the plane landed safely a short time later.

The wings were permanently bent 2–3 inches upward, and one aileron was broken and cracked in several places. A large portion of the left horizontal stabilizer had separated and severed a hydraulic line. The right horizontal stabilizer had also separated, and the auxiliary power unit (an additional jet engine that generates electrical power) had separated from its mounts in the tail section.

An even greater death-defying dive occurred on April 4, 1979, in a Boeing 727. The NTSB later determined that a leading edge slat (a wing extension used to increase lift during takeoff and landing) failed from metal fatigue. This failure caused asymmetric, but correctable, loading on the wings resulting in a rapid roll to the right and eventually an uncontrollable spiral descent. The plane dove from 39,000 feet to 5,000 feet in 63 seconds. Vertical acceleration forces increased throughout the spiral descent to a maximum of about 6 G's during the recovery. The wings may have been over-designed, or more likely, the plane's orientation prevented the wings from experiencing the full 6 G's trying to snap them off.

The plane was uncontrollable until the faulty slat separated from the wing. Numerous moving parts in the wings were damaged or missing (spoilers, fairings, aileron tabs, actuators, etc.). The plane landed safely 45 minutes after the dive.

Tail Breaks Off from Aerodynamic Forces

On November 12, 2001, just minutes after leaving New York's JFK Airport, aerodynamic forces broke off the vertical stabilizer (tail section) of American Airlines Flight 587, an Airbus 300. This crash was unique for several reasons, one being that for the first time, the structural failure of a non-metal component caused a plane crash.

Since the switch from fabric and wood, planes have been built from aluminum. A typical strength of an aerospace aluminum alloy might be 65,000 pounds per square inch. (This means a one-inch-square cross section will fail with a tensile pull of 65,000 pounds.) Carbon fibers just 3 to 5 microns in diameter can have strengths of 200,000 to 900,000 psi. However, structures cannot be made solely from such wispy fibers. Instead, tens of thousands of fibers are glued together with epoxy to form a single-ply composite. The long carbon fibers give significant strength in the direction of the fibers. When the composite is loaded perpendicular to the fibers, however, the strength is equal to the considerably weaker epoxy. To compensate, multiple layers or plies are used. Each layer has fibers in a different orientation. The end result is a component with fibers running in many different directions.

The military has used composite aircraft for a number of years in stealth technology. In addition to superior strength-to-weight ratios, composites absorb radar better than metal.

Except for the most recent Boeing 787 design with its all-composite fuselage and wings, commercial aviation has stuck its toe into composites more gingerly. Although superior strength-to-weight performance is very important for the weight-conscious aircraft industry, composites are difficult to manufacture. Also, many years of experience are key in this safety-driven industry.[15] The vertical stabilizer, providing supporting structure for the rudder, was the first major structural component in any commercial aircraft to be made solely of carbon fiber epoxy composites.

Although the rudder is used for a variety of flying adjustments, the most significant structural loading on the tail occurs during engine failure.[16] If, for example, the right engine is out, the left engine will rotate or yaw the plane clockwise. The rudder is turned into the yaw (to the left) creating a counter-rotation in the opposite direction. A left rudder actually creates a slight sideways motion of the plane (sideslip), as shown in Figure 3.14. The windblast on the tail during the sideways motion creates a rotational torque that counters the motion caused by the unbalanced thrust.

At lower speeds, there is less sideways windblast force on the tail than at high speeds. For this reason, the rudder needs to rotate at much larger angles when traveling at lower speeds and smaller angles at higher speeds. (A large rotation at high speeds would overload and break the tail.) The rudder control system limits allowable rudder rotation depending on the speed. For the Airbus 300, the rudder limits at various airspeeds are given in Table 3.1.

Table 3.1. Airspeed versus Rudder Limit for an Airbus 300

Airspeed (knots)	Rudder Limit (degrees)
0–165	30.0
220	14.5
250	9.3
270	7.0
310	5.0
350	4.0
395	3.5

WHAT IS WAKE TURBULENCE?

A plane flies because of a pressure difference on the wings creating lift. The higher pressure on the bottom of the wing (and lower pressure on the top) rolls off the wing tips in a swirling vortex. These horizontal mini-tornadoes are more commonly known as wake turbulence. The strongest vortices are produced by large planes flying slowly. Wake turbulence is normally a concern for small planes flying behind large planes. However, even large aircraft are required per FAA rules to be separated by two minutes or four nautical miles.

The broken tail on Flight 587 initially started with a relatively modest encounter with "wake turbulence." American Airlines Flight 587 took off about 1 minute and 40 seconds behind Japan Airlines Flight 47, a Boeing 747. The flight crew discussed with air traffic control the necessary spacing to avoid wake turbulence.

The wake turbulence experienced by Flight 587 was studied and not considered excessive. In addition to being separated by at least 3,800 feet vertical elevation, the two planes were at no time during flight closer than 4.3 miles. However, this distance was close enough for the wake turbulence of Flight 47 to jostle Flight 587. Attempting to make the flight more comfortable, the first officer of Flight 587 responded to the wake turbulence with excessive rudder movements.

The NTSB determined that the tail separation of Flight 587 was immediately preceded by a series of back and forth rudder movements. At the time of tail separation, the plane was headed with an 8 to 10 degree nose-left sideslip (the plane was flying sideways 8–10 degrees). This sideways flight created a lateral or sideways load on the tail. When full-rudder-left was quickly followed by full-rudder-right, the added windblast scooped out by the rudder overloaded the tail.

Moving the rudder full-left or full-right is perfectly safe, but moving it fully in either direction quickly followed by the opposite direction overloads the tail. The sequence is illustrated in Figure 3.14. The vertical stabilizer on Flight 587 broke off. An airplane cannot fly without its vertical stabilizer. All 260 passengers and crew were killed.

How can the tail fall off in a relatively simple maneuver? The certification of the tail loads was based on a far-left or far-right rudder movement. A single rudder movement by the pilot did not overload the composite tail, but the series of back and forth movements did. From interviews with other

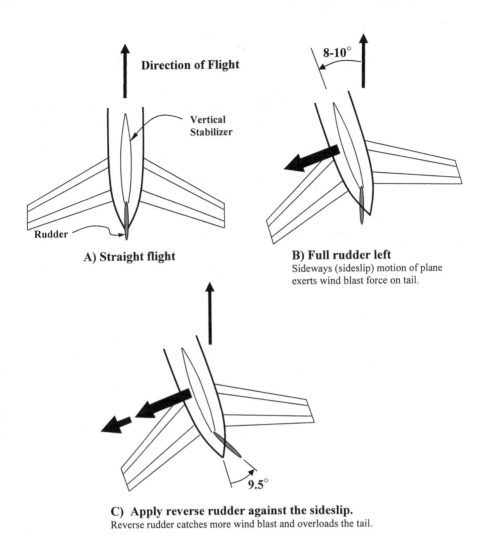

Direction of Flight

Vertical
Stabilizer

Rudder

A) Straight flight

8-10°

B) Full rudder left
Sideways (sideslip) motion of plane
exerts wind blast force on tail.

9.5°

C) Apply reverse rudder against the sideslip.
Reverse rudder catches more wind blast and overloads the tail.

Fig. 3.14. Effect of full-rudder-left followed by quick full-rudder-right overloads the tail with windblast.

pilots, the NTSB concluded that the first officer had a tendency to over-react to wake turbulence. The NTSB also concluded that pilots were not adequately trained for problems associated with repeated cyclic rudder inputs. Also, the rudder controls for this particular Airbus design had the lightest pedal forces of all aircraft reviewed. A light pedal force results in pilot inputs far too easily becoming maximum rudder movements.

Overloading the tail is exacerbated with faster rudder movements. If the rudder is rotated 8 degrees, the nose will rotate 8 degrees in the opposite direction. But if the rudder is rotated very suddenly, the plane's momentum will actually cause the nose to overshoot to perhaps 10 degrees before gently oscillating back to its equilibrium position of 8 degrees. Greater angles of sideslip (the more the plane is flying sideways) create greater lateral loading on the tail. Any overshoot increases windblast loading on the tail.

Pilots were shocked to learn how easily the tail could be broken off. They presumed that mechanical stops would prevent motions that could damage the plane. The engineers, however, suggested that they design planes for the way they are supposed to be flown—and the rudder is not supposed to be cycled quickly back and forth. Everyone agreed this point was not emphasized enough in training and procedures. The situation was summed up in an *Aviation Week & Space Technology* editorial titled "AA587: It Should Not Have Been So Shocking." The editorial explained that many airline pilots were shocked, "but the reaction of engineering test pilots amounted to: We knew that. There's a lot of test pilot lore that doesn't make it to the line pilot, and likewise the manufacturers often are not aware of some of the interesting ways in which their products are used."

In spite of what appears to be an all too easy set of circumstances, a tail had never broken off before during "normal" flight.[17]

Small Planes

In-flight breakup of small planes is more common. Typically, the inexperienced pilot becomes disoriented (for a variety of reasons) and enters an uncontrolled dive. The plane breaks up when the pilot tries to pull out of the dive and exerts excess inertial forces on the structure. The pilot can also go astray into a thunderstorm with catastrophic results.

4...

Pressure, Explosive Decompression,
and Burst Balloons

he concept of pressure and the idea of "being under pressure" have both scientific and common, non-scientific meanings. Webster's dictionary defines pressure as "the burden of physical or mental distress." The physicist, desiring a more precise meaning, has to wait for the third listed definition: "the force exerted over a surface divided by its area." Of course, a hazardous situation produced by sudden air pressure changes in an airplane may also result in a "burden of physical or mental distress" on everyone involved.

As aircraft design progressed from an open to an enclosed cockpit, a generation passed before a pressurized fuselage became part of the mix. The majority of small, general aviation airplanes, usually flying lower, remain unpressurized for simplicity. Pressurization has been a major consideration for large commercial aircraft, with early attempts proving disastrous. Equipment must be added to create pressure in the cabin, the equipment must be operated correctly, and the fuselage must be reinforced to contain the pressure. All these factors are sources of potential mechanical failure and human error. Modern systems, providing redundancy to anticipate nearly every human error and mechanical failure, will forcefully protest with alarms in the cockpit if a condition of too little or too much pressure is approached.

With too little pressure, the passengers and crew can lose consciousness and even die. Perhaps more frightening are the subtle changes that can occur inside the pilot's brain. The pilot and other members of the flight crew may act like they just returned from a three- (or perhaps four- or five-) martini lunch.

With too much pressure the airplane can respond like a balloon confronted by a pin. The analogy is accurate; this situation has happened more than a few times, though current design methods have made such a catastrophy extremely rare. However, most passengers would be alarmed to discover that every commercial aircraft flying today is in fact designed to eventually fall apart from metal fatigue. The aircraft is expected to be retired (or significantly revamped) before that happens. Additional safety is provided by periodic inspections and appropriate maintenance. Constant vigilance is required and assumed by the design engineers.

Too Much Pressure: Aloha Airlines Flight 243

Robert Schornstheimer had been promoted to captain of his own Boeing 737 just 10 months earlier, but he was no rookie behind the controls. Captain Shornstheimer had been flying 737s for 10 years as first officer with Aloha Airlines. Typical of the level of experience required for a position of such responsibility, Captain Schornstheimer had logged over 8,500 flight hours with almost 80% of them at the controls of a 737. The first officer, Madeline Tompkins, one of the few women pilots at the time had accrued over 8,000 total flight hours with about 3,500 hours in a 737. In addition to vast experience, pilots are the most examined (and reexamined), tested (and retested), certified (and recertified), and trained (and retrained) professional group anywhere.

The accident on April 28, 1988, was significantly beyond the flight crew's training and experience, and it gives insight into why airlines require extremely experienced pilots. Although this crew would be considered relatively young, flying was second nature for them and they were able to respond almost routinely to a near disaster.

Twenty minutes into the flight, the flight crew heard a loud explosion, similar to a thunderclap, somewhere behind them. This was followed by passenger screams and a loud, whooshing wind noise. The first officer's head was jerked back by the airflow and she stated that debris, including pieces of gray insulation, was floating in the cockpit. The captain observed that the cockpit entry door was missing and that there was blue sky where the first class ceiling had been. The flight crew quickly put on oxygen masks and began an emergency descent.

In spite of the plane shaking and rolling a little bit, the controls feeling springy,[1] the no. 1 engine failing, the flight crew being unable to communicate with the flight attendants, and the nose gear position indicator light failing to illuminate, the flight crew managed a normal touchdown and rollout. (Air traffic control was able to observe and communicate that the nose gear was in fact down.)

At the time of the accident, all of the passengers were seated with their seat belts on. Those passengers sitting in the window seats in rows 4 through 7 received lacerations and concussions. The most serious passenger injury occurred in window seat 5A, in which an 84-year-old woman suffered a fractured skull and other broken bones. The plane landed with several passengers unconscious.

Passengers seated in row 2 and the center seats of rows 4 through 7 experienced only lacerations. Unfortunately, a flight attendant standing in row 5 was observed by passengers being ejected from the plane. Another flight attendant, standing in row 2, was struck violently on the head by the flight deck door and suffered a concussion and head lacerations. The third flight attendant, standing in row 15, was thrown to the floor during the decompression. Although bruised and battered, she was able to crawl up and down the aisle during the emergency descent to assist passengers. The remaining passengers in rows 8 to 21 received minor lacerations and eardrum injuries. The passengers in the very rear were relatively unharmed. Of the 95 passengers and crew, the final NTSB report lists 8 serious injuries, 57 minor injuries, 29 unharmed, and 1 missing, presumed dead.

A section of fuselage approximately 18 feet along the length and 14 feet around the circumference had blown out in an explosive decompression very similar to a pinpricked balloon (Figures 4.1A and 4.1B). Additional damage to the plane included five consecutive fractured floor beams and two additional beams almost completely broken. Floor beams run across the fuselage, slightly below the centerline, and are spaced every 20 inches (Figure 4.2). When pressure was lost in the passenger cabin, the pressurized cargo hold below the floor beams blew out into the cabin. A 50-foot length of cabin floor (except at the reinforced area where the wings attach) had floor beams displaced upward to some degree.

Explosive decompression refers to a violent expansion (and loud noise) from the cabin air being released under pressure. It is not a chemical explosion, which requires ignition of a fuel vapor mist or a terrorist bomb.

A balloon (and an airplane fuselage) will expand and eventually burst with excess pressure. The fuselage, unlike the balloon, is protected by safety valves that automatically open and allow excess pressure to blow out to prevent an overpressure condition. Components of the pressurization system

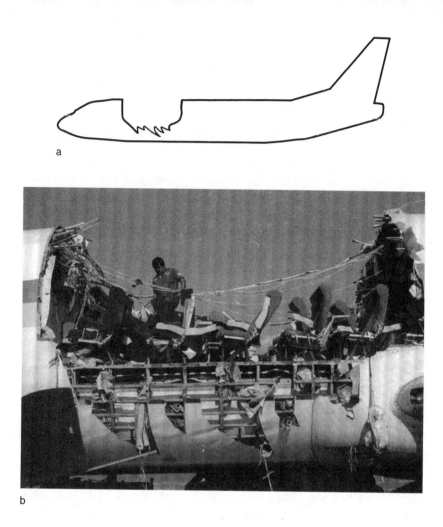

a

b

Figs. 4.1A and 4.1B. Damage to Aloha Flight 243, a Boeing 737. *Photo:* NTSB

from Flight 243 were removed and subjected to the standard test procedures for new equipment. No defects were found.

The fuselage with normal operating pressure will burst like a balloon if the fuselage's condition degrades due to metal fatigue. Metal fatigue can be briefly illustrated by bending a paper clip back and forth. Each back and forth bending of the clip represents one load cycle, comparable to a takeoff and landing pressurization load cycle. Eventually, the paper clip and fuselage will fracture if enough load cycles are applied. The study of metal fatigue is essential for aircraft design and is covered in more detail in Chapter 6.

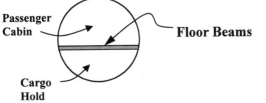

Passenger Cabin

Floor Beams

Cargo Hold

Fig. 4.2. Floor beams

Aloha Airlines flies short hops between the Hawaiian Islands. The average flight for routes serving the six main Hawaiian Islands is only 20 minutes. Planes on these routes will accumulate more takeoff and landing cycles than comparable planes for other airlines. (The accident plane had already experienced its ninth takeoff of the day, and it was only 1:25 p.m.) Manufactured in 1969, this particular 737 had 89,680 takeoff and landing cycles, at the time the second highest in the entire 737 world fleet.

To understand what went wrong with Aloha Flight 243, it is helpful to better understand pressure.

Pressure Revisited

Returning to our dictionary definition of pressure—exerting a force on a surface—the physicist requires the precision of an exact value for force and area.

Pressing on a 1-inch × 1-inch cross section (Block A, Figure 4.3) with 20 pounds of force results in a pressure underneath the block of 20 pounds divided by 1 inch^2 or 20 pounds per square inch. This value of pressure is usually referred to as 20 psi.

Pressing with the same force on a block 0.5 inches × 0.5 inches = 0.25 square inches (Block B) results in a pressure four times higher or 20 divided by 0.25 = 80 psi. When spiked heels first appeared on women's shoes, hotels could not understand why their carpeting was suddenly wearing out faster, until someone remembered the definition of pressure.

$$\text{Pressure} = \frac{\text{force}}{\text{area}}$$

Fluid pressure is more relevant to fuselage design. Fluids come in two types: liquids and gases. A liquid has a specific volume, whereas a gas will spread itself into whatever volume it is placed. Fluid pressure, in units of pounds per square inch, equals the weight of a column of fluid on a 1-square-inch surface. If a column of water 1 inch × 1 inch and 23 feet high weighs 10 pounds, the pressure on the bottom of the column is 10 psi. As the height of water increases, the pressure increases.

Using 62.4 pounds per cubic foot as the weight density of water, the weight on the bottom of any 1 in² column of water is easily calculated. Every 2.3 feet of water height results in a pressure of about 1 psi (Figure 4.4).

Water, or any liquid, is essentially incompressible. A unit volume experiences very little change in density (or volume) even at extremely high pressures, because the atoms of a liquid are virtually in contact. Gases, on the other hand, consist of widely spaced molecules (or, occasionally, widely spaced individual atoms). This spacing readily explains many of the unique properties of gases, including extremely low density, small frictional forces, and tremendous compressibility. Explosive decompression is directly related to the compressibility of air.

The difference in molecular spacing for a liquid and a gas can be estimated for the water molecule. Steam, at its boiling point, has a density about 1,665 times less than water at the same pressure and temperature. Even though the steam molecules are in constant random motion,[2] their

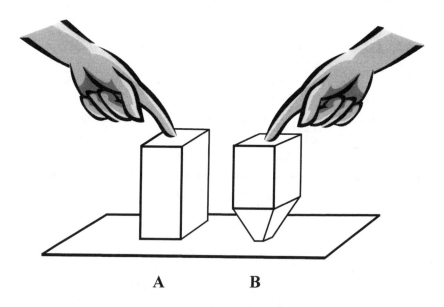

A   **B**

Fig. 4.3. Pressure is a force applied to an area. The same force applied to a smaller area results in more pressure.

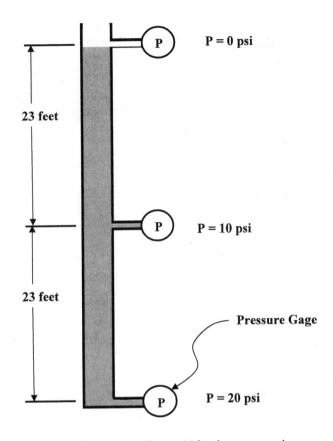

Fig. 4.4. The pressure along a 46-foot-long water column

average spacing can be approximated by assuming that one molecule exists at each corner of a cube (Figure 4.5).

To match the experimental evidence that steam is 1,665 times less dense than liquid water, the volume of the steam cube must be 1,665 times larger than the liquid water cube. The volume of the steam cube, L × L × L, must equal 1,665, or L = 11.85 (the approximate cube root of 1,665). Therefore, the steam water molecules occur (on average) with a spacing between them of 11.85 molecules.[3] This compares to one molecular spacing for liquid water.

Air Pressure

Air pressure also equals the weight of the column of air above the surface. However, unlike for water, the weight of air does not equal the volume of the column times some constant density. A unit volume of air expands as the air

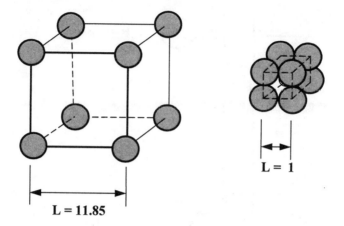

L = 11.85

L = 1

Fig. 4.5. The water molecules in steam are separated by approximately 11.85 molecular spacings. This compares to liquid water molecules separated by one molecular spacing.

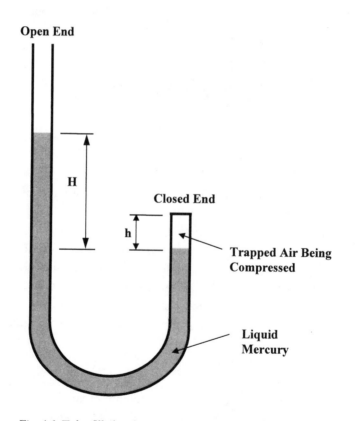

Open End

H

Closed End

h

Trapped Air Being Compressed

Liquid Mercury

Fig. 4.6. Tube filled with mercury; increase H, and h decreases.

molecules move farther apart with the decreased pressure at higher elevations. The result is continuously decreasing density with increasing elevation.

Boyle, of Boyle's Law fame, described the relationship between pressure, volume, and density with a simple experiment in the seventeenth century. Mercury is poured into a U-shaped tube with one end open and one end closed (Figure 4.6). Doubling the pressure on the trapped air (by increasing the height of the mercury column) halves the volume in the closed end. Triple the pressure, and the volume reduces to a third. Thus, the product of pressure and volume is a constant (PV = constant).

At sea level, atmospheric pressure is 14.69 psi and the density of air is 0.0762 lb/ft³ (about 800 times less than the density of water). If pressure at 5,000 feet is 12.23 psi, then Boyle's Law predicts the density at 5,000 feet to be (12.23/14.69) × 0.0762 = 0.0634 lb/ft³. Figure 4.7, accounting for the reduced density of air at lower pressures, plots the variation of air pressure with altitude.

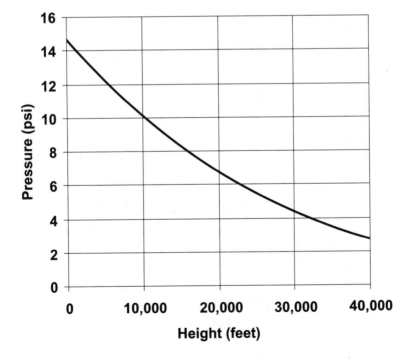

Fig. 4.7. Air pressure versus altitude

BOYLE'S LAW GIVES RELATIONSHIP BETWEEN PRESSURE AND VOLUME

Another way of stating Boyle's Law is

$$\text{initial pressure} \times \text{initial volume} = \text{final pressure} \times \text{final volume}.$$

Boyle's Law can be used to calculate the new pressure in the closed end of a tube, such as the one shown in Figure 4.6, when the volume changes due to changing pressure. If the initial volume and pressure of air in the closed end are 10 in³ and 14.7 psi and the pressure increases, thereby decreasing the trapped volume to 5 in³, the new pressure in the closed end is

$$\text{final pressure} = \frac{\text{initial pressure} \times \text{initial volume}}{\text{final pressure}}$$
$$= \frac{14.7\ (10)}{5} = 29.4\ \text{psi}.$$

If the volume of a gas doubles, its density decreases by a factor of two. Therefore, volume is inversely related to density. Boyle's Law can be rewritten in terms of density by replacing initial volume and final volume with 1/(initial density) and 1/(final density):

$$\frac{\text{initial pressure}}{\text{initial density}} = \frac{\text{final pressure}}{\text{final density}}.$$

To further illustrate the incompressibility of water versus the compressibility of air, repeat Boyle's experiment with water instead of air trapped in the closed end. The height h of the trapped water will remain virtually constant for all levels H of liquid mercury.

The air pressure at any altitude could be calculated by evaluating the weight of the column of air, similar to what was done with water, if the weight is adjusted for changing density. Assume that the density is constant along some length of column, perhaps 100 feet. The weight of the 100-foot column divided by the area of the column equals the reduction in pressure at a 100-foot elevation, Boyle's Law is used to calculate a new density at this lower pressure. By repeating the process, the pressure at any altitude can be found.

One would think that the human body is normally under a lot of pressure from the weight of the atmosphere. However, the pressure completely

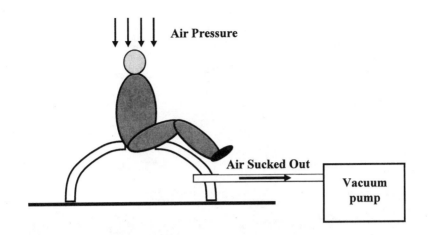

Fig. 4.8. Person sitting on an inverted bowl with a 12″ × 12″ hole. When the air is sucked out of the bowl, a force of over one ton, due to air pressure, presses the person into the bowl.

surrounds us: top, bottom, inside, and out. The net effect is negligible. To fully feel the effect of pressure, the following thought experiment could be performed.

Sit a person on a bowl with a 12 inch × 12 inch hole. Connect the bowl to a vacuum pump and evacuate all the air. The full weight of the atmosphere (a force of 14.7 psi × 12 inch × 12 inch)—equal to over a ton of force—is trying to push the person into the bowl (Figure 4.8). Unfortunately, a similar situation did happen. An airplane engine burst and knocked out a window, and as related in the following section, the person sitting next to the window, despite wearing his seat belt, was forced out of the plane by the pressure inside the cabin.

Pilot Boredom: A Source of Explosive Decompression?

On November 3, 1973, National Airlines Flight 27 was about two hours into an otherwise routine flight when the pilots of the DC-10 wondered out loud where the auto-throttle control signal came from. After speculating about the effects of interrupting certain electrical circuits on the autopilot system, the flight engineer pulled the circuit breakers to observe the results. The circuit breakers were subsequently reset. About 36 seconds after the first circuit breaker had been thrown, everyone on board was startled by a frightening explosion on the right side of the plane. In the next

Fig. 4.9. Engines on a DC-10. *Photo:* Joop Stroes

instant, the whole plane briefly received a severe buffeting and shaking. A few seconds later, there was a series of rapid, loud bangs against the right side of the fuselage.

In the cockpit, the fire warning light came on for the no. 3 engine, so the flight engineer discharged two fire extinguisher bottles in the engine. Warning lights indicated numerous other systems were out: the no. 1 and no. 3 AC generators, the AC bus (the power grid), and the DC bus. Also, low oil and hydraulic pressure were indicated in the no. 1 engine. The captain switched on the emergency power, which restored power to the no. 1 engine systems. (Figure 4.9 shows the engines on a DC-10.)

The first explosion resulted when the no. 3 engine burst. Thirty-two of the 36 fan blades exited the engine turning at 3,600 rpm (revolutions per minute). The flying blade fragments damaged a window, causing both outer and inner window panels to blow out explosively (Figures 4.10A and 4.10B). The enormous rush of air through the missing window caused the plane to buffet and shake. The passenger sitting at that window was slowly sucked out, head and shoulders first, despite having his seat belt on. It was later determined that the belt had eight inches of slack.

Figs. 4.10A and 4.10B. Damage to engine no. 3 and fuselage of Flight 27. Only six fan blades are seen attached. *Photos:* NTSB

No. 3 engine debris damaged this window.

Airframe damage from no. 3 engine debris

The window opening was 16⅛ inch × 10⅝ inch with rounded corners of 4½ inch for a total area of about 122 in². The air pressure outside at 39,000 feet is 2.85 psi. The cabin pressure is maintained about 8.1 psi higher. A crude estimate of the force trying to push the passenger out the window is 8.1 psi × 122 in², or nearly 1,000 pounds. Efforts to pull the passenger back into the plane by another passenger were unsuccessful. We have to consider the large forces involved, the difficulty of grip, and the frantic state of the person attempting the rescue.

Calculations estimated that the aircraft decompressed from 10.95 psi internal cabin pressure to 3.6 psi in 26 seconds. Decompression tests have not been done on a plane as large as a DC-10. However, tests on smaller aircraft at various cabin pressures and opening areas showed a danger of ejection at pressure differences as low as 5.2 psi.

Examination after landing revealed serious damage in many locations. In addition to losing the window next to the no. 3 engine, there were numerous other tears and punctures in the fuselage. The damaged areas ranged in size from 170 to 540 square inches (equivalent to squares 13″ × 13″ to 23″ × 23″). Additionally, the no. 1 engine oil tank and electrical wiring were damaged. The cables for the "up" control right-side elevator and "nose left" rudder trim were severed. Minor damage also occurred in the tail-mounted no. 2 engine.

The NTSB determined that the no. 3 engine accelerated to abnormally high speeds, resulting in vibrations that caused a fan-tip rub condition. The rocking of the blades in their slots allowed the blades to move forward until they sheared off the blade retainer. The blades then broke completely free, becoming unguided missiles. In fairness to the pilots, it was never clearly demonstrated that their actions resulted in the engine exceeding its operating limits. The NTSB did fault the flight crew for conducting an untested failure analysis of the autopilot system. Such an experiment, without the benefit of training and specific guidelines, should not occur during normal aircraft operations.

Numerous theories were developed during the investigation, but the precise reason for acceleration and destructive vibration was never determined. However, this did not prevent corrective measures being taken. Changes were made to hold the blades more securely in place. The fan blade retention devices were redesigned to increase the retaining forces from 18,000 to 60,000 pounds. The fan blade tip clearance was also increased to minimize the possibility of contact and any associated de-

Fig. 4.11. BAC-111 cockpit; left window blew out on Flight 5390.
Photo: Andy Fairminer

structive vibration. The improvements were incorporated into all existing engines of the same design. (Burst engines are discussed in more detail in Chapter 5.)

On June 10, 1990, British Airways Flight 5390 experienced an even more bizarre demonstration of internal cabin pressure. While climbing through 17,300 feet on departure from Birmingham, England, the left windscreen blew out of the British Aircraft Corporation BAC-111 (Figure 4.11). After hearing an explosion, some of the 81 passengers noticed that the sudden decompression had sucked the pilot out of his seat belt and halfway out a gaping hole. (After takeoff, pilots typically remove their shoulder harness and loosen their seat belt for comfort.)

The captain caught his feet in the control column. By pulling the column forward, the plane went into a steep dive. The copilot, after reaching for his oxygen mask, fought tornado winds and flying debris to regain control of the plane while two flight attendants fought for control of the dangling captain. Eventually, the flight attendants wrestled the captain out of the control column (but not into the plane) and managed to hang on to him until the plane landed safely 20 minutes later. Meanwhile, the copilot con-

Fig. 4.12. Longitudinal tear in pipe filled with water and frozen

tinued an emergency dive—the procedure for decompression—to bring the plane down to a breathable altitude.

The captain remembers having difficulty breathing in the 350 mph windblast and turning around; he remembers little else. He suffered relatively minor injuries: a broken elbow, wrist, and thumb, and frostbite on one hand. Later, it was learned that the windscreen had been replaced the night before with 84 (of the total 90) undersized bolts.

Throughout the history of pressurized flight, there have been a few passengers sucked out of planes. Except for the extraordinary circumstances described here, double-pane and plug doors have solved this problem in the modern era. However, smaller planes usually do not have these design features. In the early 1960s, the FAA studied potential ejections from small planes. They concluded that a 180-pound test dummy without a seat belt could be ejected from a 700-ft^3 cabin if sitting within 14 inches of a 900-in^2 window (or within 10 inches of a 300-in^2 window) in a plane pressurized to 8.55 psi. The significantly larger volume of compressed air (and corresponding greater air blast) inside the DC-10 and BAC-111 planes meant that ejection was considerably easier in the incidents described above.

Design for Pressure

If a pipe is filled with water and frozen, the frozen water will expand and burst the pipe with a longitudinal tear, as shown in Figure 4.12. A similar pipe will rupture with an identical pattern if overpressurized above some critical pressure value. An overpressurized cylindrical balloon will burst with the same longitudinal tear. If the experimenters are not patient enough to wait for the excess pressure to occur in the balloon, they can speed up the process with a pin. In this case, a longitudinal tear will run in both directions from the pinprick until the tear reaches the ends of the balloon.

The different geometry at the ends of the balloon will disrupt the longitudinal tear pattern. If a longer cylindrical balloon is used, the tear will increase in length, running along the length of the cylinder.

The most likely orientation for a fatigue crack due to pressure cycles in a fuselage is the same as in the balloon and pipe. The tear is perpendicular to the dominant stress pattern, known as "hoop stress." Coincidentally, the hoop stress for a can of beer is about the same as in the fuselage of a Boeing 747 or any other large commercial jet plane.

Stress is pressure inside the metal trying to rip it apart. Stress has the same units as pressure: force per unit area or psi. Fundamental to the design of all structures is a test of the material's strength. A test sample is loaded until it breaks. For example, if a rod with a cross-sectional area of 0.5 in² breaks when a 32,500-pound weight is applied, the tensile strength of the metal is said to be 32,500/0.5 = 65,000 psi, a typical value for an aerospace aluminum alloy. Metals are incredibly strong. For the same aerospace aluminum alloy, a rod with a 1-in² cross-sectional area would be able to support the weight of 20 cars if loaded in pure tension, as shown in Figure 4.13 (assuming each car weighs 3,000 pounds).[4] A similar rod of high-strength steel would support 80 cars or more. (High-strength, heat-treated steels are used in a few highly stressed aircraft components; one example is landing gear.)

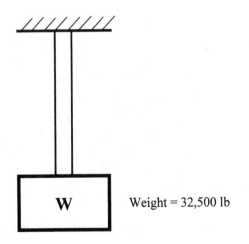

Weight = 32,500 lb

Fig. 4.13. Example of a tensile test. Secure a rod on top, and hang a weight from the bottom until the rod breaks.

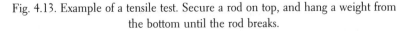

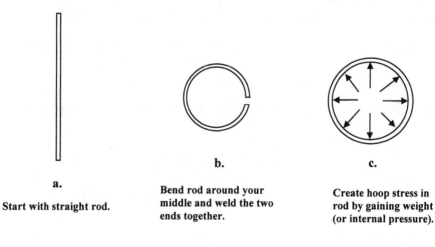

b.

a.

Start with straight rod.

Bend rod around your
middle and weld the two
ends together.

c.

Create hoop stress in
rod by gaining weight
(or internal pressure).

Fig. 4.14. Example of a hoop stress

Instead of hanging weights, it is more common to pull on the
metal to be tested with instrumented hydraulic pullers known as a tensile test machine. With this machine, the force, deflection, and stress
can be carefully measured as the breaking load is gradually applied. If
metal fatigue testing is being done, the number of cycles to failure is
recorded.

If a rod is bent into a circular hoop, wound tightly around your waist,
welded into a continuous circle, and stretched by the expansion of your
middle (let's say you gained 100 lb), a hoop stress is created in the circular rod (Figure 4.14). Stress results when the metal rod is loaded and
stretched. Now consider two rods of equal length. The first rod is stretched
by amount L. The second rod is bent into a circle (with ends welded to
create a continuous hoop) and loaded with an internal pressure. The diameter of the hoop expands until the circumference grows by amount L.
Both the straight rod and the rod bent into a hoop will have the same stress.[5]
Stress, whether due to a pressure hoop stress or a tensile pull, has the same
potential to rupture the material.

The stressed portion of a pressurized cylinder, far removed from any reinforcing ribs, end caps, or other disrupting geometry, will act like a series
of individual hoops under stress. The pressurized cylinder will break when
a critical hoop stress is reached. The critical value is an intrinsic property
of the material whether it is a latex balloon, a copper pipe, or an aluminum
alloy fuselage.

The Aloha Flight 243 Hoop Stress

The dominant stress in the fuselage of Flight 243 was the hoop stress required to restrain the internal pressure. The hoop stress went to zero when the fuselage burst open. The fuselage is also designed for a variety of maneuvering loads, but these only result in stresses when the particular maneuver is performed. The NTSB did fault the pilots for not following the appropriate emergency checklist which states, in part, that "if structural integrity is in doubt, airspeed should be limited as much as possible and high maneuvering loads should be avoided."[6] That is to say, the pilots, who performed admirably under extreme circumstances, should have been gentler with the plane!

The equation for stress in a pressurized cylinder is easily derived from fundamental principles, primarily static equilibrium. Static equilibrium requires that all forces acting on a body must sum to zero. (If the sum of the forces does not equal zero, the result is acceleration, as described by Newton's Second Law, force = mass × acceleration.) For the case of a pressurized cylinder, the pressure forces trying to split the cylinder must be balanced and in equilibrium with the internal stresses in the pipe wall.

The axial stress in a pipe is particularly easy to derive with static equilibrium. The axial stress is perpendicular to the hoop stress (Figure 4.15).

Consider a pressurized section of pipe with welded caps on each end. If the weld on one cap is completely severed, and a finger holds the cap in place, as shown in Figure 4.16, the finger must supply a force that reacts against the internal pressure, which tries to blow the cap off. Using the jargon of the physicist or engineer, the cap must be in static equilibrium.

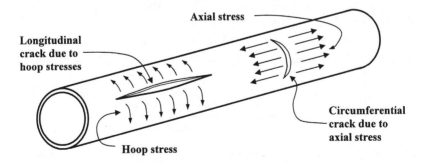

Fig. 4.15. Orientation of hoop stresses and axial stress and resulting fracture orientation in a pressurized cylinder

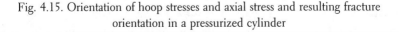

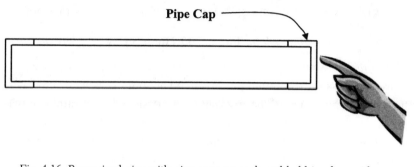

Pipe Cap

Fig. 4.16. Pressurized pipe with pipe caps on each end held in place with an
external force

The sum of the forces must equal zero. The pressure forces pushing to the
right must be in static equilibrium with the force supplied by the finger
pushing to the left:

Force in finger = Pipe cross-sectional area × Internal pressure.

For those who like to check the arithmetic with specific values, assume the
pipe has an inside diameter of 1 inch and an internal pressure of 100 psi.
The inside cross-sectional area of the inside diameter will be 0.785 in².
The required external force to hold the caps in place equals 0.785 in² ×
100 psi = 78.5 pounds.

If the pipe cap's weld is intact, it will supply the force previously provided by the finger. The internal pressure or stress in the pipe weld equals
the force divided by the cross-sectional area of the weld. For a pipe wall
thickness of 0.1 inches, the cross-sectional area of the pipe wall is 0.345
in². The stress in the weld will equal 78.5 pounds/0.345 in² = 227.5 psi.
This is an axial stress perpendicular to the hoop stress.

If the above analysis is repeated with thickness and diameter used as
variables instead of specific values, the equation for axial stress could be
derived as

$$\text{axial stress} = \frac{\text{pressure} \times \text{diameter}}{4 \times \text{thickness}}.$$

Pressure acts perpendicular to all surfaces, as shown in Figure 4.17a. However, the vertical component of pressure force on the top half of the cylinder will be cancelled out by the pressure force on the bottom half. The net
effect is horizontal force acting to the left on the projected area, as shown

in Figure 4.17b. The internal hoop stresses must react against the pressure force.

The hoop stress that results from internal pressure trying to expand the diameter of the pipe can also be derived from static equilibrium. If the internal pressure force (= internal pressure × projected area) trying to rip the pipe apart is equated to the internal stresses trying to hold it together, an equation for hoop stress can be derived.

$$\text{hoop stress} = \frac{\text{pressure} \times \text{diameter}}{2 \times \text{thickness}}.$$

This simple equation illustrates certain commonsense notions about hoop stress in a pressurized cylinder. Hoop stress will increase with

- increased pressure;
- increased diameter; (With increased diameter, the pressure acts over a larger area, resulting in a bigger pressure force trying to rip the cylinder apart.)
- decreased pipe wall or fuselage skin thickness. (For example, if corrosion thinned the pipe wall or fuselage skin, the safe pressure rating would have to be reduced.)

Also, the hoop stress is twice the axial stress, which explains why pipes, balloons, and fuselages will rupture, as shown in Figure 4.15.[7]

Modern jetliners are designed for an internal pressure of 7.5 to 9.0 psi. The stresses for common aircraft (and a can of pop) appear in Table 4.1.

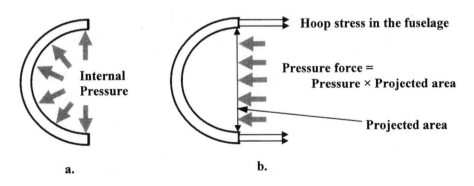

Internal Pressure

Hoop stress in the fuselage

**Pressure force =
Pressure × Projected area**

Projected area

a.

b.

Fig. 4.17. Hoop stresses are in equilibrium with the pressure forces trying to separate the cylinder.

Table 4.1. Examples of Fuselage Hoop Stress

	Internal Pressure (psi)	Thickness (in)	Radius (in)	Hoop Stress (psi)
737	7.5	0.036	74.0	15,400
777	8.6	0.070	122.0	15,000
DC-10	8.7	0.071	118.5	14,500
747	9.0	0.063	128.0	18,300
Pop can*	50.0	0.004	1.3	16,250

*Pressures vary depending on amount of CO_2 injected and on temperature. These values were experimentally determined by engineering students.

These fuselage hoop stress values should be compared with the tensile strength of 2024-T3 aluminum sheet (a common fuselage alloy), which is 65,000 psi. Of course, fatigue and fracture add additional, more complicated criteria for design stresses. (Metal fatigue occurs at stresses lower than the single-load tensile strength value.)

Pop cans seem to be designed to the same stress level as a 747. Presumably, a pop can is not as critical an application (and doesn't experience cyclic loading) and could operate at a higher stress level. There are other practical considerations. If the pop can's wall thickness was reduced any more, the aluminum sheet would be difficult to handle and subject to manufacturing defects. The possibility of manufacturing defects is also a very important issue for jetliners, because defects can initiate fatigue cracks.

The History of Pressurized Flight

The benefits of flying at 20,000 feet or higher were recognized in the early 1920s. A bomber flying at higher altitudes could elude enemy aircraft. Commercial aircraft could fly above turbulent and dangerous weather systems, increasing passenger safety and comfort. Although the military was experimenting as early as the 1920s with high-altitude flying, the equipment was very primitive. The Lockheed XC-35, first delivered to the U.S. Army Air Corps in 1937, is considered the first successful pressurized cabin. This plane, capable of flying at 25,000 feet, was intended to be an experimental flying laboratory testing high-altitude flying procedures and equipment. Of course, people cannot breathe properly at this altitude (hypoxia is discussed later in this chapter), so new equipment was required to maintain air pressure in the fuselage at a breathable level.

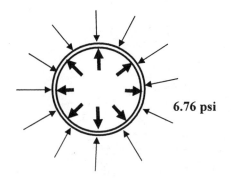

6.76 psi

Fig. 4.18. Boeing's Stratoliner was designed for 10.92 psi internal pressure and 6.76 psi external pressure for a net differential of 10.92 − 6.76 = 4.16 psi.

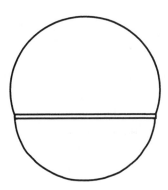

Fig. 4.19. Many modern jet aircraft often use a double-lobe design with floor beams in between. The equation for hoop stress will work in each lobe.

The Boeing B-307 Stratoliner, introduced in 1938, was the first commercial aircraft with a pressurized cabin. The Stratoliner was designed to carry 33 passengers and cruise at 20,000 feet. The cabin pressure of the Stratoliner was maintained at a pressure equivalent to 8,000 feet altitude (10.92 psia).[8] When flying at 20,000 feet, the pressure inside the fuselage (10.92 psia) was greater than the outside atmospheric pressure (6.76 psia) (Figure 4.18). The Stratoliner fuselage was effectively designed as if it was a pressure vessel with an internal pressure of 10.92 − 6.76 = 4.16 psig.

The Stratoliner was the first airplane to have a flight engineer as a member of the crew. The engineer was responsible for maintaining pressurization, power settings, and other subsystems, leaving the pilot free to concentrate on flying the aircraft.

Before pressurized cabins, the fuselage cross section was oblong, or even rectangular, in shape. With pressurized cabins, the cross section became a circle, the most efficient shape for internal pressure (Figure 4.19). This

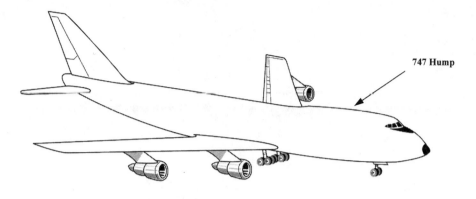

Fig. 4.20. Hump on 747 has flat sections that increase the hoop stress.
Graphic: NASA

simple rule of nature concerning internal pressure and circles was violated with the hump on the 747 (Figure 4.20). Boeing achieved the distinctive hump on top of the 747 using flat panel sections. However, the stresses in the flat panels were higher than in circular ones, and a premature fatigue problem occurred. The result was costly repairs, but no disasters.

The introduction of the de Havilland Comet was a significant event in the history of aviation. It was the first commercial jet aircraft, and it was designed to fly at 40,000 feet. The ability to fly at this altitude resulted in the highest fuselage internal pressure ever used. This new, higher pressure, coupled with a lack of design knowledge about metal fatigue, caused two planes to be destroyed in mid-air with explosive decompression. The design changes since the Comet disasters explain why the problem has been significantly solved, and why Aloha Flight 243 managed to hang together and safely land without being completely destroyed. Additional discussion, provided later, is required to explain why explosive decompression caused as much damage as it did to Flight 243.

The de Havilland Comet and Explosive Decompression

Sir Geoffrey de Havilland (1882–1965) was one of Britain's earliest and most innovative aviation pioneers. He had constructed and flown his own aircraft by 1910. In 1920, he established the de Havilland Aircraft Company.

One of de Havilland's more famous innovative planes was the WWII Mosquito. In 1938, de Havilland proposed an all-wood bomber or reconnaissance aircraft that would be so fast it could be unarmed. The

British Air Ministry was generally hostile to the plan and initially turned it down.

With persistence the plane was built. De Havilland correctly surmised that aluminum and skilled metal workers would be very scarce commodities during the war. There would be, however, many experienced carpenters and piano, cabinet, and furniture makers available. (Howard Hughes's famous Spruce Goose was also made out of wood during WWII for the same reasons.)

When the prototype Mosquito achieved a top speed of 392 mph (20 mph faster than the Spitfire), the Air Ministry was finally won over. Hermann Goering, commander of the German Luftwaffe, summed it up best in January 1943 when his public address was interrupted by a low-level attack by a squadron of Mosquitoes. He said, "It makes me furious when I see the Mosquito, I turn green and yellow with envy! The British, who can afford aluminum better than we can, knock together a beautiful wooden aircraft that every piano factory over there is building."[9]

The Mosquito was the fastest aircraft in the Royal Air Force's Bomber Command until May 1951 and flew many famous and unique missions in WWII.

De Havilland also took the lead in jet construction, building the engines for the Meteor, England's first operational jet aircraft (first flown in March 1943). The Vampire, England's second jet aircraft, was completely designed by de Havilland in 1942.

Since WWII meant the suspension of work on commercial aircraft, the airlines of the early 1950's were little changed from the late 1930's. For example, the DC-7, Douglas's largest and last piston engine airplane, was a stretched version of the DC-6, a fifteen-year-old design. The DC-7s 3,400 horsepower engines had 72 noisy, pulsating pistons (4 engines with 18 pistons each) that produced unpleasant vibrations in the passenger compartment.

Because of their experience with jet engines and jet-powered planes during WWII (the only company experienced in both), de Havilland was confident they could design a passenger jet plane that would far surpass existing technology. They began in September of 1946, and design, development, and flight tests lasted until the first paying customers took off on May 2, 1952.

The Comet 1, flying at significantly higher altitude and speed than existing piston passenger planes, was an instant sensation. The jet engines were also quieter and smoother than piston engines. Passenger comfort was further increased by flying above the weather. Orders poured in for the

Comet and the new, faster, and higher-capacity Comet 2. In September of 1952, de Havilland announced the development of the Comet 3 for transatlantic service. By May of 1953, de Havilland had orders for 50 Comets and was negotiating 100 more. They had completely overtaken existing technology and were on their way to dominating the commercial aviation market.

The cabin of a Comet maintained a pressure equivalent to 8,000 feet elevation when the aircraft was flying at 40,000 feet. The aircraft had an internal pressure 8.25 psi greater than the external pressure—about 50% more than any previous design. Although the Comet was extensively tested structurally, the fatigue testing that was performed turned out to be flawed. Metal fatigue was not well understood at that time.

A series of crashes soon followed. On March 3, 1953, a Comet crashed during takeoff in Pakistan. All passengers and crew died in the crash and ensuing fire. The Indian investigation concluded that the accident resulted from excessive nose-up attitude during takeoff causing a partly stalled condition and excess drag.

On May 2, 1953, a Comet took off from Calcutta and disappeared suddenly in a thunderstorm. The disappearance was rapid, occurring during communications with the Delhi airport. A public inquiry conducted by an Indian High Court judge concluded that the aircraft encountered severe gusts in a thunderstorm, leading to loss of control and in-flight breakup during an attempted recovery. The wreckage was flown to Britain's Royal Aircraft Establishment for further investigation; they confirmed the Indian Court's findings.

On January 10, 1954, a Comet took off from Rome. Forty-four miles southeast of the island of Elba, after climbing through 26,000 feet, the pilot got cut off in mid-sentence shortly before the plane disappeared over the Mediterranean. The mid-sentence interruption is an important clue to the suddenness of the disaster. Based on contemporary understanding of fatigue and fracture, it is believed that a fatigue crack slowly grew to a critical size. Once the critical size is reached, the entire fuselage rapidly tears in a few milliseconds with an explosive decompression very similar to what happens when a balloon is popped (see balloon experiments later in this chapter).

The Elba Comet had only 1,290 pressure cycles (takeoffs and landings). Metal fatigue was not even considered initially as a failure mode, because two sections of the fuselage had been pressure tested for 18,000 cycles during design and development.

After examining a number of possible explanations for the Elba accident, a large number of modifications were completed. In-flight fire was considered to be the most likely cause, and many possible sources of ignition were investigated. Armor plating was fitted around the engine to guard against a lost turbine blade penetrating the fuselage. Eleven weeks after the accident, with no explanation emerging from the investigation and still no sign of metal fatigue in the ongoing fatigue tests, it was decided to resume flying the Comets.

British Overseas Airways Corporation (BOAC)[10] resumed Comet service on March 23, 1954. Fewer than three weeks later, a nearly identical breakup occurred over the Mediterranean, the fourth fatal accident in under 13 months. This accident occurred after only 900 pressure cycles. Four days later, the Minister of Transport and Civil Aviation announced to the House of Commons the withdrawal of the Certificate of Airworthiness for all Comets. A few days later, Prime Minister Winston Churchill declared that "the cost of solving the Comet mystery must be reckoned neither in money nor manpower." The investigation became more intense.

BOAC suspended all Comet flights pending a major review with de Havilland and the British Ministry of Transport and Civil Aviation. In view of what was at stake, three Royal Navy salvage ships were called in to assist with the recovery effort. Because of the trouble explaining the crashes, it was decided on April 18, 1954, to destructively fatigue test one of the existing Comets, a fourth-production Comet with 1,230 flight cycles. The test was done in a water tank for safety. A complex loading cycle to simulate a typical flight involving transfer of loads from the landing gear to the wings, pressurization, depressurization, and finally transfer back to the landing gear was developed to approximate a three-hour flight in 10 minutes. The test rig (Figure 4.21) operated continuously.

As more salvaged pieces were put together from the earlier Elba crash (the second Mediterranean crash was in much deeper water), the sequence

Fig. 4.21. Aerial view of test rig used for cyclic loading of Comet fuselage

Fig. 4.22. Fatigue crack in fourth production Comet tested in a water tank

of breakup became increasingly clear. Pieces of the cabin were found embedded in the tail section. The cabin failed first and violently hurled internal cabin pieces against the tail.

At the end of June, the test fuselage failed in the water tank after 1,230 actual flight cycles and an additional 1,830 simulated cycles for a total of 3,060 cycles. An eight-foot-long fatigue crack was found originating from a rivet hole at the corner of the escape hatch cutout (Figure 4.22). Additional fatigue cracks were found initiating from rivet holes around the cutout for the antenna on top of the cabin.

After studying the potential trajectory path of the fuselage in the third Comet breakup (the January 1954 crash over the Mediterranean), a large section of the roof, including the cutout "window" for the antenna was finally recovered. At a rivet hole near the cutout, the distinctive markings of metal fatigue were found. At this point, the mystery was considered solved.

The fatigue crack shown in Figure 4.22 was associated with the stress concentrations of the window cutout. A stress of almost 45,700 psi was measured experimentally on the test fuselage in the water tank. This compares to a value of only 28,000 psi predicted by de Havilland designers. At the time, it was not common design practice to consider every detailed notch and potential stress riser in the fuselage. De Havilland did not evaluate the fatal peak stresses associated with a rivet hole near a window.

De Havilland had designed the Comet fuselage for 2.5 times the operating pressure of 8.25 psi and tested the fuselage to two times the operating pressure, or 2P. This pressure exceeded the current practice, as specified by the British Civil Airworthiness Requirements, which calls for a proof test of 1.33P and a design safety factor of two against rupture (i.e., rupture is expected at 2P). It was widely believed that metal fatigue would not occur if the fuselage passed a proof test of two times the operating pressure. Later, metal fatigue became more prominent in studies originating with the Royal Air Force. After the Comet crashes in Pakistan and India in early 1953, there had been concern about fatigue of the fuselage. So, in July 1953, de Havilland had initiated fatigue testing of a forward part of the cabin. By September, the test section had received 18,000 pressure cycles of 8.25 psi. The tests ended when a fatigue crack was initiated in a corner of the window. This compared favorably to the design life of 10 years and 10,000 cycles. Based on this testing, therefore, everyone was convinced—at least until the later tests were conducted in 1954—that metal fatigue was not the problem.

How did the Comet fuselage, tested to over 18,000 cycles, experience three failures with less than 3,000 cycles? There are several possible answers:

1. Metal fatigue by its very nature is highly statistical with much scatter in the data. A factor of nine to one is sometimes quoted. In other words, if one section is tested to 18,000 cycles, then it is expected that tests on many identical sections would result in one of them failing in as few $18,000/9 = 2,000$ cycles. The proper way to do this type of testing is to test many sections and do statistical analysis on the results. However, that approach is too expensive for a multimillion-dollar fuselage. The modern approach still tests only one or two planes, but hundreds of small-scale and component tests are also conducted. Added safety is ensured by performing structural inspections after only a small percent of the expected design life is used up.

2. The fuselage section tested to 18,000 cycles was first given a proof test to 16.5 psi. This one-time overload would "yield" the aluminum near the rivet hole, resulting in compressive residual stress. This compressive residual stress would superimpose and reduce any subsequent pressure stresses during the fatigue test. The one-time proof test of twice the operating pressure effectively invalidated any subsequent fatigue testing by making the test fuselage more fatigue resistant than the aircraft actually used in flight, which did not have this overpressurization.

By 1954, the public's confidence in the Comet was gone. Propeller planes took back commercial aviation until the Boeing 707 made its first commercial flight in October of 1958. Learning from the Comet's experience, the 707 had 15 months of structural testing including 50,000 pressure cycles on a section of fuselage inside a water tank. The pressurized fuselage was also guillotine tested, demonstrating damage tolerance to, for example, a thrown turbine blade ripping through the fuselage.

As a result of the Comet disasters, full-scale fatigue testing was initiated in all future aircraft. More emphasis was given to design for fatigue. This ultimately led to "fail-safe" and "damage-tolerance" design practices in which the fuselage is designed to be safe even if a 40-inch-long crack exists.

Exactly What Explodes During Explosive Decompression?
The Energy of Compressed Gas

An airplane fuselage filled with air and pressurized to 8.25 psi above normal atmospheric pressure, or 8.25 + 14.7 = 22.95 psia pressure relative to a vacuum, will explode if the pressure is rapidly released. The same fuselage pressurized with water will not explode. Pressurized gas is dangerous. Pressurized water is not. The compressed gas contains a great deal of stored energy whereas the compressed water contains very little energy. There are several ways to explain this. One way considers the amount of compression that occurs and evaluates how much energy is required from a piston pump to obtain this compression.

The volume of a Boeing 747, for example, is about 59,000 ft^3. Boyle's Law can be used to calculate the volume of air at normal atmospheric pressure that will have a pressure of 22.95 psia absolute when compressed to 59,000 ft^3.

$$\text{Final volume} = \frac{\text{initial pressure}}{\text{final pressure}} \text{ initial volume.}$$

$$\text{The required volume of air} = \frac{14.7 + 8.25}{14.7} \; 59,000 \text{ ft}^3 = 92,100 \text{ ft}^3.$$

The pressure testing of the 747 on the ground is equivalent to starting with a cylinder filled with air at atmospheric pressure and initial volume of 92,100 ft^3 and compressing it to 59,000 ft^3 at 22.95 psia (Figure 4.23). Imagine the amount of energy required to obtain this compression with a pis-

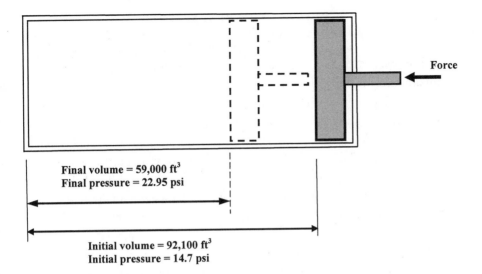

Final volume = 59,000 ft³
Final pressure = 22.95 psi

Initial volume = 92,100 ft³
Initial pressure = 14.7 psi

Fig. 4.23. The pressurization of a 747 is equivalent to compressing air.

ton pump that displaces 1 ft³ on every stroke. This pump would have to stroke $92,100 - 59,000 = 33,100$ times.

Water is usually described as incompressible, but in fact is slightly compressible. The same 747 filled with water at 22.95 psia results in 1.65 ft³ of compression. In other words, 1.65 ft³ of water will have to be added to keep the fuselage completely filled after pressurization. The same piston pump would have to stroke only 1.65 times to compress the water in the 747 fuselage and increase the atmospheric air by 8.25 psi. This is 20,000 times less than the air pump.

In both cases, the two pumps boost the pressure of the water and air from standard atmospheric pressure to a value that is 8.25 psi greater. Clearly, the forces involved are the same. If the two pumps receive the required energy by burning gasoline, more gasoline will be used by the pump that has to operate 20,000 times longer.

From the principle of conservation of energy, the work done by the pump does not disappear but becomes energy stored in the compressed fluid. The energy stored in the compressed air is 20,000 times greater than the energy stored in the compressed water.

This phenomenon explains why the pressure testing of the Comet was done with incompressible water instead of compressible air. It also explains why full-scale pressure testing initially done on all commercial planes after the Comet was done with water. Later, as they became more confident in their analysis of metal fatigue, they switched to testing with air. Boilers, nuclear reactors, and other pressure vessels are routinely tested with water and are rarely tested with air for similar safety reasons.

The difference between pressurized water and pressurized air is also demonstrated by bursting balloons. Several 4-inch-long, 0.75-inch-wide cylindrical balloons were filled with air until they expanded to roughly 12 inches long and 2.75 inches diameter. The balloons were burst with a pin. The experiment was repeated with balloons filled with water. The balloons filled with air demonstrated more internal energy than the water-filled balloons. The air-filled balloons demonstrated

- acoustic energy by making a loud "pop" versus a muffled splash;
- fracture energy by tearing and fragmenting more (typically three fragments for the air balloon versus two for the water balloon);
- kinetic energy by the pieces of balloon flying 90 to 120 inches from the burst site versus the water balloon collapsing in place. (If a metal container filled with pressurized gas were to fragment for whatever reason, dangerous missiles would result.)

There is also significant strain energy in the stretched latex. This strain energy actually drives the tearing. However, since the air and water balloons were expanded to roughly the same size, the strain energy is approximately the same in both cases. The only difference is the internal energy of the air and water.

The explosive decompression blast from the balloon can also be felt on skin. If a hand is placed slightly above the pinprick of a water balloon, the hand will not get wet. If an identical balloon is filled with water and then emptied through its nozzle, a slight water residue will remain inside the balloon. If this balloon is then filled with air and pricked, an air-water mist blast will be noticeable on a hand held near the pinprick.

The fracture pattern in the balloon is always the same. The tear runs longitudinally along the length of the cylinder (lined up with the maximum hoop stress) until it reaches the ends (Figure 4.24). The tear then runs along the sphere-cylinder seam, the new location of greatest stress.

The sphere and cylinder each have their own natural stress states when pressurized. Each will deform a different amount and develop different stresses with internal pressure. When the spherical end is connected to the cylinder, additional stresses are created as each shape tries to force the other

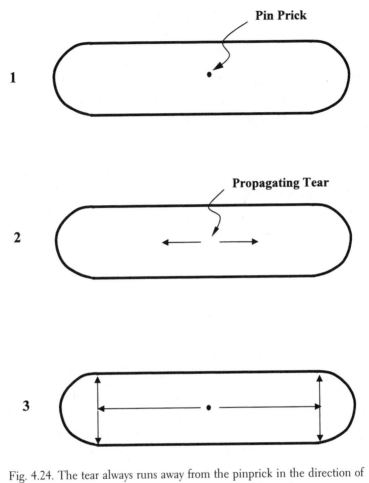

Fig. 4.24. The tear always runs away from the pinprick in the direction of maximum hoop stress until it reaches the ends.

shape to its deformed state. This results in the largest stress occurring where the two different geometries are connected, exactly where a balloon tears at the two ends. The end tears in the air-filled balloon always ran a full 360 degrees, tearing off the two ends and blasting them 90 to 120 inches. The water-filled balloon would try to tear in the same pattern, but the two ends never completely tore off.

The longitudinal tear pattern can be altered if an external force creates stresses greater than the hoop stress. Hold the nozzle end of a balloon and pull on the opposite end trying to elongate the cylinder. If the ends of the balloon are pulled hard enough, the axial stress will become larger than the hoop stress and the fracture will change from a longitudinal to a circumferential orientation. If the ends are twisted, a spiral tear will form. In a fuselage, these additional loads would be analogous to maneuvering and wing loads. The additional stresses from any additional loading must be added to the pressure stresses.

The previous discussion compares the difference in stored energy between compressed gas and compressed water. The actual amount of energy stored can be calculated by using the physicist's definition of work.

A Physicist's Definition of Work and Energy

Work refers to a force acting over a distance. The physicist defines work as force × distance.

For example, the work to lift a 10-pound weight 2 feet is

$$\text{work} = 10 \text{ lb} \times 2 \text{ ft} = 20 \text{ ft lb of work.}$$

To lift the same block 4 feet requires twice as much work.

Energy is the ability to do work. From the conservation of energy, work and energy are equivalent. One hundred ft lb of energy must be expended (or spent) to accomplish 100 ft lb of work. It takes energy to lift the weight. (The energy can come from a variety of sources.) The weight contains more energy the higher it is lifted. This energy is called gravitational potential energy and equals weight × height, the work done to lift the weight. If the weight is allowed to fall under the acceleration of gravity, conservation of energy requires the potential energy of height be converted into the kinetic energy of motion. The weight falls faster and faster as it loses height. Its potential energy is converted to kinetic energy.

For the pressurized fuselage, it takes work to compress the gas. Conservation of energy requires that this work reappear as energy stored in the compressed air available for popping a balloon or exploding a fuselage. The energy contained in compressed gas can be determined from the definition of work.

As explained earlier, the pressure testing of a 747 is equivalent to using a piston to compress air in a cylinder from 92,000 cubic feet down to 59,000 cubic feet. After this compression, the air will now have a pressure 8.25 psi greater than atmospheric pressure, or 22.95 psia.

The energy in the compressed gas can be calculated from the work equation.

$$\text{work} = \text{force} \times \text{distance} = \frac{\text{force}}{\text{area}} \times \text{area} \times \text{distance}$$
$$= \text{pressure} \times \text{area} \times \text{distance.}$$

This calculation cannot be made directly because of the compressibility of air. The pressure continuously changes as the volume changes. An incremental calculation can be done in which the pressure is adjusted after each increment of compression.

$$\Delta\text{Work} = \text{pressure} \times \text{area} \times \Delta\text{distance}.$$

The Greek symbol "Δ" or delta, is used to indicate an increment of some quantity, in this case an increment of work and an increment of distance.

A few simple calculations show the energy contained in a pressurized 747 to be about one billion inch pounds. One billion inch pounds of energy are equivalent to raising an 800,000-pound 747 about 108 feet (1,296 inches) and dropping it on your foot. (The energy associated with a hard landing of an MD-10 is discussed in Chapter 2.) When the 747 is raised 1,296 inches, it will contain about one billion inch pounds of gravitational

---

ENERGY OF COMPRESSED GAS: THE CALCULATION

To pressure test a Boeing 747, 92,000 cubic feet of air must be compressed to 59,000 cubic feet. To calculate the energy in compressed gas, we must do an incremental calculation that adjusts for changing pressure.

If the piston compresses the volume from 92,000 ft³ to 90,000 ft³, Boyle's Law predicts the new pressure will be

$$\text{final pressure} = \frac{\text{initial volume}}{\text{final volume}} \times \text{initial pressure} = 15.026 \text{ psi.}$$

Substituting the *average* pressure, $(14.7 + 15.026)/2 = 14.863$ psi, and the increment of volume change, volume = 2,000 ft³ = area × Δdistance, into the work equation gives 14.863 lb/in² × 2,000 ft³ × 1,728 in³/ft³ = 51,366,000 in lb of work for this increment of compression.

The process can be repeated for the next increment of compression (from 90,000 ft³ to 88,000 ft³, in which case the numbers are average pressure = 15.1967 psi, and new increment of work = 52,519,968 in lb.

A spreadsheet can be set up to repeat this incremental calculation of work until the compression equals 59,000 ft³, in which case the total work is found to be 1,038,300,000 in lb.

potential energy. Right before it hits your foot, it will contain that same amount of energy as kinetic energy.

Pressurizing a fuselage with a dangerous amount of compressed gas energy is not an unsafe event. It happens tens of thousands of times each day. The energy release rate and how it is released determines a safe versus a dangerous situation. If the energy in the compressed gas is released gradually through a small, stable hole, a long blast of air blowing through the hole will safely dissipate the stored energy. If the energy released becomes focused on a propagating crack in the fuselage and the metal starts to tear and rupture faster than the compressed gas is being dissipated, a dangerous situation exists.

How does the metal start to tear? Metal fatigue will cause small cracks to grow with every pressure cycle. Initially this growth is very small: 10^{-6} to 10^{-5} inches per cycle are realistic values for fatigue crack growth. (Actual values will vary with many factors.) This process can be demonstrated by bending any piece of sheet metal. For example, take a sheet of aluminum about 1 inch wide, 5 inches long, and 0.007 inches thick cut from a mini-blind slat. Use scissors to cut a crack "starter" of a few hundredths of an inch long. The scissors' cut mimics a stress notch that might be from a rivet hole or a manufacturing nick or defect. After numerous bending cycles (that mimic pressurization of the fuselage), notice that a fatigue crack can be seen growing across the width of the sheet metal.

Fatigue crack growth is a very common phenomenon in aircraft and must be managed to maintain a safe structure. Inspection for metal fatigue is a fundamental part of normal maintenance of the aircraft. The inspection procedures are approved and certified by the FAA. The fatigue crack must be found before it reaches a critical stage resulting in, for example, the wing falling off, fuselage breakup, or failure of hundreds of other critical components.

As the fatigue crack grows, various scenarios can play out.

1. The crack is detected by normal maintenance and repaired before it affects the plane's ability to pressurize.
2. The crack grows slowly until a noticeable leak or slow decompression occurs. If the pressurization system cannot keep up with the leak, oxygen masks are automatically deployed, and the plane lands safely at the closest airport.

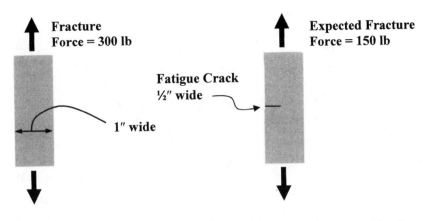

Fracture Force = 300 lb

Expected Fracture Force = 150 lb

Fatigue Crack ½" wide

1" wide

Fig. 4.25. Mini-blind slat breaks with 300 lb tensile force. Experimental hand force required for fracture of slat with pre-existing ½" fatigue crack is estimated to be a few pounds.

3. The crack reaches a critical length and begins to grow rapidly, even catastrophically until something interrupts and stops the crack growth. The crack may not stop growing until the plane fragments. As the crack grows, the plane simultaneously loses pressure as the air inside blasts out through the crack. The reduced pressure reduces the stresses driving the crack growth. At this point, there is a race to see if the plane safely decompresses or, as in the case of the Comet, bursts like a balloon.

For the balloon, the pinhole is immediately unstable. The balloon is not "damage tolerant" of a pinhole. In other words, the critical length of an unstable flaw in a balloon is the size of a pinpoint. For the fuselage, the fatigue crack is initially stable. The crack has to grow to an unstable length (longer than 40″) after many pressure cycles for the fuselage to potentially break up.

The concept of critical crack length can be illustrated with a simple experiment (Figure 4.25). A small length of a mini-blind slat was destructively tested until it broke with a tensile force of 300 pounds. Starting with a full-width blind, cut the fatigue starter and bend the blind repeatedly until a crack grows halfway across. The remaining section is expected to have half the strength, or to break at 150 pounds. Instead the remaining section easily rips apart with only a few pounds of tensile pull. This is a remarkable result. Half the slat's width is removed, but its strength is reduced by a factor of 40 or 50.

The fatigue crack in the Comet grew with every pressure cycle until it reached a critical length, at which point the whole plane rapidly broke up.

The concept of critical crack length is better understood with the following experiment. With a new strip of aluminum mini-blind, grow a fatigue crack about 20% across the width, and then try to pull the strip apart. When the strip does not pull apart, continue testing by increasing the increment of crack growth followed by a tensile pull. Repeat crack growth and tensile pulling until the strip rips apart. You will see that there is a well-defined critical crack length that distinguishes "does not break" from "easily breaks," with very little transition between the two.[11]

What explains this phenomenon? There is some effect of stress concentration. The sharp notch concentrates stress and creates a higher local stress value. However, it is more complicated than simple stress concentration. The stress does rise at the crack tip, but if higher stresses were the only consideration, a stronger material would better resist crack propagation. Yet the exact opposite happens: a stronger material has less resistance to rapid fracture and tearing.

It turns out to be an energy mechanism. The crack absorbs energy from the stressed structure as it grows and the structure gives up energy. If the rate of energy given up by the structure is greater than the rate of energy required to propagate the crack, the crack may grow in an unstable manner.

Energy absorption versus strength requirements is counterintuitive. Common sense tells us that greater strength is more protective and better. However, there are exceptions. Consider two tubes of cardboard 10 inches long and 2 inches in diameter. One is filled with bubble gum and the other with steel. Place one end of each tube against your head and strike the other end hard with a hammer. Which one protects the head better, the stronger steel or the energy-absorbing bubble gum? The concept of energy-absorbing material being superior is revisited in Chapter 8.

Aerospace engineers are constantly searching for more tear- or fracture-resistant materials. This property is also known as increased "toughness." Unfortunately, as materials get stronger, with rare exception they become less tough or have decreased ability to resist tearing. These are conflicting goals. A certain minimum strength is required in the fuselage skin to contain the pressure and resist the hoop stress. If a stronger material is used, a thinner fuselage skin can be used. This saves weight, a very good thing for an airplane. (Airlines can fly more passengers and make more money with any weight reduction, or manufacturers can add more safety-enhancing equipment.) However, the stronger material is less tear resistant, so instead of the fuselage being stable with a 40-inch crack, it may fly apart with a 20-inch crack. This potential to tear directly converts into operating costs. The airplane has to be taken out of service frequently to inspect for structural fatigue cracking and maintain an adequate level of safety.

The process of fatigue crack growth and fracture is well understood by now, and the engineers are not defenseless in their methodology to prevent destruction of the plane. Modern design methods greatly improve the likelihood of safe decompression. For example, the crack can grow until it is stopped by a crack arrestor, essentially a reinforcing band on the fuselage designed for this purpose.

"Fracture mechanics" is used by the engineer to predict the rate of fatigue crack growth and the critical length of the crack when unstable, rapid crack growth occurs. Careful determination of these two parameters is used to set inspection intervals. The aerospace industry uses the phrase "damage tolerance." A damage-tolerant design considers

- prediction of the most likely fatigue initiation sites;
- prediction of crack growth rates;
- prediction of the critical crack size at which crack growth becomes unstable and dangerous;
- analysis of required inspection intervals;
- required accuracy of inspection methods.

The inspection intervals and procedures are carefully coordinated with a damage-tolerant design. The inspection plan is approved and certified by the FAA.

Size Matters

The decompression rate depends on the cabin volume, the size of the opening in the fuselage, the pressure difference between the inside and outside, and—in the case of metal fatigue—the speed of crack propagation. Larger cabins, with a larger volume of compressed air, take longer to decompress. Larger openings or leaks in the fuselage will result in faster decompression rates.

There are three cases commonly referred to:

1. Explosive decompression, a very dangerous situation
2. Rapid decompression—get out the emergency breathing equipment
3. Slow decompression, which may go unnoticed

Explosive decompression is very rapid. It occurs in less than a second with a loud explosion-like pop. The pop is caused by a pressure shock wave propagating at the speed of sound. If the fuselage suddenly ruptures, the energy stored in the compressed air is released rapidly in a massive pressure shock wave. This energy is also available to propagate fatigue cracks.

During an explosive or rapid decompression, the rapid rush of air from the cabin will sweep out anything not secured. Dirt and dust will affect vision for several seconds. Fogging will also result, adding to the confusion in the cockpit and cabin. (A sudden decrease in pressure and temperature changes the amount of water vapor that can be held in the air. The excess water vapor appears as fog, condensed droplets suspended in the air.)

The U.S. Air Force has studied the effects of very rapid decompression on the body. During decompression, gases expand within body cavities. The abdomen may expand with the diaphragm being displaced upward by the expansion of trapped gas in the stomach. If severe enough, it can result in unconsciousness and shock. Usually, abdominal distress can be relieved by the passage of excess gas.

Because of the delicate nature of lung tissue, the lungs are potentially the most vulnerable part of the body during a rapid decompression. The Air Force reports no serious injuries resulting from rapid decompression with open airways, even while wearing an oxygen mask. However, disastrous, even fatal consequences can result if the breath is forcibly held with the lungs full of air. Under this condition, the lungs become over-extended by the high pressure, causing tearing and rupture of lung tissue and capillaries. The air trapped in the lungs can also be forced into the circulation system by way of ruptured blood vessels, resulting in massive air bubbles moving through the body and lodging in vital organs.

Lung damage is considered nearly impossible in a commercial aircraft because of the large volume of air inside the plane. To damage lung tissue, the plane must decompress in about 0.2 seconds. Recall that the DC-10 with the missing window (partially filled with a passenger, at least briefly) took 26 seconds to decompress. Even Aloha Flight 243, with a gaping 18-foot × 15-foot hole, did not decompress fast enough to cause internal injuries. However, lung injuries may be possible in a smaller, private jet and it is certainly possible in a military fighter plane.

Size was definitely a contributing factor in the Payne Stewart[12] accident. Because of the massive damage during impact, the NTSB was unable to conclusively determine why the cabin depressurized or why the flight crew failed to deploy emergency oxygen in a timely manner. The 265 cubic feet of the Learjet cabin (Table 4.2) certainly reduced the response time available to the pilots

On October 25, 1999, at approximately 9:19 a.m., Payne Stewart and five others took off from Orlando in a Learjet bound for Dallas with 5,300 pounds of fuel. At 9:27 a.m. and 23,300 feet, the first officer acknowledged clearance from air traffic control to climb to 39,000 feet. At 9:33 a.m., the Jacksonville controller instructed the flight crew to change radio frequencies but failed to get any response. The plane failed to level off at 39,000 feet and eventually reached a maximum altitude of 48,900 feet. The Learjet was considered out of control from that point on, and it flew for almost four hours by autopilot to the crash site near Aberdeen, South Dakota.

An Air Force F-16 first made visual contact with the Learjet at 10:52 a.m. and observed no external damage, but the pilot could not see inside the cabin, presumably because of frost inside the windows. In theory, the fighters could have tried to tip or nudge the wings of the plane to change its course, but it is not clear if the Learjet's autopilot would simply have automatically corrected for this disturbance. The Air Force briefly considered shooting the plane down until it was determined that the plane would run out of fuel in a sparsely populated area. A South Dakota highway patrol officer observed the crash. As the Learjet ran out of gas, it plunged into the ground at a 90 degree angle, resulting in a severe impact at approximately 1:20 p.m.

Table 4.2. Examples of Cabin Volumes

Plane	Cabin Volume (ft^3)	Passengers
747	59,000	416–568
777	46,000	305–440
DC-9	5,840	80–115
Learjet	265	6–8

The NTSB considered this accident to be two separate events: (1) loss of cabin pressure, and (2) failure of the flight crew to receive supplemental oxygen in a timely manner.

The flow control valve, which supplies air to the cabin for pressurization and heating, was found in the closed position. Assuming no other mechanical breaches in the fuselage, this would cause depressurization of the cabin to a dangerous level in about two and a half minutes.

An additional source of pressurized and heated air is supplied to the windshield to prevent icing, even if a mechanical breach occurs elsewhere in the cabin. This windshield defogger air is also designed to provide backup breathing air. Frosting of the windows seems to indicate an interruption of the inflow bleed air.

The following from the NTSB report[13] gives partial insight into complexities of the control panel settings for the ventilation system.

> In the event of a loss of normal pressurization, windshield defogger air can be routed into the cabin as an emergency source of pressurization by ensuring that the IN NORMAL/OUT DEFOG knob is in the IN NORMAL position, setting the WSHLD HEAT switch to AUTO, and setting the CABIN AIR switch to OFF (closing the flow control valve). Pressurization will then be maintained automatically. If pressurization is not maintained in the AUTO position, cabin altitude can be maintained by manually controlling the OUTFLOW VALVE using the UP/DN control switch, located on the pressurization module.

As part of normal ventilation, pressurized air exits from the outflow valve, thus preventing carbon dioxide buildup.

Closure of the flow control valve and interruption of bleed air inflow could occur from pilot error (wrong control panel settings) or from mechanical failure. The severity of the crash prevented precise determination. Many components and control systems were destroyed beyond recovery. For example, only fragments of passenger oxygen masks were recovered making it impossible to determine if they had been activated.

The NTSB determined the probable cause as incapacitation of the flight crew members as a result of their failure to receive supplemental oxygen following a loss of cabin pressurization. Investigators were unable to explain why emergency oxygen was not used or why the cabin depressurized.

The families sued Learjet, claiming a cracked component caused the outlet airflow valve to break away from the plane's frame, resulting in the decompression. After six months of testimony, the jury returned a not guilty verdict on June 8, 2005. They found no negligence in the design or manufacture of the plane. When the families' attorney was asked why he thought the jurors ruled the way they did, he replied, "there was a lot of science in this case, a lot of technical testimony."[14]

As discussed in Chapter 1, NTSB reports are not directly admissible in court. Their standard of establishing "probable cause" is not a legal standard. The publicly available NTSB investigation is, of course, available to the expert witnesses on both sides of any lawsuit and is expected to indirectly make its way into the courtroom. At least one party will likely have an incentive to dispute the NTSB findings with their own experts.

RAPID DECOMPRESSION OCCURS when the air leaks out faster than the cabin compressors can maintain pressure. There is no explosion. The passengers are in danger of hypoxia, but there is plenty of time for them to deploy their emergency oxygen masks.

In the case of slow decompression, the problem can be so gradual that it is difficult for people to notice while everyone slowly passes out. However, multiple alarm systems have to fail before this situation presents an actual safety hazard. Slow decompression could be due to a slow leak, improper settings of the cabin pressure controls, or any of a number of simple maintenance problems.

A more common sequence of events, which does not affect the plane's ability to land safely and is in many ways a testament to the success of modern fatigue design methods, is as follows:

1. A large crack occurs. This may or may not be accompanied by a loud bang, depending on the rate of decompression. Depending on the amount of energy released, there may be considerable damage.
2. The crew is either unable to maintain cabin pressure, or they are not even aware there is a problem.
3. The plane lands safely proving again that fail-safe design, redundant systems, and proper inspection solve most problems.

One of the longest, documented fuselage fatigue cracks was found in a DC-9 on March 4, 1996, during normal maintenance. Surprisingly, the 38.75-inch crack had not affected the plane's ability to maintain pressure on the previous flight. The plane had 82,325 cycles. The plane was repaired and quickly returned to service. Meanwhile the FAA mandated inspection in that area, of all DC-9s with follow-up inspection required every six months.

Other examples of in-flight fuselage cracks include the following:

- On December 5, 2004, a 747 made an unscheduled landing in Alaska after experiencing a rapid decompression at 30,000 feet. A 12-inch tear was found in a pressure bulkhead. The plane had 27,243 cycles (high for a 747, which is designed for fewer cycles).
- On June 13, 2000, a 737 made an unscheduled landing in Brazil after a rapid decompression at 29,000 feet. A 28-inch fuselage crack was found. The plane had 78,198 cycles.
- On August 23, 1995, the crew of an L-1011 reported hearing a loud bang at 33,000 feet. A 12-foot-long circumferential crack was found in a pressure bulkhead. The plane had 25,691 cycles (high for an L-1011). Due to the cost of repairs, the plane was retired by Delta and purchased by Lockheed for fatigue testing.

There are many more examples of cracking found during normal inspections.

Cargo Doors: Another Source of Explosive Decompression

Plug doors have increased the safety of passenger doors in commercial airlines for many years. A plug door wedges into the frame of a fuselage, similar to a sink plug wedging into a drain. This design can make opening the door complicated. In one common configuration, the door must first open into the cabin a short distance before it rotates 90 degrees to swing outside of the fuselage. Cabin pressure prevents opening from the inside and provides a better seal, even at low altitudes. For example, Boeing reports that at 1,500 feet in a 747, the cabin differential pressure is about 0.3 psi. This gives a pressure thrust of 900 pounds force that must be overcome by anyone trying to open the door.[15] This pressure increases significantly at higher altitudes.

In a 1981 incident, the handle of a 747 door was seen to rotate 60 degrees from the normal closed position shortly after takeoff. With some difficulty, two cabin crew members were able to keep the handle closed. If

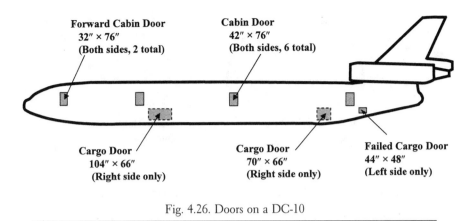

Forward Cabin Door
32" × 76"
(Both sides, 2 total)

Cabin Door
42" × 76"
(Both sides, 6 total)

Cargo Door
104" × 66"
(Right side only)

Cargo Door
70" × 66"
(Right side only)

Failed Cargo Door
44" × 48"
(Left side only)

Fig. 4.26. Doors on a DC-10

they let go, a loud sound of rushing air was heard. Engineers back at the airport instructed them to let go and see where it stopped. They refused. The captain told them that the door, being a plug design, would not open. However, the cabin crew became increasingly alarmed and were instrumental in the decision to return to the airport for an emergency landing without incident.

The initial movement of a plug door must be inward into the fuselage. (Depending on the design, many plug doors tilt and then swing out of the cabin and out of the way.) A much larger cargo door cannot move inward because of space limitation. Most cargo doors have latching and locking mechanisms, which have been the source of a few accidents.

The worst accident with decompression from a cargo door occurred on March 3, 1974, near Paris in a DC-10. From the cockpit voice recorder, investigators determined that at 9,000 feet there was a muffled explosion and a sound of rushing air indicating a sudden decompression. The cargo door (Figure 4.26) had not been properly latched and locked prior to take-off, and the flight deck aft cargo door warning light was out because of a maladjusted warning switch. The sudden blowout of the cargo door resulted in a rapid loss of pressure in the cargo hold. The higher pressure in the cabin above the cargo hold damaged the floor in between the two. This floor support structure and the cabin floor collapsed downward into the cargo compartment, sucking out two rows of passengers still strapped to their seats (Figure 4.27).

The aircraft structurally survived all of these events. However, the control cables from the cockpit to the control actuators in the tail were routed through the cabin floor beams. Collapse of the floor severed the control

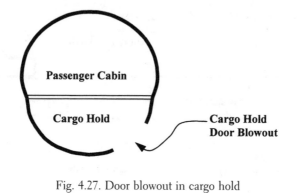

Fig. 4.27. Door blowout in cargo hold

cables and made pitch control of the airplane impossible, resulting in the plane crashing approximately 10 minutes after takeoff. There were 346 fatalities.

After this accident, the FAA required all wide-body transports to withstand cargo deck decompressions associated with openings, for whatever reason, of at least 20 square feet. Airline operators were given the choice of strengthening the cabin floor to prevent collapse or of enlarging the venting between the cargo hold and passenger cabin with blowout panels. Sufficient venting will prevent large pressure differentials between the cargo and passenger cabins. McDonnell Douglas incorporated both methods, and the floor collapse problem never occurred again.

The cargo door closing procedure is a multi-step process involving closing, latching, and locking. A door warning system is supposed to indicate if this has not occurred. The external lock handle cannot be stowed and a vent door will remain open if the door is not properly closed, latched, and locked. Proper stowing of the external lock handle and a closed vent door serve as a final visual indication that the sequence has occurred correctly.

The vent door, added to planes in 1970, is another redundant safety feature. A vent door is a small, flapper like door built into the cargo door. The vent door connects directly to the latch of the cargo door. If the latching linkages do not properly latch, the pilot will see with a visual check that the vent door is open. If the pilot incorrectly takes off with the cargo door not properly latched and the vent door open, the cargo hold cannot be pressurized. The cabin pressure would eventually blow out the vent panels in the floor—not a good situation, but also not an explosive decompression.

Somehow, all of these safety features were defeated in the 1974 DC-10 crash. Investigators determined that the door-mounted hooks had failed to engage the fixed pins properly and that the additional heavy force used to close the external latch handle had resulted in bent internal rods and tubes without locking the door securely. The vent door incorrectly closed because it was activated by the manual latch handle instead of the latching pins.

Numerous mechanical and electronic changes were made to the door closing sequence. Unfortunately, not all door problems have been solved.

Another door incident occurred on February 24, 1989, this time in a 747. Again, an improperly closed 110-inch × 99-inch (roughly 9-foot × 8-foot) cargo door blew out at about 22,000 feet and a differential pressure of 6.5 psi. Most of the blowout panels were activated in the floor, apparently preventing catastrophic collapse of the floor.

Additional damage to the airplane included an approximate 10-foot × 15-foot hole in the right side of the fuselage. A 13-foot × 15-foot section of fuselage skin also separated from the airplane above the cargo door. The floor beams adjacent to and inboard of the cargo door fractured and buckled downward. Because of the floor damage, the connection between the seats and the floor broke, and seats were ejected through the gaping hole with the loss of nine passengers. No metal fatigue or corrosion was found in the fuselage to suggest an alternative explanation for the extensive damage.

The flying debris also knocked out two of the four engines. In spite of all this damage, an emergency landing was successfully completed. The plane was considered repairable. Although this tragic loss of life is unacceptable, the damage-tolerant robustness of the structure is remarkable.

Too Little Pressure: Hypoxia

Hypoxia can be defined as insufficient oxygen being supplied to tissues for their physiological needs, despite an adequate supply of blood.

Hypoxia is not caused by a lack of oxygen per se. If hypoxia was caused by a lack of oxygen, the body's ventilation system could compensate with an accelerated breathing rate. (The body will, in fact, compensate for mild hypoxia with faster breathing.) Hypoxia is actually caused by an insufficient oxygen pressure to adequately drive the diffusion mechanisms of oxygen across the tissue in the lungs. This can be explained by Dalton's Gas Law, which states that "the total pressure of any mixture of gases is the sum of the individual pressures." In other words, if 21% of air is oxygen, the partial pressure due to oxygen is 0.21×14.69 psi $= 3.085$ psi. Think of gas pressure being the kinetic energy of atoms bouncing off the walls of a container. If 21% of the atoms are oxygen, then 21% of the pressure is from the oxygen molecules.

By the time the air gets to the lungs, the body has saturated the air with water vapor, which contributes its own partial pressure. Because of the added water vapor, the partial pressure of oxygen in the lungs is reduced to approximately 14% or 0.14×14.69 psi $= 2.056$ psi. The hemoglobin in the blood returning from the tissues carries oxygen at a partial pressure of about 0.77 psi. The difference between 2.056 psi and 0.77 psi is the differential pressure driving the diffusion of oxygen through lung tissue. The oxygen will diffuse across a semi-permeable membrane (in this case the alveola-capillary junction) from an area of high oxygen concentration to an area of lower concentration. Simultaneously, carbon monoxide will diffuse across the same lung tissue driven by a concentration gradient in the opposite direction.

Since the partial pressure in the blood returning to the lungs depends on body mechanisms (chemistry, diffusion rates, etc.) and is more or less the same at any altitude, there is very little tolerance for lowering the oxygen partial pressure in the lungs and still maintaining proper diffusion of oxygen in the lungs.

Federal aviation rules require that pilots have supplemental oxygen when flying longer than 30 minutes above 12,500 feet, and for all flight above 14,000 feet. FAA Advisory Circular 61-107 provides the estimates shown in Table 4.3 of useful consciousness at various altitudes.

The U.S. Air Force Flight Surgeon's Guide states that intellectual impairment is an early symptom of hypoxia that makes it difficult for individuals to comprehend their own disability. Thinking is slowed, calculations are unreliable, memory is faulty (especially of the immediate past), and judgment is poor. There may be a release of basic personality

Table 4.3. Times of Useful Consciousness
at Various Altitudes

Altitude (ft)	Sitting Quietly	Moderate Activity
22,000	10 min	5 min
25,000	5 min	3 min
30,000	1 min	45 sec
35,000	45 sec	30 sec
40,000	25 sec	18 sec

traits and emotions, as with alcoholic intoxication, including euphoria, elation, pugnaciousness, overconfidence, and moroseness.

Pilots often train in a decompression chamber to experience and identify their individual symptoms of hypoxia. A typical test involves the test subjects being given a checklist of tasks to perform, similar to following an emergency checklist. The pilots tend to exhibit one of two types of behavior: page flicking or fixating. The page flickers will flip mindlessly through the checklists, while the fixators will stare at one page. (Keep in mind that these are professional pilots who are not being surprised with a decompression and are determined to focus as long as possible.) Because of these behaviors, the first step of the procedure when a decompression occurs always calls for the pilots to put on their oxygen masks before doing any additional thinking.

To maintain sufficient partial pressure of oxygen and support proper breathing, cabin pressure in all commercial flights is maintained at a pressure equivalent to an altitude of 8,000 feet. If the cabin pressure reaches the equivalent of 10,000 feet, a warning alarm sounds automatically. If the cabin pressure goes up to 12,000 feet,[16] the pilots are supposed to don their oxygen masks before attempting to debug the problem. At 14,000 feet of cabin altitude pressure, the passenger's oxygen masks automatically deploy. If the flight crew cannot control the loss of cabin pressure, the procedure calls for a rapid rate of descent until the plane reaches around 10,000 feet of actual altitude. This procedure is practiced by all pilots once a year in a flight simulator. Of course, the simulation is done without the reality of an explosive decompression which might include splitting ear pain, condensation of water vapor into a thick mist, and the general confusion of papers and other loose items blowing around the cockpit.

Hypoxic Flight Crew

In the following story, the general pandemonium of an explosive decompression did not occur. The story illustrates a sneakier enemy: gradual decompression and the slow onset of hypoxia with some people drifting in and out of hypoxic stupor. "Some people" is not supposed to include the pilots, though. This incident aboard Flight 406 occurred May 12, 1996, in a Boeing 727.

Upon reaching a cruising altitude of 33,000 feet, the cabin altitude warning alarm sounded. The captain noticed that the right air conditioning pack was off. Along with the flight engineer, he attempted to reinstate the pack without using a checklist. The cabin pressure continued to climb to 14,000 feet, at which time the warning lights illuminated and the oxygen masks deployed in the cabin. While attempting to correct the cabin altitude, the flight engineer inadvertently opened the outflow valve, which resulted in a rapid loss of cabin pressure.

The lead flight attendant was in the cockpit serving meals, and the captain asked her to see if the oxygen masks deployed in the passenger cabin. She observed the masks were down, leaned against the cabin door, and promptly slumped to the floor.

The first officer, who had only 10 hours of flight time in the airplane, initiated an emergency descent. He later stated that when questioning the captain, he was not getting a normal response. He then looked at the captain and noticed his head tilted to the left with his oxygen mask only partially on his face. Simultaneously, the flight engineer got out of his seat to assist the unconscious flight attendant. The flight engineer's oxygen mask was pulled from his face, and the next thing he remembers is the plane being below 10,000 feet. By then, the lead attendant had recovered and repositioned the oxygen masks on the captain and flight engineer. They recovered in seconds and an emergency landing was made shortly after without incident. The airplane was inspected and flight tested the next day. The airplane's pressurization system functioned with no anomalies.

The NTSB stated the probable cause as "the failure of the captain and flight engineers to utilize a checklist to troubleshoot a pressurization system problem, and the flight engineer's improper control of the pressurization system which resulted in an inadvertent opening of the outflow valve and subsequent airplane decompression."[17]

To better understand the story, an explanation of the ventilation system is required.

The pressurization system in a jet aircraft is analogous to maintaining pressure in a balloon with a hole in it. The engines pump pressurized "bleed air" into the fuselage, the temperature of which is controlled by air conditioning packs. The air circulates through the cabin and into the cargo hold and exits via the cargo outflow valve.

Constant air circulation is required to avoid carbon dioxide buildup. Federal regulations require a ventilation system that provides 10 cubic feet per minute of fresh air per person at 8,000 feet pressure altitude. Older model airplanes, such as the DC-9, the B-727, and half of the DC-10s, provide 100% fresh air to the aircraft cabin. Newer models of jet aircraft, including B-737, 747, and 757, some DC-10s, and A-300, 320, and 310, provide up to 50% recirculated air. The recycled air system uses less engine compressed air and conserves fuel. The effectiveness of these filtration systems is sometimes the focus of debate on cabin air quality.

The World Health Organization (WHO) states,

> The quality of aircraft cabin air is carefully controlled. Exchange with outside air and filtration of recirculated cabin air provide a total change of air 20–30 times per hour. This level of ventilation is much greater than that in any building and ensures that contaminant levels are kept low. Modern aircraft recirculate up to 50% of cabin air. The recirculated air is passed through HEPA (high-efficiency particulate air) filters, which trap particulate material, bacteria, fungi and most viruses. Consequently, recirculated cabin air is very clean.[18]

Boeing asserts that cabin air is replaced every two to three minutes, depending on the airplane's size. The greatest health risk is sitting in close proximity to potentially sick passengers.

The cargo outflow valve operates to adjust the rate of air escape to maintain the desired cabin pressure. With air escaping at a regulated rate, more air must continually be blown into the cabin to maintain pressurization. The fuselage is a pressure vessel with constant inflows and outflows. This process is automated to adjust for different internal pressures required as the aircraft changes altitude (the ear popping experience). The equivalent altitude in the plane slowly goes from ground level to 8,000 feet as the internal cabin pressure slowly decreases from 14.7 psia to between 10.7 and 12.6 psia, depending on the altitude. (Recall that cabin pressure is limited by the structural strength of the fuselage and is designed to be 8.25 psi greater than the outside pressure.)

The inlet air actually originates inside the jet engine compressors. The compressors are used to provide sufficient pressure to blow air into, through, and out of the pressurized cabin (Figure 4.28). The air inlet enters the cabin and circulates through the cargo hold before it exits out the cargo outflow valve.

On the ground, the cargo outflow valve is initially wide open. As soon as the plane lifts off, the valve starts to close and the cabin begins to pressurize. The airplane may be climbing at thousands of feet per minute, but the cargo outflow valve (regulated by pressure sensors) automatically ad-

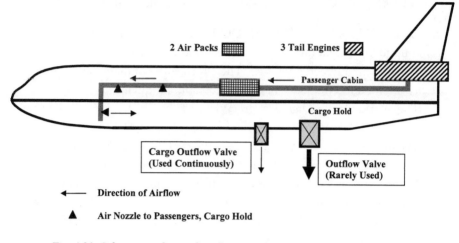

Fig. 4.28. Schematic of aircraft airflow circulation system for a Boeing 727

justs to raise the cabin pressure perhaps hundreds of feet per minute, a rate equivalent to driving up a mountain.

The air comes out of the engine compressors heated and must be cooled by blowing cold outside air past the heated compressor air. This heat exchange takes place in an "air pack." The air pack must accommodate extremes in temperature, from 95°F summer air temperatures on the runway to −50°F at 40,000 feet. The air pack is, therefore, a system that cools, heats, mixes, compresses, and expands various air streams as needed to provide the necessary air temperatures.

The Boeing 727 has three engines providing inflow compressor air to two air packs. This provides flexibility for a number of potential operating conditions. Bypassing compressor air to the air pack reduces engine power. There are a variety of reasons why the pilot might want to switch around which engine supplies air to the air packs. For example, if an engine is experiencing problems, it does not need the additional loading associated with providing compressor air to the air pack.

There may be rare circumstances in which the plane should take off with only one air pack turned on. With a particularly heavy takeoff weight the pilot might want the added power of shutting off one air pack. Also, an air pack might be shut off if it overheats—a fire on an airplane must be avoided at all costs.

The captain and flight engineer on Flight 406 stated that both air packs were on for takeoff—the normal situation. However, the incident on this flight started with an alarm and the captain noticing that one air pack was off. As with most critical systems, having two air packs provides increased reliability.

However, one air pack will not blow enough air into the cabin to keep it sufficiently pressurized at an altitude of 40,000 feet. The 727 is certified to fly at only 25,000 feet if just one air pack is operating. Both air packs should have been operating on Flight 406. One air pack did not provide sufficient air pressure, so the cabin altitude warning horn sounded at about 32,000 feet. While the captain located the silencer button, he noticed that the right air pack toggle was in the off position. The most likely explanation is that this air pack was never turned on in the first place, which is consistent with no mechanical equipment problems being found by the NTSB investigation.

As explained to me by a former 727 pilot, it is important to follow the emergency checklist. The emergency checklist for depressurization is, in part, to

1. put on oxygen mask and keep on till the situation stabilizes;
2. close cargo outflow valve;
3. initiate an emergency descent to 5,000 feet.

The captain on Flight 406 was slow to put on his mask and became unconscious. The flight officer became unconscious after removing his mask to assist the flight attendant. The first officer donned his mask, maintained consciousness, and ultimately initiated the emergency descent. In the ensuing confusion and onset of hypoxia, it appears that the rarely used outflow valve was inadvertently opened, leading to rapid decompression.

A depressurization is normally associated with a mechanical breach in the fuselage. This usually occurs because of metal fatigue or mechanical damage (e.g., a burst engine, a door or window failure). Basically, the extra hole in the fuselage permits more air to leak out than is being supplied by the air packs. The procedure in this situation calls for the normal operating holes in the fuselage—in other words, the cargo outflow valve—to be closed.

Correctly following items (1) or (2) above would have greatly stabilized a quickly worsening situation. The flight crew failed to do either. Instead, in the confusion, the flight engineer inadvertently opened the manual outflow valve.

In addition to the cargo outflow valve, the manual outflow valve is a second outlet valve for emergency decompression that the flight crew can initiate on purpose. For example, if smoke or other toxic fumes are circulating in the cabin, the crew can open this valve to quickly flush the smoke out (assuming a fire is not making smoke faster than it can be removed). Again, a pilot I spoke with (who states he experienced one decompression in 25 years) said that this valve is not to be used except in a variety of unusual circumstances. It was a major error for the Flight 406 crew to open the manual outflow valve.

Eventually, the first officer did catch up to the correct procedure. The flight crew has to solve the breathing problem, above all else, by flying to an altitude where breathing can occur without the assistance of any mechanical devices.

Passengers should, of course, follow instructions. Specifically, they must put on their own oxygen mask before assisting children. If a child passes out for a few seconds, the adult can easily revive him or her. However, if the adult passes out and the flight attendant has to find the unconscious passenger among a few hundred people, the long delay could become a serious medical problem.

Glancing into the cockpit (for me, as a novice) always results in the same commonsense notion: there are too many dials and switches on the control panel. It's certainly true for a hypoxic flight crew.

Helios Flight HCY522

On August 14, 2005, Helios Flight HCY522, a Boeing 737, left Cyprus at 6:07 a.m. local time. According to the flight data recorder, the cabin altitude warning alarm sounded at 12,040 feet. At 6:14 a.m. and an altitude of 15,996 feet, the captain contacted the company's ground engineer to report problems with the takeoff configuration warning alarm and cooling equipment.

Because of conflicting information (the flight crew was debugging the wrong problem), a somewhat confusing conversation between the captain and the ground engineer ended at 6:20 a.m. Shortly afterwards, the operation's dispatcher called the flight crew without receiving a response.

At 6:23 a.m., the autopilot leveled the plane off at 34,000 feet.

At 8:23 a.m., after numerous attempts to contact the aircraft, the Greek Air Force intercepted the plane. The F-16 pilots reported that the first officer appeared to be unconscious, the pilot appeared to be missing from the cockpit, and one or two other people seemed to have entered the cockpit. At 8:49 a.m., the left engine flamed out, followed by the right engine 10 minutes later.

Impact occurred at 9:03 a.m.

The Greek investigators indicated that, because of previous pressurization problems with this plane, the maintenance crew performed a pressure check of the cabin the night before the flight and left the pressurization system controls on "manual," a setting almost never used. If set on manual, the cabin cannot be pressurized unless the outflow valve is manually shut.

The low-cabin-pressure alarm did sound at 14,000 feet, but unfortunately it sounds similar to other warning alarms. Since the pressurization switch is almost never set on manual, and the post-takeoff checklist was ambiguous about the pressurization control settings, the flight crew became hypoxic trying to diagnose the wrong problem. The plane continued to climb to 34,000 feet and flew for nearly three hours on autopilot before running out of fuel and crashing with 121 fatalities.

Although the German pilot and the Cypriot copilot made obvious errors while trying to debug the problem in English, Boeing was also faulted. The Greek investigators recommended that Boeing "clarify" the instructions in the aircraft maintenance manual concerning placement of the pressure controls after any maintenance. They also recommended that Boeing "consider enhancing the design of the Preflight checklist" to include a specific item "instructing the flight crew to set the pressurization selector to Auto."[19]

Aloha Flight 243 Revisited: Fuselage Design

The explosive decompression of Aloha Flight 243 resulted from a structural failure of the Boeing 737 fuselage (see Figure 4.1).

The fuselage of a 737 has an extensive network of reinforcing ribs. The circumferential ribs occur every 20 inches along the length of the fuselage; the longitudinal ribs are spaced every 9.5 inches around the circumference.[20]

Figure 4.29 shows a section of the 133-foot-long, 148-inch-diameter 737 fuselage. The spacing of the ribs is shown approximately to scale. A side view of the fuselage (Figure 4.30) shows the aspect ratio of the circumferential and longitudinal ribs, as well as the stringer spacing. Figure 4.31 shows the interior of a 737 with its reinforcing ribs and floor beams exposed.

This type of construction is known as a semi-monocoque fuselage, which is a reinforced monocoque or shell construction. The main advantage of

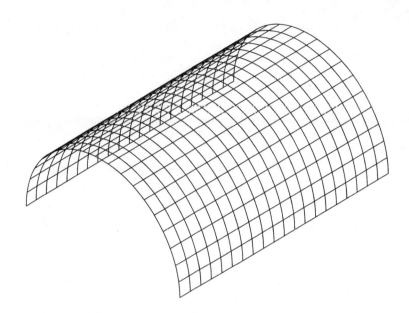

Fig. 4.29. Section of Boeing 737 fuselage showing circumferential and longitudinal reinforcing ribs. Spacing is shown to approximate scale. *Image:* Lowell Stanlake

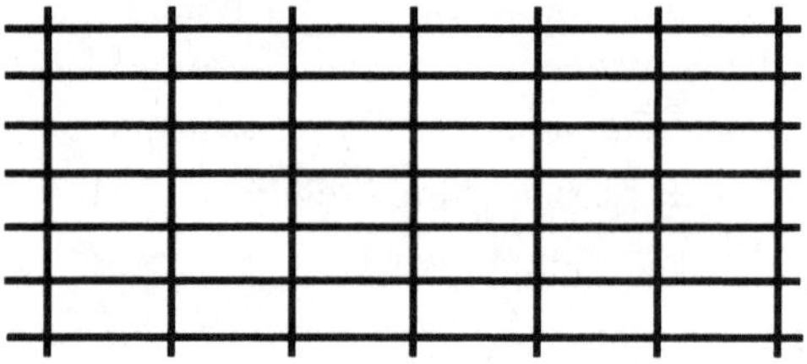

Fig. 4.30. Side view of fuselage showing rib spacing. Shown are 6 × 6 = 36 "bays" or a 120″ × 57″ section of Boeing 737 fuselage.

the semi-monocoque fuselage is that the structural integrity does not depend on a few members. A semi-monocoque fuselage may withstand considerable damage and still hold together. The network of ribs illustrates the fail-safe and damage-tolerance design principles. Because of the redundant reinforcement, the loss of one or two reinforcing ribs is not expected to be critical to the fuselage as a whole. Damage can result from impact debris (a burst engine), corrosion, fatigue, or some combination of the three.

The skin of the 737 fuselage is 0.036 inches thick, more than enough to contain the internal pressure. Surprisingly, the primary purpose of the

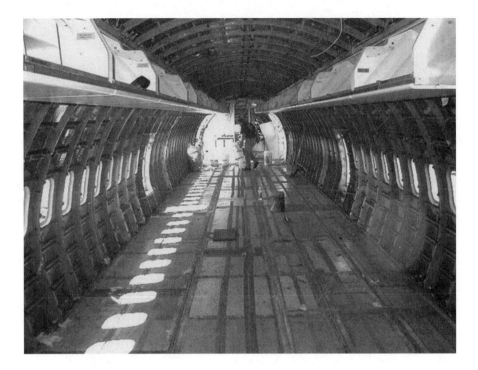

Fig. 4.31. Interior of 737 with exposed reinforcing ribs and floor beams
Photo: Larry Reeves, Artificial Reef Society of British Columbia

reinforcing ribs is not pressure containment. The ribs are designed to withstand maneuvering loads on the fuselage. For example, a one-engine operating scenario results in a massive torque on the fuselage.

Torque

Torque (from the Latin word meaning to twist) is a force applied to the end of a lever arm. A 20-pound force on the end of a 10-inch wrench supplies a torque of 10 inches × 20 pounds = 200 inch pounds on a bolt head (Figure 4.32).

$$\text{Torque} = \text{force} \times \text{length}.$$

The engine of a Boeing 737 supplies about 22,000 pounds of thrust and hangs on the wing about 17 feet (204 inches) from the center of the fuselage. This results in a torque of 22,000 pounds × 204 inches = 4,488,000

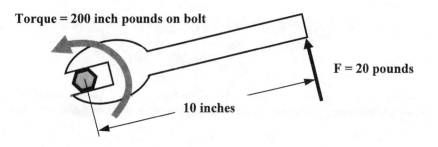

Torque = 200 inch pounds on bolt

F = 20 pounds

10 inches

Fig. 4.32. Force on a wrench that "torques" a bolt

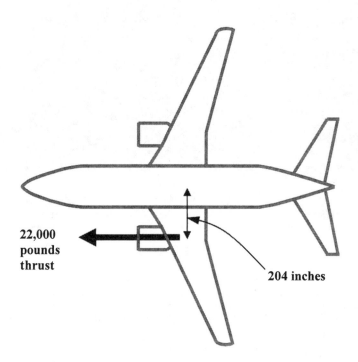

22,000 pounds thrust

204 inches

Fig. 4.33. The thrust of a single engine creates a torque on the fuselage.

inch pounds on the center of the fuselage (Figure 4.33). Of course, the engine on the other wing supplies an equal but opposite torque at the same point on the fuselage. The net torque on the fuselage is zero.[21] But what if one engine is out?

All commercial jets are perfectly safe to cruise, land, and even take off with one engine inoperative. Additional engines are a redundant safety feature. With one engine out, the tail provides a counter torque (Figure 4.34); otherwise, the plane would spin like a top.

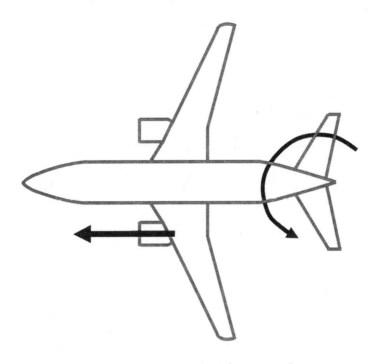

Fig. 4.34. When only one engine operates, the tail must provide an opposite torque to prevent the plane from spinning.

The torque from the engine and the torque from the tail can be envisioned by attaching a shaft to the center of the tail and another shaft to the center of the fuselage at the wing attachment (Figure 4.35). Each shaft could be gripped by super-strong hands and twisted with 4,488,000 inch pounds of torque. The shaft corresponding to the engine torque is twisted clockwise, and the tail shaft is twisted counterclockwise. The end result: the fuselage acts like a box with two large torques or bending moments, one at each end (Figure 4.36).

This type of loading on the fuselage results in the exaggerated distortion shown in Figure 4.37. The top edge compresses and the bottom edge elongates. Also shown in Figure 4.37 is the neutral axis at the center of the fuselage. The neutral axis marks the demarcation line above which the fuselage compresses and below which it elongates.

The distinction between compressive loading and elongation loading (Figure 4.38) is important and should be understood, because they cause entirely different failure mechanisms. Normally, failure is understood to be tearing or fracture associated with excess tensile stress. Fracture occurs

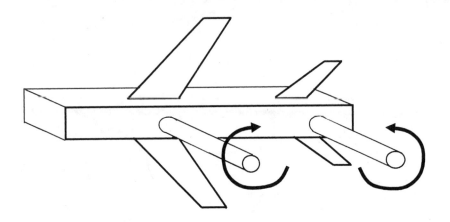

Fig. 4.35. Consider the fuselage as a box with two shafts connected to it. The movement from one engine operating is equivalent to grabbing the shaft and twisting it with a torque. The rudder supplies a balancing torque.

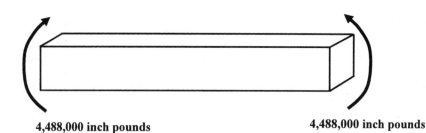

4,488,000 inch pounds　　　　　**4,488,000 inch pounds**

Fig. 4.36. One-engine operation results in a torque on the fuselage with a counter torque from the tail.

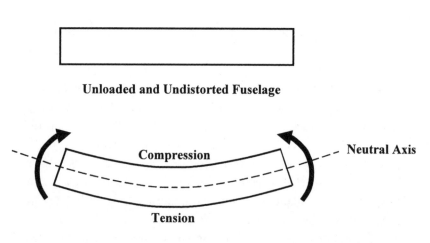

Unloaded and Undistorted Fuselage

Compression　　　　　**Neutral Axis**

Tension

Fig. 4.37. Exaggerated distortion of a fuselage loaded with two torques. The top edge is compressed and the bottom edge elongates.

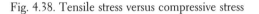

Tensile Stress

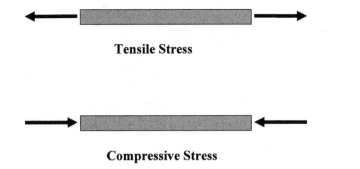

Compressive Stress

Fig. 4.38. Tensile stress versus compressive stress

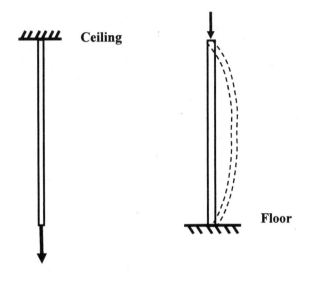

Fig. 4.39. Yardstick supports a large load when loaded in tension, but it buckles with a fraction of the load when loaded in compression.

when the strength of the material is exceeded in a single load. Fatigue crack growth is also associated with tensile stress. The fatigue crack opens and extends with a tensile stress. However, the fatigue crack closes and does not propagate with a compressive stress. Compressive stresses do not cause failure due to fracture or fatigue; rather, they result in failure by "buckling." Buckling can be demonstrated with a yardstick. A yardstick, if attached to the ceiling, can easily support a person's weight. But, leaning just slightly on the top of a yardstick sitting on the floor will quickly cause bowing and buckling (Figure 4.39).

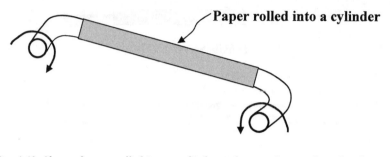

Paper rolled into a cylinder

Fig. 4.40. Sheet of paper rolled into a cylinder and torqued on each end with PVC elbows used as hand grips

The yardstick is an example of gross buckling, in that the entire structure buckles. The fuselage would buckle like a yardstick if for some reason the entire fuselage was loaded in compression. More likely, however, the fuselage buckles locally, where only that part of the fuselage in compression buckles.

A "test" fuselage made from a sheet of paper rolled into a cylinder can be torqued to demonstrate the structural problem of torsion on a fuselage. Actually, any tube will do, including the cardboard center from a roll of paper towels. Bending the cylinder with both hands correctly demonstrates the problem. However, the tube appears to crush under the hands, distracting from the mechanism of collapse caused by torsion and specifically the compressive stresses resulting from the torsion. I will describe a slightly more elaborate experiment that eliminates hand crushing.

I obtained two PVC elbows from a hardware store and attached a threaded coupling to each elbow to serve as hand grips. I then rolled a sheet of paper into a tube and taped the elbows to each end of the paper cylinder, as shown in Figure 4.40. The two elbows are torqued, resulting in compression on the top edge of the cylinder. The torqued paper fuselage simulation does not fail in tension, nor does it tear. Instead, it bulges inward or buckles on the compression side. Flimsy structures, especially aerospace structures, can fail by buckling with compressive stresses.[22]

To reinforce the fuselage when torqued, longitudinal reinforcing ribs known as stringers are spaced every 9.5 inches around the circumference and run the length of the fuselage. The force in a single rib can, in principle, be easily calculated, but in practice the calculation rapidly becomes tedious. When the entire fuselage is loaded by an external torque, some of the reinforcing ribs will be in tension and some of them will be in compression. Those that are in compression are susceptible to buckling just like the yardstick.

The actual longitudinal reinforcing rib of a 737 is a 4-inch-wide strip of sheet metal bent into a hat-shaped cross section (see Figure 4.50). The cross-sectional area of the longitudinal rib is 4 in × 0.036 in = 0.14 in². Evaluating a circular rod with the same area for buckling illustrates the rapid degradation of strength with increased length. As the rod becomes longer and more flimsy, it buckles at a lower load. For example, a 10-inch-long circular rod buckles with a compressive load of 1,500 pounds, a 20-inch-long rod buckles with 376 pounds, and a 50-inch-long rod buckles with only 60 pounds.

Figure 4.41 illustrates the decrease of compressive strength of the circular rod as it becomes longer. Compare this to the same rod's tensile strength of well over 9,000 pounds (assuming the rod is aluminum with a

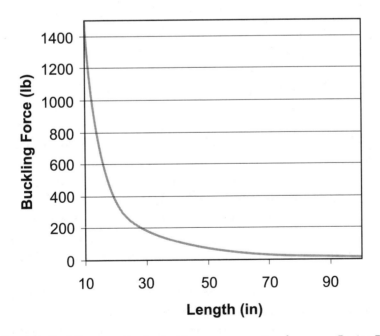

Fig. 4.41. Buckling strength of rod with same cross-sectional area as a Boeing 737 horizontal reinforcing rib of various lengths

tensile strength of 65,000 psi). This tensile strength remains constant for any length. Because of this weakness in compressive loading, circumferential ribs, known as frames, are added every 20 inches to give the longitudinal stringer ribs lateral support. This 20-inch spacing is a standard for most large commercial planes.

The network of frames and stringers (see Figures 4.29 and 4.31) are required to sustain a variety of maneuvering load conditions. These design loads were well established before the problem of metal fatigue became understood. The design process has two steps:

1. Design the reinforcing ribs for structural loads on the fuselage.
2. Verify that the network of ribs provides adequate safety for metal fatigue and adjust as needed.

Design Improvements since the Comet Disasters

Two de Havilland Comets were destroyed because of metal fatigue and explosive decompression. Since then, significant progress has been made in design methods for fatigue and fracture. Very costly and lengthy fatigue testing is now part of the design cycle. Fatigue problems are discovered during tests before they become an in-flight disaster.

Having the benefit of the Comet's experience with metal fatigue (Boeing traded design methods for pylon-mounted engines in exchange for de Havilland's test data on fatigue), the 707 had 15 months of structural testing, including 50,000 pressure cycles on a fuselage section inside a water tank.

Typical modern design methods require a fatigue crack to be stable up to a length of 40 inches. This means that a fatigue crack has to grow longer than 40 inches before it is expected to propagate rapidly and destroy the plane, as occurred with the Comets. The worst-case scenario assumes the circumferential rib is also cracked with metal fatigue, as shown in Figure 4.42. This damage is based, in part, on the hypothesized damage made by uncontained engine fragments from a burst engine (see Burst Engines, Chapter 5).

These design requirements do not mean the plane is certified to fly with fatigue cracks. The plane is simply designed to be safe and will not experience an explosive decompression if a fatigue crack grows to a length of 40 inches. Inspection for fatigue cracking provides an additional margin of safety. A fatigue crack is expected to be found and repaired long before it reaches 40 inches.

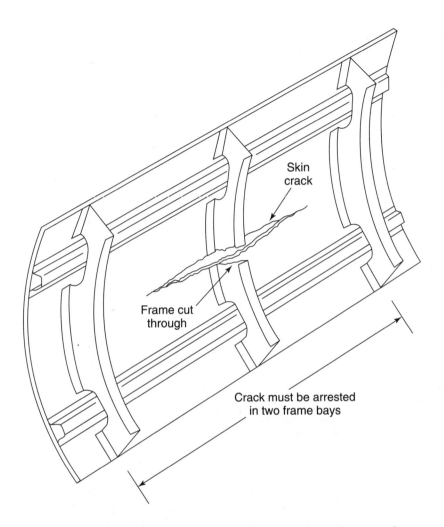

Fig. 4.42. Standard design practice requires fatigue crack growth to stop after growing two rib spacings or 40 inches total length. *Source*: FAA, Report DOT/FAA/CT-93/69.II

Controlled Decompression Failure Mode

The propagating fatigue crack is engineered to "turn the corner" and propagate perpendicular to the origin fatigue crack, as shown in Figures 4.43 and 4.44. (The fatigue crack initially grows perpendicular to the hoop stress; see earlier discussion on hoop stress.) This produces a flap that opens—the so-called controlled decompression failure mode. The flap acts like a large safety valve. The pressure blows out reducing the hoop stress driving the crack growth, and explosive decompression is prevented.

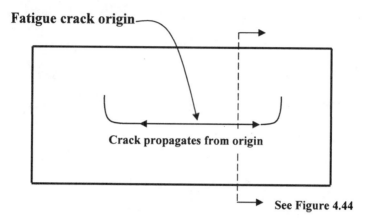

Fatigue crack origin

Crack propagates from origin

See Figure 4.44

Fig. 4.43. Propagating fatigue crack turns the corner, producing a flap that opens. The flap relieves the internal pressure and reduces the hoop stress.

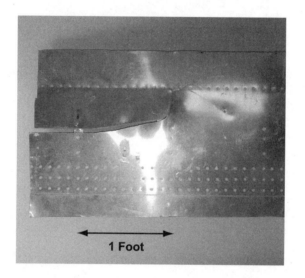

1 Foot

Fig. 4.44. Example of fatigue crack turning the corner to create a flap

Special test panels similar to a 737 fuselage were constructed to study fatigue crack growth and detection. The section shown in Figure 4.44 was cut from such a panel. Also shown in Figure 4.44 is a top row of rivets that attaches an underlying longitudinal hat-shaped reinforcing rib and three rows of rivets that make up a lap joint (explained later).

Balloon Test

A tear in a balloon can be made to turn the corner, as described above, by adding strips of tape. Two complete circumferential loops of tape are added to the cylindrical balloon and the balloon is pricked between the two tape loops (Figure 4.45). The tape acts like tear straps, which are described below.

In the case of this balloon experiment, the tear turns the corner, but it does not stop. The tear turns and runs 360 degrees around the circumference of the balloon. The balloon fragments into three pieces. Similar fragmentation of an aircraft would result in complete in-flight destruction, as with the Comet disasters. (In some tests, depending on the geometry of the balloon tested, the spacing of the tape was important. If the tape was spaced too far apart, the tear would run through the tape without turning the corner.)

Aloha Flight 243: Fuselage Design Details
Tear Straps on the Boeing 737

Several design features have been developed since the Comet disasters to make the fuselage structure more resistant to metal fatigue. These new design details were incorporated into the Aloha Boeing 737 fuselage.

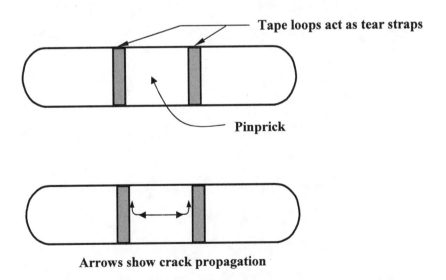

Tape loops act as tear straps

Pinprick

Arrows show crack propagation

Fig. 4.45. Tear straps in a balloon result in same tear patterns as in a fuselage. However, the balloon fragments into three pieces when the tears turn and run 360 degrees.

One of these new details is tear straps. A tear strap, a reinforcing strip attached to the fuselage skin (Figure 4.46), is designed to arrest the fatigue crack or make the crack turn the corner and form a flap. Tear straps act like reinforcing Band-Aids attached to the fuselage skin to arrest fatigue crack growth.

The skin panels are fabricated from two aluminum sheets, each 0.036 inches thick, glued together with an epoxy adhesive. The inner sheet is masked. Unwanted excess material on the inner layer is chemically removed, leaving a waffle pattern of reinforcing strips or tear straps (Figure 4.47). The tear straps are at 10-inch intervals both longitudinally and circumferentially. This corresponds to a tear strap under every circumferential frame, in between each frame, and under every longitudinal rib.

Lap Joints on the Boeing 737

The fuselage structure consists primarily of skin, frames (circumferential ribs), and stringers (longitudinal ribs). Skin panels are joined longitudinally at lap joints where the sheet metal of the upper skin panel overlaps the

Fig. 4.46. Close up of shear clip detail on Boeing 737 test panel. Underneath the shear clip is a tear strap.

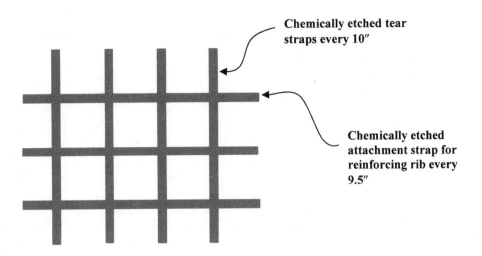

Chemically etched tear
straps every 10"

Chemically etched
attachment strap for
reinforcing rib every
9.5"

Fig. 4.47. Two aluminum sheets are adhesively bonded to form the fuselage skin. A
waffle pattern of tear straps is formed by chemically etching from one side.

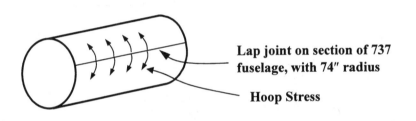

Lap joint on section of 737
fuselage, with 74" radius

Hoop Stress

Fig. 4.48. The lap joint lines up with the maximum hoop stress.

sheet metal of the lower skin panel (like shingles on a roof) by about three
inches (Figure 4.48). The overlapped area in a 737 is adhesively bonded
and riveted with three rows of rivets. The three rows of rivets are seen in
Figure 4.49.

The center row of rivets secures a longitudinal reinforcing rib to the lap
joint. Each skin panel is about 18 feet long. The adhesive and rivets were
independently capable of withstanding the static hoop stress. Disbonding
of the adhesive adds loading to the rivets, making them more susceptible
to fatigue.

In a lap joint connected with multiple rows of rivets, the first and last
rows take most of the hoop stress. As shown in Figure 4.50, the hoop stress
to the left is mostly supported by the first row of rivets on the left. (Correspondingly, the hoop stress to the right is mostly supported by the first row
of rivets on the right.)

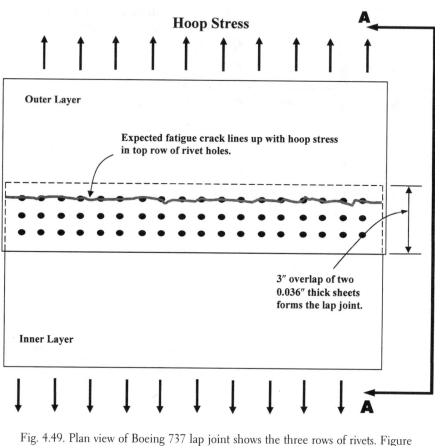

Hoop Stress

Outer Layer

Expected fatigue crack lines up with hoop stress in top row of rivet holes.

3″ overlap of two 0.036″ thick sheets forms the lap joint.

Inner Layer

Fig. 4.49. Plan view of Boeing 737 lap joint shows the three rows of rivets. Figure 4.50 shows section A-A from the side.

In addition to the first and last row of rivets supporting most of the hoop stress, the outer layer of fuselage skin contains a knife-edge in the rivet hole, as shown in Figure 4.51. This knife-edge serves as a stress riser to initiate fatigue cracks. The net effect is that the top row of rivets on the outer layer is most susceptible to fatigue cracking. Also shown in Figure 4.51 are rivet head designs that eliminate the knife-edge. These rivits were recommended by the FAA when problems with the glued lap joint became known.[23]

The lap joint design was changed on the 737 because of disbanding of the glued joint. On early production runs (planes 1 through 291, including Aloha Flight 243), the doubler sheet was chemically milled away at the lap joint locations (Figure 4.52). For later aircraft, the doubler sheet was

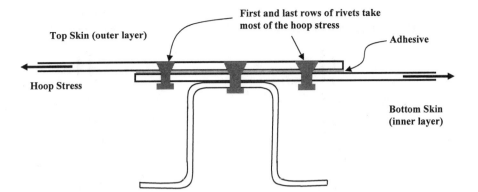

Fig. 4.50. Side view of longitudinal lap joint with attached hat-shaped longitudinal reinforcing rib (section A-A in Figure 4.49). Joining is by adhesive bonding and three rows of rivets.

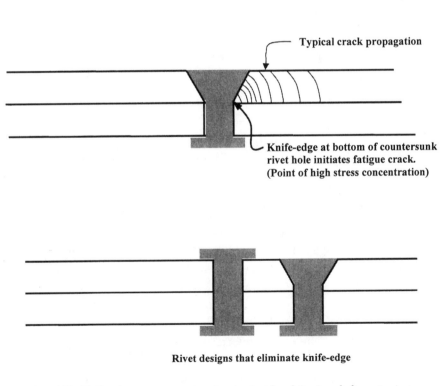

Fig. 4.51. Knife-edge stress concentration in rivet head (top) and alternate rivet designs that eliminate knife-edge

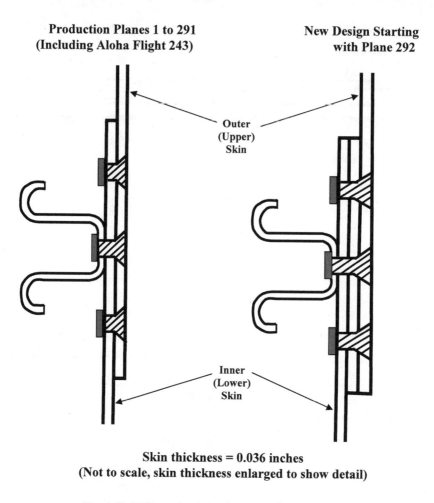

Outer
(Upper)
Skin

Inner
(Lower)
Skin

Skin thickness = 0.036 inches
(Not to scale, skin thickness enlarged to show detail)

Fig. 4.52. Different lap joint designs in the Boeing 737

retained on the outer panel of each lap joint to provide a third layer of 0.036-inch sheet metal in the critical lap joint.

In addition to the three rows of rivets, the lap joint in the Aloha Flight 243 airplane was also glued. In fact, the hoop stress was designed to be totally supported by the glue, with the rivets as a backup system. The rivets concentrate the forces and stresses with increased susceptibility to fatigue. (The epoxy-bonded surfaces transfer loading more uniformly than rivets.) Without the force concentration of a rivet, thinner material can be used in the fuselage skins with no degradation of fatigue life. The lap joint with adhesive is a better joint, if disbonding does not occur. If disbonding does occur, the rivets will still support the hoop stress, but the overloaded rivets will develop fatigue cracking in the fuselage skin.

The cleaning and etching process used for gluing the skin panels of the early 737 planes did not provide a uniform surface oxide for bonding. The service history also showed a degraded bond quality from condensation. Once in service, moisture could enter the joint in the region of disbonding and corrosion could occur. The moisture and corrosion contributed to further disbonding of the joint due to accumulation of oxides and water wicking into the joints.

In planes produced after 1972, the glued adhesive lap joint was replaced by a riveted joint with increased thickness (Figure 4.52). The 737 flown for Aloha Flight 243 was delivered in 1969.

Fatigue Testing of the Boeing 737

The 737 was designed for an economic service life of 20 years and 75,000 cycles. The fail-safe design criteria established by the manufacturer required that the fuselage be able to withstand a 40-inch crack without suffering catastrophic failure. The placement of the tear straps every 10 inches in both longitudinal and circumferential directions was designed to redirect any running crack in a direction perpendicular to the crack. This redirection of the crack was expected to cause the fuselage skin to flap open, releasing the internal pressure in a controlled manner.

Boeing demonstrated the ability of the fuselage to fail safely within two bays during certification testing. A guillotine test produced a 40-inch separation in a pressurized section of fuselage test panel. As expected, the tear redirected itself upon reaching a tear strap.

The complete fuselage of the 727 (predecessor of the 737) was fatigue tested for 60,000 cycles (one economic design life). For the 737, a representative section of the fuselage was fatigue tested for 150,000 cycles (two economic design lives). No fatigue cracking or disbonding was observed during this testing. However, no consideration was given to the joining of adjacent rivet hole fatigue cracks, the possibility of disbonding of the lap joint, or the effect of corrosion on the lap joint.

In 1986 (before the Flight 243 accident), Boeing acquired a 737 with just over 59,000 takeoff cycles. An additional 70,000 simulated takeoff, pressurization, and landing cycles were applied by Boeing engineers. Skin cracking was observed at 79,000 cycles and again at 89,000 cycles. Growing from both sides of the rivet holes, the fatigue cracks ranged in size from

0.37 to 0.67 inches from tip to tip. With additional pressure cycling, the cracks joined up to form a 32-inch crack at about 100,000 cycles. Testing continued till 100,673 cycles when the crack reached almost 40 inches, the skin flapped open (as shown in Figures 4.43 and 4.44), and a controlled pressure release occurred, as expected. However, adhesive disbonding did not occur during this testing.

What Went Wrong With Aloha Flight 243?

The fuselage remaining after the accident did not contain the origin of the failure. Most of the failed panels were lost at sea and never recovered. Attached portions of the remaining fuselage were examined for fatigue cracking. In one lap joint, fatigue was found in the knife-edge of the rivet holes of 16 consecutive rivets. The longest fatigue crack was 0.18 inches. Disbonding of the lap joint and tear straps was also found in this area.

Disbonding occurs in random locations, overloading nearby rivets and causing fatigue cracking in the rivet holes. The rivet hole cracks expand and link up as shown in Figure 4.53. Eventually, the linked fatigue cracks reach a critical length and the fuselage rapidly unzips.

The first officer of Flight 243 arrived at 5:00 a.m. for the preflight inspection, required before the first flight of the day. He parked on the lighted park-

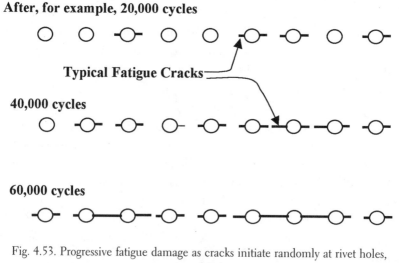

Fig. 4.53. Progressive fatigue damage as cracks initiate randomly at rivet holes, grow, and link up

ing area and inspected the exterior of the aircraft in predawn darkness. He found nothing unusual. No other visual checks were performed before the other eight flights of the day—none were required per FAA rules. A passenger reported (after the accident) noticing, during boarding, a skin crack running through a top row of lap joint rivets. The NTSB believes the top rivet row was visibly cracked before departure. It is unclear if the first officer missed the fatigue crack seen by the passenger, or if the fatigue crack appeared after the 5:00 a.m. inspection and before the eighth flight of the day at 1:35 p.m. Both are possible.

The NTSB concluded that disbonding and joining of multiple rivet hole cracks prevented the controlled decompression and fail-safe flapping mechanism from occurring. The multiple rivet fatigue cracks lined up and represented an unexpected weakness. Disbonded tear straps added to the weakness.

The NTSB believes that Aloha Airlines had sufficient information regarding lap joint problems to have implemented a maintenance program to detect and repair any lap joint damage, because

- Aloha's 737s were high-cycle planes, including the number 1 and number 2 (the accident plane) high-cycle planes in the world fleet;
- Aloha operated in a harsh, corrosive environment (salt water atmosphere);
- Aloha had discovered a 7.5-inch lap joint crack in another airplane;
- Boeing had issued a service bulletin covering inspection and repair of lap joints and addressing possible disbonding, corrosion, and fatigue of lap joints;
- the FAA had issued a similar Airworthiness Directive, which recommended the replacement of countersunk rivets with protruding head rivets. This replacement would require drilling out and eliminating the knife-edge. However, this was not done because of ambiguity as to which rivets needed replacing.

The NTSB faulted the Aloha Airlines maintenance program for not detecting the disbonding and fatigue damage.

Although this accident was preventable, the linkup of multiple fatigue damage did represent a partial defeat of the fail-safe design philosophy. The fact that the plane did not break up completely does prove the mettle of fail-safe design. Nonetheless, the Aloha Flight 243 accident is an unacceptable situation.

Aloha Airlines' Other 737s

Following the accident, the NTSB visually inspected the exterior of Aloha Airlines' other 737s. Much evidence of corrosion was found in the lap joints, including swelling and bulging of the skin, pulled and popped rivets, blistering, and scaling. Three other high-cycle Aloha 737s were pulled for "heavy" inspection and maintenance. These planes had 90,051 (the highest of any 737 in the world fleet), 85,409, and 68,954 cycles. After finding extensive fatigue cracking and disbonding, Aloha decided the two oldest planes were beyond economic repair. They were sold for parts and scrap. The third plane was out of service for a year receiving repairs.

High-cycle planes, if properly maintained, are not inherently unsafe. However, if fatigue cracking, disbonding of lap joints, and corrosion are not repaired in a timely manner, they will lead to accelerated fatigue cracking and eventual failure.

Industry Response to Aloha Flight 243

The day after the Aloha Airlines accident, the FAA issued an Airworthiness Directive applicable to all Boeing 737 planes with more than 55,000 pressure cycles. The directive required flight at reduced altitude (23,000 feet) until a visual inspection of specific lap joints could be completed.

A week later, the FAA expanded the required inspection to all 737s with more than 30,000 pressure cycles. Inspection of additional lap joints was required, and more sensitive inspection methods were added. (See eddy current discussion in Chapter 6.)

Within the next seven months, the FAA expanded the amount of required inspection a third and a fourth time. Repairs of specific lap joints were phased in depending on the number of cycles on the plane.

The national response to this problem was impressive. Major research programs were initiated and millions of dollars spent on research. In 1988, Congress passed the Aviation Safety Research Act of 1988. This act increased the scope of the FAA mission to include research into the causes, effects, and mitigation of fatigue and environmental degradation of aircraft structures. The first response to the accident was an industry-wide review of design and maintenance programs. It was concluded that with structural modifications and proper maintenance, the service life of planes could be safely extended. As a longer-range approach, the industry established the

Airworthiness Assurance Working Group (AAWG) and the FAA established the National Aging Aircraft Program and the National Aging Aircraft Research Program.

In 1993, the AAWG report "Structural Fatigue Evaluation for Aging Airplanes" identified 14 structural details to be monitored and evaluated. Meanwhile, the FAA, working with NASA and the USAF, identified three technical areas for research: fatigue and fracture, nondestructive inspection, and flight loads.

From this research, new analyses and inspection methods were developed. Eventually, the FAA phased in new rules requiring more damage-tolerance design methods and more inspection of aging planes. Eventually, new rules required full-scale fatigue testing to demonstrate that "widespread fatigue damage" would not accumulate in rivet holes or elsewhere.

The Aloha Flight 243 explosive decompression could have been avoided with appropriate maintenance. The near disaster served as a wakeup call. The national response was increased vigilance at all levels: researchers, designers, airlines, and regulators.

China Airlines Flight 611

The problems and solutions of explosive decompression are understood well enough by now to make any future occurrences extremely unlikely, even in aging aircraft. However, it can still occur due to human error.

On May 25, 2002, the worst-ever explosive decompression disaster took place. Radar data showed China Airlines Flight 611, a 747, experienced an in-flight structural breakup at 35,000 feet. Radar operators watched the plane break into four distinct pieces before disappearing from the screens.

The size of the debris field, roughly 1 mile × 3 miles, was consistent with an in-flight breakup. Recovery of the cockpit voice recorder showed no anomalies; in fact, the last recorded sound was the pilot humming a popular tune.

The fuel tanks and doors were found substantially intact, ruling out fuel tank explosion or blowout of the doors. There was no evidence of a bomb. Explosive damage leaves telltale patterns.

The plane had been delivered in 1979. A tail strike occurred during a hard landing in 1980. A temporary repair was completed the next day by installing two reinforcing doublers made of 0.062-inch-thick aluminum, essentially a temporary patch. At that time, it was specified that a permanent repair was to be made within four months.

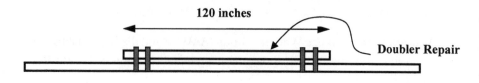

Fig. 4.54. A repair patch covered the 747 tail strike damage.

The plane was grounded three-and-a-half months later for a fuselage bottom repair. Investigators were never able to find any engineering process documentation on this repair, nor did Boeing find any record indicating completion of the permanent repair.

A 120-inch-long, 22-inch-wide, 0.10-inch-thick repair patch was installed (Figure 4.54). The repair patch was riveted with two rows of rivets on each side. The airplane had accumulated 21,398 cycles since new and 20,631 since the time of the repair.

The tail strike region and repair patch were recovered from the debris field. Evidence of scratches in an area 120 × 20 inches was observed. The fuselage skin thickness had been reduced with some attempt being made to polish out the scratches. Long scratches as deep as 0.009 inches were found on recovered sections of the tail strike region. Multiple fatigue cracks were found in the scratches, including one over 15 inches long. Although the investigators were unable to determine the length of cracking prior to the accident flight, analysis suggests a crack of at least 71 inches in length— a crack long enough to cause unstable crack growth and structural separation of the fuselage.

Fatigue cracks initiated out of the tail strike scratches and grew for 22 years. To "fix" a scratch, Boeing permits removal of skin damage up to 20% of the original thickness for scratches less than 10.2 inches long and 15% for longer scratches. Otherwise, the scratch must be cut out and removed.

The scratches on the Flight 611 airplane should have been cut out. The Boeing repair manual called for cutting out scratches this widespread and adding replacement material. The repair should have been covered with a patch that exceeded the cutout by at least three rows of rivets. Boeing approval (no documentation was ever found) is required for a repair this extensive.

Common sense would indicate that a patch twice the thickness of the original skin would make the fuselage twice as strong. Not so, writes Steve Swift in the 1996 spring edition of *Flight Safety Australia*. A stronger patch can, in fact, make the structure weaker. The load must be correctly transferred from the fuselage skin to the patch. If there are multiple rows of rivets connecting the patch to the skin, a thicker patch disrupts the load sharing and concentrates even more loading on the first row of rivets. The end result could be a less fatigue-resistant structure.

Every aircraft suffers cracks, corrosion, and occasional scrapes with service vehicles, and it all gets patched up. A hundred patches on an aircraft is not unusual.

Although a not entirely satisfying explanation, correct repair technique and adequate inspection would have prevented the Flight 611 disaster. More recently, the FAA has mandated new inspection requirements for repair patches.

Incidentally, China Airlines has one of the worst safety records in commercial aviation, with nine fatal crashes between 1970 and 2002.

More Balloon Testing

Numerous balloon tests have been described in this chapter. I will use one last test to demonstrate how appropriate structural reinforcement will convert a structure susceptible to explosive decompression when damaged to a safer condition of slow decompression. I cut duct tape to a width of about 0.5 inches and wrapped it circumferentially around a cylindrical balloon expanded to a 2.75-inch-diameter. I also simulated two reinforcing circumferential ribs with duct tape.

If the tape, or ribs, are far enough apart, they don't seem to have any effect. The balloon still pops, though the tearing may be less. If the reinforcements are too far apart, they are incapable of arresting the tear and the tear passes right through the tape. The same effect is seen in a fuselage. The running crack will gain speed, and the longer and faster it is, the harder it is to arrest. The fatigue cracks must be arrested before they become too long and start moving with excessive energy.

If the duct tape frames are placed close enough, the tear can be forced to change from a longitudinal to a circumferential tear, albeit a jagged circumferential tear. In this case, the loop stress is being reduced to the point where the axial stress interacts with the tearing.

If, in addition to the circumferential frames, longitudinal ribs of duct tape (i.e., stringers) are also added, the balloon will eventually become fail-safe. The balloon does not pop, but rather, it quickly decompresses with a hiss.[24] The balloon does not tear past the tape. The spacing of the tape seems to be very important. The balloons used (different balloons may give

ATOMIC BONDING: LATEX BALLOON VERSUS ALUMINUM FUSELAGE

Numerous principles about fuselages have been illustrated with a balloon. There is, of course, a fundamental difference between rubber and aluminum. The balloon, upon pressurization, will expand 300–400%, and it will snap back violently during decompression. The aluminum fuselage will also expand and snap back, but by a considerably smaller amount. The Boeing 737 fuselage, in between the reinforcing ribs, will expand about 0.2 inches when pressurized and snap back by this amount when depressurized.

The aluminum atoms form a tightly packed crystal. Loading the crystal stretches the individual atomic bonds. Upon removal of the load, they rebound like tiny springs. The long molecular polymer molecules in the balloon are analogous to a strand of spaghetti or string.

The backbone of all polymer molecules is a repeating carbon-carbon bond. (Different polymers have different atoms or molecules hanging off the carbon atoms.) The number of carbon-carbon bonds can be tens of thousands or more.

Although the carbon-carbon bond remains constant at an angle of 109.5 degrees, the long chain of carbon bonds is relatively flexible and free to rotate out of the plane of the page. This results in severe kinking of the long and skinny molecule, analogous to a rolled up ball of string (Figure 4.55). When the two ends of the balled up molecule are pulled on under a load, the string elongates. All unreinforced polymers will stretch considerably longer than metals will because of this effect. In the case of some rubbers, the stretch can be several hundred percent of the original length.

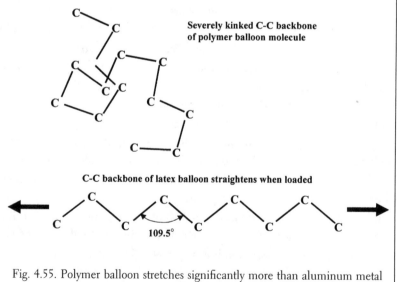

Severely kinked C-C backbone
of polymer balloon molecule

C-C backbone of latex balloon straightens when loaded

Fig. 4.55. Polymer balloon stretches significantly more than aluminum metal crystals because of the nature of polymer bonds.

different results because of thicker rubber, higher-strength rubber, manufacturing defects, or a different diameter) failed safe with about ⅝ inch between circumferential ribs and ½ inch between longitudinal ribs. The
failure of the balloon then changed from explosive decompression to slow
decompression. This is a very important distinction for a plane at 35,000
feet.

In the spirit of full disclosure of the experimental results, and it is unclear if it matters, I used two complete circumferential wraps of duct tape.
Also, I wrapped the balloon tightly enough to see a slight squeeze in it.

5...

Jet Propulsion, Burst Engines,

and Reliability

United Airlines Flight 232

This chapter tells the story of the crash of United Flight 232 and the national research effort that resulted from it. Also explained are jet engine operation and design and the forces that caused the engine on Flight 232 to fly apart.

IN YET ANOTHER EXAMPLE of why commercial pilots occasionally need every one of their thousands of hours of experience,[1] Captain Al Haynes, on July 19, 1989, crash-landed a DC-10 in Sioux City, Iowa, without any hydraulic controls. This means the rudder, ailerons, flaps, elevators, and all other flight control surfaces were unresponsive.

Upon landing, the right wing tip struck the ground, breaking off the wing and spilling fuel. The plane flipped upside down, burst into an enormous fireball, and broke into multiple sections as it cartwheeled and skidded more than 1,200 yards down the runway. Miraculously, 185 of the 296 occupants survived. The flight crew were universally hailed as heroes.

Shortly after the accident, an NTSB official reported the problem with the plane's hydraulics and further stated he could not explain how the flight crew had retained even limited control from the cockpit. Other experts marveled that "it was quite a feat keeping the plane in the state of Iowa, much less getting it to the end of the runway."[2]

As explained by Captain Haynes after the fact, any unusual occurrence results in cockpit confusion for about 30 seconds—the time it takes to look up the appropriate procedure and checklist in the flight manual. Imagine their dismay as they discovered that there was no procedure, emergency or otherwise, for flying the plane without hydraulic controls.

Another option for the flight crew was to call United Airlines' maintenance headquarters for assistance from their experts who are on call for any flight emergencies. Reviewing the transcript of that conversation, one wants to scream, How many times do they have to say they lost all hydraulics? There simply are no contingency plans for flying an airplane without hydraulics because (1) with system redundancy, it is considered virtually impossible to lose all hydraulics or, at least, it is beyond belief, and (2) if it does happen, it is impossible to fly the plane. This explains the response of United's maintenance expert who asked a repeated variation of the question, You lost what?

United Flight 232 left Denver bound for Chicago at 2:09 p.m. and reached a cruising altitude of 37,000 feet. The first indication of a problem was a loud bang followed by a lurch and shuddering of the plane. Flight attendants were knocked off balance and some to the floor. The passengers remained calm.

Meanwhile, in the cockpit, the flight crew noticed from the instrument panel that the tail-mounted engine had failed. After shutting down the damaged engine, the first officer announced, "I can't control the airplane." The plane was turning right in spite of being given inputs to turn left. As Captain Haynes told the story in May 1991: "I said the dumbest thing I've ever said in my life, I said, 'I got it.' I didn't have it for very long. Because we immediately determined that we could not control the airplane."[3]

Apparently, the tail engine (Figure 5.1) had burst, damaging all hydraulic controls in the airplane. The damage resulted in hydraulic fluid leaking out of all three hydraulic systems. Besides lack of control, the flight crew had two other problems: the plane was slowly pitching up and down like a slow-motion roller coaster, and it was rolling to the right.

When the power to the engines is reduced, there is less speed and lift. The nose drops but airspeed picks up as the falling plane begins to accelerate from gravity. The airplane tries to fly its "trim" speed. With increased speed and lift, the nose lifts, speed decreases, and the cycle repeats with the plane pitching up and down. This pitching motion is a natural tendency of all planes and occurs whenever the plane's airspeed or pitch is disturbed from its equilibrium condition. Pilots reflexively make minute adjustments to the control surfaces to prevent these oscillations. However, pilots have no training or experience making these adjustments with en-

Fig. 5.1. Tail engine of DC-10, similar to the one that burst on Flight 232.
Photo: Anders Olsen

gine thrust only. Under great duress, the flight crew had to relearn how to fly the plane.

Many of the required new flying skills were counterintuitive. For example, when the nose starts to go up, the plane slows down, approaching a stall. Flying rule number one is to avoid stalling at all costs. A stall occurs when the angle of the wings relative to the air stream is such that lift is insufficient to support the plane's weight. When a plane stalls, it falls nose down, dives, and potentially crashes.

When the nose pitches up, the throttle must be reduced to dampen the pitching motion, an exactly opposite response to avoiding a stall. The throttle had to be reduced, but not too much.

The damaged tail resulted in asymmetric airflow. In addition to pitching, the plane tried constantly to roll to the right. Without engine adjustments, the plane would roll over and enter a dive. Three times the plane rolled (the wings tilted) up to 38 degrees and was close to flipping over.

About 13 minutes into the crisis, the captain was notified that Denny Fitch, an off-duty United Airlines DC-10 training pilot, was aboard and had offered assistance. Fitch was quickly invited into the cockpit.

Captain Haynes picks up the story:

It was sort of funny listening to [and] reading the transcript because [Fitch is] about fifteen minutes behind us now, and he's trying to catch up, and everything he says to do we've already done. And after about five minutes—that's 20 minutes into this operation—he says, "We're in trouble!" We thought: that's an amazing observation, Denny. And we kid him about it. . . . When he [finally] found out that he didn't have any knowledge for us, he says, "Now, what can I do?" I said, you can take these throttles, and try to help us. . . . So what he did, he stood between us . . . and he took one throttle in each hand, and now he could manipulate the throttles together. . . . And we said, give us a right bank, bring the wing up, that's too much bank, try to stop the altitude. He'd try to respond. And after a few minutes of doing this, everything we'd do with the yoke, he could correspond with the throttles. So it was a synchronized thing between the three of us. . . . So that's how we operated the airplane, and that's how we got it on the ground.[4]

They still had very limited controls and could only turn right. The plane was still pitching up and down, albeit at a manageable rate. After a series of five right turns (some made on purpose to reduce the altitude and some the airplane did on its own), the plane lined up with the Sioux City Airport for an emergency landing.

The last instructions from the captain to the head flight attendant were "you'll get the command signal to evacuate, but I really have my doubts you'll see us standing up, honey. Good luck, sweetheart."

To land safely, the plane must be flying straight and level with forward and downward velocities within safe limits. A DC-10 normally lands, depending on wind, weight, and other conditions, at about 160 mph forward velocity and 3 to 5 feet per second sink rate.

During takeoff and landing, the airplane's velocity is relatively low. (Lower airspeed corresponds to less airflow over the wings and less lift.) To increase lift during these slower speeds, the wing area is increased by adding moving parts on the wings' leading and trailing edges. The part on the leading edge is called a slat, while the part on the trailing edge is called a flap. The flaps and slats move along metal tracks built into the wings. The large flap area also increases air resistance or drag. This helps the airplane slow down for landing. Because of the hydraulic damage on Flight 232, the flaps and slats were not extended.

United Flight 232 was attempting to land, without any meaningful controls, at almost 250 mph and a sink rate of over 30 feet per second. (Chapter 2 discusses sink rates for other accidents.)

The Crash Landing of Flight 232

The first impact, from the main landing gear, gouged an 18-inch-deep hole in the foot-thick concrete runway.

The center section (rows 9–30) contained the majority of passengers and came to rest, inverted, about 3,700 feet from initial impact. This section was eventually destroyed by the post-crash fire. Thirty-three of the 35 passengers who died from smoke asphyxia were in rows 22 to 30. The ceiling collapse was greatest in this area and complicated evacuation for the fire victims. Twelve of the 33 fire victims also had impact injuries, which further slowed their evacuation. The remaining two fire victims, seated outside rows 22–30, were elderly and had difficulty exiting. The fire burned for two hours.

The tail section and about 10 seats broke off reasonably intact and with limited serious injury. Except for the tail section, the cabin behind the wings was substantially destroyed by impact. First class was similarly destroyed (Figure 5.2). With four major fuselage breaks, the plane separated into three pieces. Not all the breaks resulted in separation.

Fuselage breaks indicate high-impact forces and are usually associated with serious injury. Fuselage breaks also disrupt the seat-floor attachments. Detached seats release and injure passengers. This is similar to an automotive accident when seat belts are unused and passengers are ejected from the car. External flames from fuel leaks can also enter the cabin through breaks in the fuselage.

The cockpit was totally demolished and crushed around the flight crew. However, the shoulder harnesses and lap belts remained intact and restrained the three pilots (and Denny Fitch), who all survived—albeit with serious injuries. (The first officer had broken ribs and a broken pelvis.) It was over 35 minutes before they were extracted with the aid of a forklift. Part of the delay involved disbelief about anyone surviving inside such wreckage.

There were four infants on board, roughly 1 to 2 years old. Their parents were instructed to place the babies on the floor and hold them in place after assuming the crash position (head between knees). Two babies re-

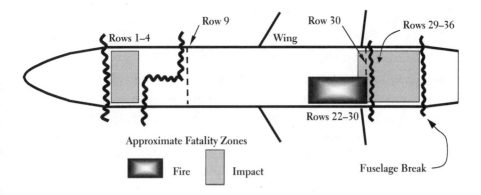

Fig. 5.2. Approximate fuselage breaks in Flight 232 and impact and fire fatalities concentration zones

mained in place, and two slid away during the impact. One was safely rescued by another passenger who heard cries after exiting the plane. He reentered the burning fuselage and retrieved the child. The fourth baby, in row 22, died of smoke inhalation. The other three babies were in rows 11,12, and 14.

Flight 232 Simulator Studies

A simulator was programmed with the aerodynamic characteristics of the tail-damaged DC-10. Forty-five simulated flights were flown by qualified DC-10 line captains, training captains, and test pilots to determine if flight crews could be trained to safely land a DC-10 without hydraulic flight controls.

It was quickly decided that there were too many variables to control. Landing speed, touchdown point, direction, altitude, and vertical velocity could all be controlled separately, but it was virtually impossible to control all of them simultaneously. The tendency of the plane to pitch up and down and roll to the right added to the control problems. The NTSB concluded that the plane could not be safely landed even with additional training. The NTSB further stated "that under the circumstances the UAL flight crew performance was highly commendable and greatly exceeded reasonable expectations."[5]

To understand why all hydraulic systems failed and why everyone found that remarkable (if not unbelievable) requires a review of the hydraulic system.

The Hydraulic Control Systems

Beginning with the Wright brothers, movement of flight control surfaces was accomplished from the cockpit with cables. As technology developed and airspeeds increased, the magnitude of the forces required became problematic. The conventional cable system evolved into two separate circuits. The first circuit, a mechanical system of cables directly linked to the cockpit controls, opens and closes ports in servo valves. Servo valves pressurize (and depressurize as needed) tubing in a second hydraulic circuit that directly moves the flight control surfaces.

Hydraulic systems use an incompressible fluid, usually oil, to transmit forces from one location to another. The concept is based on Pascal's Principle which states that when pressure is applied to a confined incompressible liquid, this pressure is transmitted, without loss, throughout the entire system and to the container walls. The simplest example is inserting a cork in a bottle completely filled with water. The pressure on the cork is force divided by the area of the cork. Assuming there is no friction in the cork, this pressure is uniformly transmitted to all surfaces of the container. (The bottom of the container has a slightly higher pressure because of the weight of the column of water acting on the bottom.)

A common application of Pascal's Principle is the hydraulic lift used to raise a car in an auto repair shop. A small force is applied to a small piston, which creates the same pressure on a large piston, but with a much larger force (force = pressure × area). Conservation of energy requires the larger piston to move a shorter distance in proportion to the ratio of the area of the small piston divided by the area of the large piston.

The airplane's hydraulic circuit consists of pumps, piping, valves, and actuators. Hydraulic pressure created by a pump moves through 0.5-inch tubing until it reaches an actuator. An actuator is typically a piston inside a cylinder (Figure 5.3). Hydraulic pressure on an aircraft usually operates at 3,000 pounds per square inch, resulting in tremendous force magnification of the pilot's manual input. A 2-inch-diameter piston would create a cylinder force of almost 9,500 pounds.

The servo valves can reverse the flows and motion of the actuator or hold the piston in one location. This would correspond to moving the flaps or slats in and out or holding them in one position.

The actual design of a particular circuit and component is quite complicated with many additional requirements. Mechanical stops are added

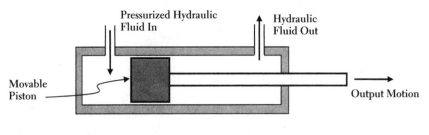

Fig. 5.3. Hydraulic actuating cylinder

to limit motion that can damage the plane. For example, in Chapter 3, I described how excess rudder movement broke the tail off an Airbus 300. That particular rudder was designed to rotate ±30 degrees at air speeds below 165 knots and ±3 degrees at speeds over 395 knots, with additional limits for in-between speeds. Added machinery simulates the "touch" of the increased pilot input forces required at higher speeds, and this leads to rather complex hydraulic circuit designs with their own "logic."

The latest designs replace the mechanical cable circuit with "fly-by-wire" electronic circuits. Hydraulic actuators are still necessary to create the large forces required to move the flight control surfaces. Increasingly, computers monitor hundreds of sensors around the aircraft and enforce the "flight envelope" to prevent instability and structural damage. For example, if the nose pitches up too high, the wings can stall and lose lift; if the wings bank at too high an angle, the plane can flip over; and if a turn is too tight resulting in excessive centrifugal forces, the wings can break off or the tail can break, as already described in Chapter 3.

The latest Boeing and Airbus designs approach this control problem with two different philosophies. The Airbus has "hard" protection features in which the computer prevents the pilot from exceeding the flight envelope. Boeing believes the flight crew, not the computer, should have final say in an emergency situation. The Boeing computer provides increasing audio and visual alarms while increasing the required yoke forces that make it progressively more difficult, but not impossible, to exceed the limits.

As an example, for bank angle protection in the Boeing 777, a warning light comes on at 30 degrees and an audio warning occurs at 35 degrees. At 60 degrees, the yoke force increases from 10–15 pounds to 60–70 pounds. The plane will still bank farther, but only with very high pilot input forces. Upon yoke release, the plane returns to stability with a 30 degree bank. With an Airbus, the bank angle is restricted to 67 degrees or 45 degrees, depending on the plane's angle of attack (angle of the wings relative to the airflow).

A hydraulic pump is required to generate the fluid pressure in the system. Smaller pressure would require bigger actuating cylinders, resulting in a weight penalty.[6] Reliability of the pump and leak free hydraulic tubing is critical for safe operation of the plane. For these reasons, considerable redundancy is built into the flight control hydraulic system.

In the DC-10, there are three different, unconnected hydraulic systems. The lines are physically separated to minimize vulnerability to structural damage. If the fluid leaks in one of the systems, it does not affect the other two. Each hydraulic system is powered by a primary and a reserve pump driven by a different engine.

The systems are also connected mechanically through an electric pump so that, if the pressure in one system is lost, the pressure in another system will immediately start a motor that will drive a pump in the failed system. If, for example, engine no. 1 shuts down and stops the pump for hydraulic system no. 1, a second pump operating from engine no. 2 takes over to pressurize system no. 1.

As a last resort, there is an air-driven generator, essentially a wind turbine that drops out of the fuselage and runs a motor to provide hydraulic system pressure. This device is automatically activated by a variety of circumstances such as total fuel exhaustion and shutdown of all three engines.

For safety and redundancy, every flight control surface is operated by two different hydraulic systems. Although some of the flight control surfaces might be inoperable, the plane can be safely flown with two of the three hydraulic systems compromised.

The probability of total hydraulic failure was estimated at something like one chance in a billion. Believe it or not, this is a meaningful calculation. Instead of three hydraulic lines, consider three coins. The probability of the first coin being heads is $\frac{1}{2}$. The probability of two coins both being heads is $\frac{1}{2} \times \frac{1}{2} = \frac{1}{4}$. (Given two coins, heads-heads is one of four possible outcomes.) The probability of three coins all being heads is $\frac{1}{2} \times \frac{1}{2} \times \frac{1}{2} = \frac{1}{8}$.

The frequency of hydraulic failure is well understood from historical data. If a hydraulic failure occurs once every 1,000 flights, then the probability of all three hydraulic lines failing simultaneously is $\frac{1}{1,000} \times \frac{1}{1,000} \times \frac{1}{1,000}$ or one chance in a billion. There is one obvious flaw in this logic: one event causing failure of all three systems was not considered.

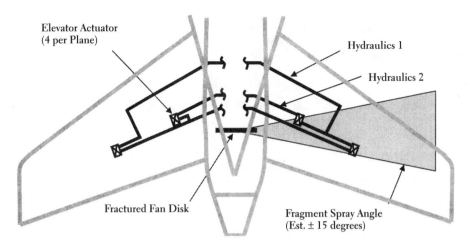

Fig. 5.4. Hydraulic lines in the tail section of a DC-10, and an approximate fragment spread angle from the damaged fan disk. *Image:* FAA

Everyone was confident that complete loss of all flight controls was impossible, which explains the initial confusion from United's ground personnel and Denny Fitch. They simply could not understand how there could be a total hydraulic failure. Flight crews are required to manage two failed hydraulic systems periodically in flight simulators as part of normal, ongoing training, but McDonnell Douglas, the FAA, and United Airlines considered failure of all hydraulics so remote that no procedures were ever developed for this situation.

Nothing works if hydraulic fluid is drained from all three systems. The designers did not consider this possibility. In all fairness to the engineers, the list of possible failure modes that they design for is more or less a list of historical failures. If it has never happened before, it does not get on the list until somebody thinks of it—usually after an accident.

The tubing for the three hydraulic systems is widely spaced, except in the tail where there are four control surfaces: left and right inboard and outboard elevators. The hydraulic tubing in the tail section is lined up with the tail engine fan disk that flew apart on Flight 232 ripping out all three hydraulic systems. The estimated trajectories by the engine fan disk fragments are shown in Figure 5.4. Small fragments had a spread angle of roughly ±15 degrees (the angle the fragments sprayed out of the engine). A heavier and more damaging fragment—one-third of the fan disk (explained in the next section)—had a spread angle of ±3 degrees.

Fig. 5.5. A child throwing bricks from a wagon illustrates conservation of momentum.

To understand why the engine fan disk burst, some background on jet engine operation, design, and manufacture is helpful.

Jet Propulsion

Several fundamental principles of physics can be used to explain how a jet engine works.

One such principle is Newton's Third Law: For every action there is an equal and opposite reaction. Usually, this is explained with the motion of a balloon. The air shooting out of the balloon is the action, and the motion of the balloon in the opposite direction is the reaction.

The balloon and the jet engine also illustrate conservation of momentum, where momentum = mass × velocity. Since the momentum of the balloon-air system is conserved (neither created nor destroyed) and remains equal to zero, an equation can be written to describe the motion of the balloon. The rearward momentum of the air molecules exiting the balloon equals the forward momentum of the balloon.

(Mass of air) × (velocity of air)
$$= \text{(mass of jet engine)} \times \text{(velocity of engine)}.$$

A child tossing bricks from a wagon also illustrates conservation of momentum. The wagon moves forward as the bricks are tossed out the back (Figure 5.5). Because the momentum of the brick-wagon system remains zero, the backward momentum of the brick must equal the forward momentum of the wagon. Knowing any three of the following variables easily allows calculation of the fourth using the conservation of momentum equation as follows:

(mass of wagon) × (velocity of wagon)
$$= \text{(mass of brick)} \times \text{(velocity of brick)}.$$

The mass of the plane is analogous to the wagon. A continuous stream of bricks is similar to the jet engine's exhaust.

Unbalanced pressure inside the balloon is another way to look at engine thrust. When the balloon nozzle is pinched, the pressure forces of the air on the inside of the balloon are balanced in all directions, resulting in zero propulsion. If the balloon's nozzle is open, the gas cannot push against the opening. An unbalanced force occurs opposite the nozzle, resulting in a forward engine force or "thrust" (Figure 5.6).

A balloon is a pressurized container that quickly loses pressure from air leakage. A jet engine is a container filled with turbo-machinery designed to suck in air and maintain constant pressure. The pressure is maintained with combustion of a fuel-air mixture. The maximum pressure in the combustion chamber, depending on the design, is about 20 to 42 atmospheres (300–600 psi).

When the fuel-air mixture ignites, the rapid expansion of combustion gases pressurizes the engine and blasts exhaust gas out the back. The inlet and outlet pressures of the jet engine are important parameters monitored by the pilot as indicators of engine performance and thrust.

The combustion process requires massive quantities of air. The front end of the jet engine, known as the compressor, provides the air. The compressor forces air into the higher pressure combustor by increasing

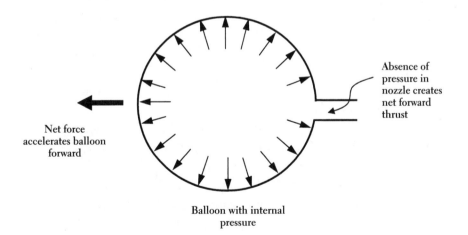

Absence of pressure in nozzle creates net forward thrust

Net force accelerates balloon forward

Balloon with internal pressure

Fig. 5.6. Pressure cancels everywhere except opposite the open nozzle. The unbalanced pressure creates forward net thrust = internal pressure × area of nozzle.

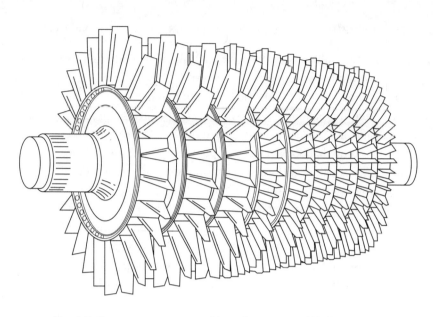

Fig. 5.7. Compressor rotor assembly with nine sets of blades or stages

the pressure (and temperature) of the air. Depending on the particular engine, the air is compressed to between ¹⁄₁₀ and ¹⁄₁₅ of its original volume outside the engine. The compression is accomplished by a rotating shaft containing many blades set at an angle (Figure 5.7). Each blade is something like a miniature propeller blade. The effect is similar to many fan blades rotating and blasting air into the engine. Usually, there will be as many as 10 to 15 stages of compression, corresponding to the 10 to 15 sets of blades.

Compressed air is blasted into the combustor. After fuel is added to the air and burned, the combustion gases expand, increase the pressure, and blast out the engine exhaust.

What rotates the compressor shaft? A set of blades on the other end of the shaft, past the combustion section, rotates this shaft. The angle of the blades is set such that combustion gases moving past will cause rotation, similar to a windmill, of the turbine (Figure 5.8). The compressor section acts like a fan blowing air into the engine. The turbine section acts like a windmill that rotates when combustion air blasts over it.

As air moves through the compressor, the velocity, temperature, and pressure of the air increase. Because of tremendous changes in air pressure and temperature (and corresponding density changes) occurring across the compressor, it is more efficient for different stages of the compressor to operate at different speeds. The most common solution to this design problem is to have two compressors operating at different speeds. This is accomplished with two shafts and two turbines, as shown in Figure 5.9.

Each rotor has its own "redline" limits or revolutions per minute (rpm) not to be exceeded, as it can result in damage to the engine. Eventually, excessive rpm will rip the rotating parts of the engine apart with centrifugal forces, just as swinging a weight on a string over your head with sufficient rpm will eventually break the string, a concept relevant to the engine failure of Flight 232. Typical rotational speeds for the low-pressure compressor and low-pressure turbine are about 3,600 rpm. Typical rotational speeds for the high-pressure turbine and high-pressure compressor are about 10,000 rpm.

Turbofan: Modified Jet Engine Design

Early jet engines were very inefficient. The exhaust gases came blasting out near the speed of sound. This kinetic energy (the energy of motion) of the hot gas is wasted. Returning to the wagon-brick system, once the brick has left the child's hand, it has obeyed the conservation of momentum and propelled the wagon forward. However the brick now has considerable kinetic energy. It would be useful to somehow capture some of that energy. One possibility is to have the brick strike an angled blade attached to a

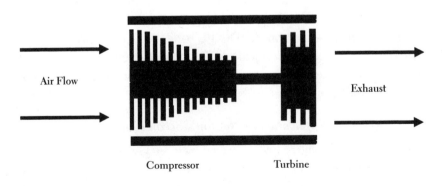

Air Flow

Exhaust

Compressor   Turbine

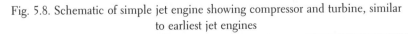

Fig. 5.8. Schematic of simple jet engine showing compressor and turbine, similar to earliest jet engines

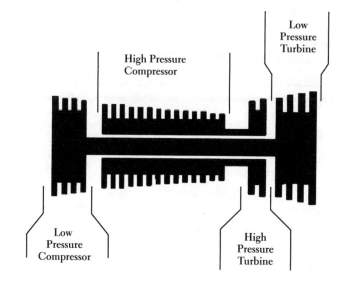

High Pressure
Compressor

Low
Pressure
Turbine

Low
Pressure
Compressor

High
Pressure
Turbine

Fig. 5.9. Most modern jet engines have two different turbines,
two different shafts, and two compressors.

shaft. The shaft begins to rotate. On the other end of the shaft is a large
fan that also rotates, providing additional thrust for the wagon. Although
this awkward contraption could in theory work, it is much easier to imagine the implementation of this concept in an actual jet engine. A second
turbine is added and rotated by the exhaust stream and the turbine rotates
a shaft. On the other end of the shaft, in front of the engine, is a large fan
(Figure 5.10). The fan acts like a propeller, except the fan blades are inside a duct attached to the engine (Figure 5.11).

In large engines used on the DC-10, B747, B757, B767, A300, A310,
and others, as much as three-quarters of the thrust delivered by the engine
is developed by the fan. An added benefit is noise reduction. The exhaust
gas is now slower and quieter.

Application of Newton's Second Law

United Airlines Flight 232 was powered by three General Electric CF6-6
engines, each with 40,000 pounds of thrust. The acceleration of the
DC-10-10, with a takeoff weight of 369,268 pounds, can be calculated with
Newton's Second Law:

$$\text{force} = \text{mass} \times \text{acceleration.}$$

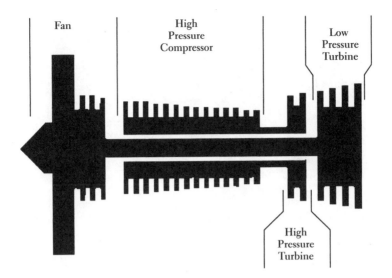

Fan

High
Pressure
Compressor

Low
Pressure
Turbine

High
Pressure
Turbine

Fig. 5.10. A fan is added in front of the engine to increase thrust.

Fig. 5.11. General Electric CF6-6 engine used in Flight 232. Note large fan
section in front.

To make the calculation correctly, the distinction between mass and weight needs to be considered. Since weight is a force on a mass from gravity,[7] Newton's Law can be used to define mass.

$$\text{Weight} = \text{mass} \times \text{acceleration from gravity, so}$$
$$369{,}268 \text{ lbs} = \text{mass} \times 32.2 \text{ ft/sec}^2, \text{ or}$$
$$\text{mass of DC-10} = \frac{369{,}268 \text{ lb sec}^2}{32.2 \text{ ft}}.$$

With mass defined, we can now calculate acceleration as

$$\text{acceleration of DC-10} = \frac{\text{force}}{\text{mass}} = \frac{32.2 \text{ ft} \times 3 \times 40{,}000 \text{ lb}}{369{,}268 \text{ lb sec}^2}$$
$$= 10.46 \frac{\text{ft}}{\text{sec}^2}.$$

At this acceleration, it would take the plane a little more than 30 seconds to reach a typical takeoff speed of 215 mph. An actual value would be somewhat longer to overcome air drag and other frictions.

The 40,000-pound thrust of the CF6-6, certified for flight in 1970, should be compared to the first American-made jet engine, also made by General Electric in 1942, and its thrust of 1,250 pounds. More recently, the General Electric GE90-115B, certified in 2004 for the newer, long-range version of the Boeing 777, is a record breaker for the largest sustained thrust of 122,965 pounds. Those three thrust numbers summarize continuous engineering design improvements in jet engine technology.

The basic jet engine concept is quite simple.[8] Simplicity adds reliability compared to a piston engine. Performance optimization requires incredible feats of engineering design.

Fan Disk Design and Manufacture

The knowledge base for a single engine component, just one block of metal, would support two or three university engineering courses. Even then, each course would be a superficial treatment based on available public information. The really good stuff is held close to the vest and labeled proprietary. The engine designers will tell you, justifiably so, that they spent

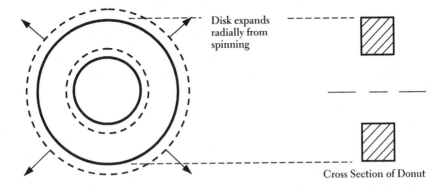

Disk expands radially from spinning

Cross Section of Donut

Fig. 5.12. Simplified fan disk represented as a disk with a center hole. When rotating, centrifugal forces expand the disk radially to the dashed lines (greatly exaggerated).

millions of dollars figuring it out and are not going to give it away to the competition.

With access to the corporate secrets, the story becomes much more complex. Everything starts to interact: metallurgy, casting, forging, machining, stress analysis, fatigue, fracture, propulsion, thermodynamics, heat transfer, fluid flow, and computer simulations. Massive test programs for these listed items also interact. Most of an engineering degree, and advanced degrees, could be taught as background material and connected to just a single component of the engine. If an entire engineering degree focused on one engine part, the new graduate would be very useful to the jet engine company but would still be considered a rookie.

Our purpose here is to outline some of the more important design requirements based on the fundamental physics involved.

The internal parts of the engine must be designed for many different operating conditions. The primary loading on all rotating components comes from the centrifugal forces trying to expand the fan disk radially, as shown in Figure 5.12.

The radial expansion is similar to the expansion that occurs in a pressurized pipe, fuselage, or cylindrical balloon and results in hoop or circumferential stresses, as described in Chapter 4.

The outside diameter, or outermost region of the fan disk, supports the least spinning mass and has the lowest stresses. The inside diameter supports the most spinning mass and, therefore, has the highest stresses. This is similar to the weights on a string, depicted in Figure 5.13. If the weights are all equal and swung over the head, the innermost piece of string supports the most spinning mass and has the highest tensile stress trying to rip it apart.

The force the swinging weights exert on the string is called the centrifugal force. The equal but opposite force the string exerts on the weights is called the centripetal force. If a half-pound weight on the end of a 27-inch-long string (approximately equivalent to the center of gravity of a fan blade on the Flight 232 engine) is swung over the head in a circle at two revolutions per second, the tension in the string is

$$\text{tension} = \text{mass} \times \text{radius} \times (\text{angular velocity})^2.$$

With appropriate conversions,[9]

$$\text{tension} = \frac{0.5 \text{ lb sec}^2 \times \text{ft } 27 \text{ in}}{32.2 \text{ ft} \times 12 \text{ in}} \left(\frac{2 \text{ revolutions}}{\text{second}} \right)^2 \times \left(\frac{2\pi \text{ radians}}{\text{revolution}} \right)^2$$
$$= 5.5 \text{ lb.}$$

The CF6 fan blade weighs about 10 pounds and rotates at 3,490 revolutions per minute or rpm. Swinging a 10-pound fan blade on the end of a 27-inch string at 3,490 rpm, the tension in the string is about 93,300 pounds. Note that there are 38 fan blades trying to rip the disk apart for a total centrifugal radial force of over 3.5 million pounds. This force is applied uniformly every 360/38 = 9.47 degrees, as shown in Figure 5.14.

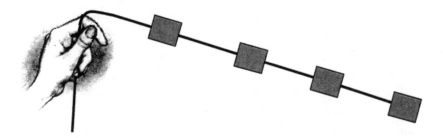

Fig. 5.13. Rotation of weights on a string results in tensile stress in the string. The innermost piece of string supports the most weight and, therefore, has the highest tensile stress.

Fig. 5.14. Over 3.5 million pounds of centrifugal force distributed circumferentially tries to rip apart the fan disk.

As shown in Figure 5.15, the actual cross-sectional shape of the disk is more complicated than a donut. However, the most highly stressed region remains on the inside diameter. The disk arm, which bolts the disk to the shaft, provides some local reinforcement. This moves the region with highest stress toward the front of the bore, which is where the disk started to fracture on Flight 232. Figures 5.16 and 5.17 show photographs of a CF6 fan disk.

There are two main design requirements for all rotating components in a jet engine: (1) a minimum excess rotational speed that causes the component to break, called a burst margin, and (2) a minimum number of takeoff-landing cycles that the component must tolerate. A burst margin is required to address the possibility of the engine controls breaking down or of the fan shaft failing (i.e., the fan shaft breaks, so there is no load on the engine, but the turbine still accelerates all components with excess rotational speed).

Design for Burst

To provide an adequate margin against burst during normal service, engine certification testing requires the engine to operate at 120% of design rpm for at least two minutes. Depending on the design, the engine may spin itself apart shortly above this required certified test condition. Actual failure depends on the detailed design of a particular engine. This means that during normal operations the engine is operating near $1/1.20 = 0.83$ or 83% of its burst speed.

The engine is relatively close to spinning itself apart at all times, a remarkable concept to comprehend. As is the case for all things that fly,

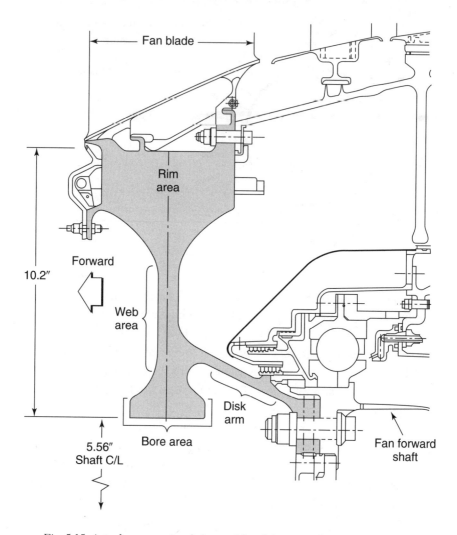

Fig. 5.15. Actual cross-sectional shape of fan disk is complex. Fatigue crack in Flight 232 initiated in the region with the highest stress, the forward section of the bore. The disk arm bolts to the fan shaft. *Image:* NTSB

weight is very important, so all aerospace components operate relatively close to their failure point. This is not meant to be an indictment of aircraft safety, but compared to ordinary items: airplanes do operate on the ragged edge. They are safe, statistically speaking, because of billions of dollars of design, analysis, and testing; unparalleled follow-up maintenance and inspection; and extensive crash investigation.

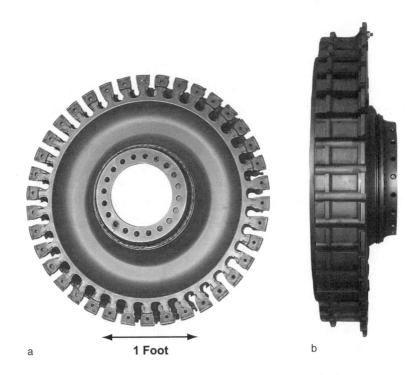

a   **1 Foot**   b

Fig. 5.16. Titanium fan disk from CF6 engine, face and side views

Fig. 5.17. Fan disk with one of 38 fan blades

Design for Metal Fatigue

The fan disk also has to be designed for metal fatigue, previously described as bending a paper clip back and forth until it breaks. Metal fatigue is extremely important to the design of all airplane components and was the cause of numerous fuselage breaks described in Chapter 4. (Metal fatigue is described in terms of atomic structure in Chapter 6.)

For a paper clip, one fatigue cycle might be bending the wire plus 60 degrees, then minus 60 degrees, and finally returning the wire to the start position. (Paper clip fatigue experiments are described in Chapter 6.) For the disk, a single fatigue cycle is one takeoff and landing where the disk spins up to its design speed and is fully loaded by centrifugal force. With enough cycles, all fan disks and all other rotating components in the engine will eventually fail due to metal fatigue.

Once a fatigue crack initiates, it will slowly grow or propagate with each takeoff cycle until it grows to a critical length and the disk flies apart. The NTSB investigators determined that the fatigue crack in Flight 232 was most likely growing from a casting defect on the very first takeoff cycle in 1972. It grew for 17 years until the crack reached a critical size, roughly 1.5 inches long and 0.25 inches deep, when the fan disk broke apart.

General Electric engineers determined, based on calculations of mechanical stresses and on component testing until failure, that the average life of the disk should be 54,000 takeoff cycles. Because fatigue data has tremendous statistical scatter and is sensitive to many manufacturing factors, the FAA certified the engine for 54,000/3 = 18,000 cycles. Therefore, after 18,000 cycles, the fan disk will be taken out of the engine and replaced, regardless of any evidence of fatigue cracking. This design concept is known as "safe-life design." The component is designed and operated within its safe-life expectation.

The fan disks were retired and saved after 18,000 takeoffs. General Electric was in the process of starting the extremely lengthy justification process to recertify the disks for 20,000 cycles. After the crash of Flight 232, however, this plan was put on indefinite hold.

Additional safety is provided by frequent inspections for fatigue cracks. The fan disk on Flight 232 had been inspected six times before the fatal flight.

Manufacturing Flight 232's Fan Disk

The fan blades, fan disk, and other colder parts of the compressor are manufactured from titanium.

Titanium was a laboratory curiosity until practical commercial processes were developed for its production in the 1940s. In the 1950s, the U.S. Air Force sponsored extensive research to optimize the alloy recipe for military applications. The main attraction of titanium is its exceptional strength-to-weight ratio at temperatures between 400°F and 1,000°F. (Aluminum begins to lose strength at 350°F and steel weighs more.) Portions of the jet engine that operate above 1,000°F are made out of "super alloys" consisting of various combinations of nickel, iron, chrome, and cobalt.

There are various steps in the manufacture of titanium fan disks. In the first step, raw materials are melted into what is called an "ingot." Titanium "sponge," essentially processed ore, is melted into a 28-inch-diameter, 7,000-pound ingot inside an electric arc furnace. The melting is done to obtain a uniform crystal structure on a microscopic level. Ingot formation is similar to a welding rod depositing a molten puddle of metal. The ingot is then forged or hammered into a new shape called a "billet," cut into a disk, and then forged again closer to the final shape. The forged blanks are then machined into the final shape. After discarding the top and bottom, the 7,000-pound ingot becomes a 6,200-pound billet, which is then cut into 700-pound forging blanks.

Molten titanium is very reactive with air, specifically nitrogen and oxygen. To prevent these reactions, the molten titanium is processed in a vacuum. It is very difficult and expensive to process molten titanium at 3,000°F and seal it against tiny air leaks. The only process used in the 1970s, and still in common use today, is called vacuum arc remelting, or VAR (Figure 5.18).

In VAR, the mold is actually a water-cooled copper electrode. The mold has to withstand high temperatures and conduct massive amounts of electricity, a significant design feat in its own right. In the case of Flight 232's engine, it is believed that the copper mold shifted because of thermal expansion and leaked air. This small air leak became Flight 232's fan disk defect.

When air leaks into the mold, the molten titanium reacts with the oxygen and nitrogen to form what is called a "hard-alpha" inclusion or defect. There is no chemical definition of hard alpha. Hard-alpha anomalies have

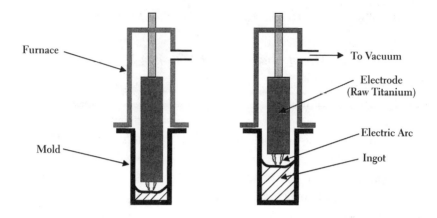

Fig. 5.18. Vacuum arc remelting process. A puddle of molten metal accumulates from the melting electrode and freezes to become the ingot.

been measured with 3.4 to 14.8% nitrogen and with up to 2.5% oxygen. Hard-alpha defects can also occur because of minute contaminants in the titanium feedstock.

There are many other types of manufacturing defects that can occur. Hard-alpha defects happen to be one of the more interesting and the focus of millions of dollars of research.

Titanium Crystals

During solidification, all molten metals form crystals. A crystal is like an orderly stack of oranges, with each orange being a titanium atom. Understanding the crystal structure of titanium helps with the explanation of many of the subsequent manufacturing steps and the nature of the hard-alpha defect.

There are several different crystal configurations important to the metallurgist (and many others important to the geologist). Titanium, upon heating, transforms from hexagonal close packed (HCP) to body centered cubic (BCC). The different crystals correspond to stacking the oranges in different patterns, as shown in Figure 5.19.

The process is completely reversible upon cooling. The crystal structure on a microscopic level, called the metal's microstructure, is very sensitive to heating and cooling rates. Engineers go to great lengths to optimize these properties, because they greatly affect mechanical strength and resistance to fatigue cracks.

The crystalline transformation temperatures are very important markers during the manufacturing sequence. The billet is formed above the crystalline transformation temperature. The finished billet is heat-treated again above this temperature to get a uniform crystal structure. After final forging, the titanium is heated above the HCP-to-BCC transformation temperature yet again, rapidly cooled or quenched, and then partially reheated. All this is done to manipulate and optimize the crystals formed and their mechanical properties.

A metal, if hit hard enough with a hammer, will deform slightly under the hammerhead because planes of atoms slide inside the crystal, a process known as "slip." A ceramic will respond differently. Complex ceramic crystals cannot slip when stressed. Instead of deforming, a ceramic will shatter if struck hard enough with a hammer. If the vacuum leaks during processing of the molten titanium, unwelcome nitrogen and oxygen atoms, trapped in the titanium crystal, will prevent slip, resulting in glasslike brittleness similar to a ceramic. The hard-alpha defect will crack first or, in the case of Flight 232, fall out and create a cavity with higher stresses. In both cases, metal fatigue crack is initiated at a much lower cycle life than would otherwise be expected.

Casting versus Forging

The atoms pack more efficiently in crystals than in an amorphous (noncrystalline) liquid. Because of this increased packing efficiency, a cubic inch of molten metal actually shrinks during solidification. Because it is impossible to cool such a large block of metal uniformly, there will be

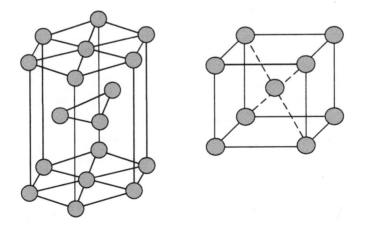

Fig. 5.19. Hexagonal close packed (HCP) and body centered cubic (BCC) crystals

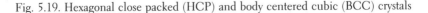

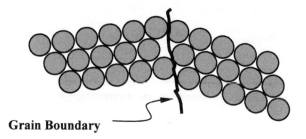

Grain Boundary

Fig. 5.20. Crystals with different orientation meet at an interface called a grain boundary.

nonuniform cooling and shrinkage. This shrinkage appears as internal porosity, cracks, and other voids.

When solidification begins, crystals initiate and grow at random sites and in random orientations. When one crystal grows into another one with a different direction, the pattern is disrupted. The disrupted interface is known as a grain boundary (Figure 5.20).

The crystal inside a grain boundary is known as a grain. Casting tends to form large grains. A fracture path can align within one large casting grain and propagate more easily. The many, many grain boundaries associated with small grains act as mini barriers to fracture. The fracture has to keep changing direction and ends up taking a more tortuous path as it hops into a new grain with a different orientation.

Because of large grains, voids, and porosity, a casting is considered inferior to a hot worked forging. (For example, all components in an automotive drive train are forged. The engine block, which operates with lower stress, is cast.)

Forging, a process known to blacksmiths for roughly 2,000 years, involves heating a metal to more than half its melting temperature[10] and beating on it with a hammer. This breaks up the coarse grains from casting into smaller grains and also welds shut the internal porosity and voids.

The 28-inch-diameter ingot is beaten or forged into a 16-inch billet with heat and massive hydraulic hammer blows. The billet is cut into eight forging blanks for additional forging before final machining.

First Inspection

The first of many inspections occurs after formation of the billet.

The billet is inspected for defects ultrasonically in a process similar to that used by doctors. A special piezoelectric crystal is used to convert electrical energy into mechanical energy. The crystal vibrates when it receives an electrical signal. The vibrations generate high-frequency sound waves that enter the billet. The 2 to 5 megahertz (1 MHz = 1 million cycles per second) sound waves bounce off the opposite surface. The piezoelectric sensor also works in reverse. The reflected sound wave vibrates the crystal and generates a measurable electrical signal. Any differences indicate an internal defect.

When accelerated, a piezoelectric crystal will also generate an electrical signal. This electrical signal can be used to measure acceleration and is the basis of a special sensor called an accelerometer. Accelerometers are used to measure impact accelerations during crash testing (see Chapter 8).

Detection of hard-alpha defects with ultrasonics can be difficult. The hard alpha, caused by excess nitrogen and oxygen, is only slightly different in composition from the surrounding material. The smaller nitrogen and oxygen atoms sit interstitially (between) the titanium atoms and have little effect on ultrasonic properties of the base material. The regions of excess nitrogen gradually blend into the titanium making any ultrasonic measurement somewhat diffuse.

An increased nitrogen content results in increased brittleness. The brittleness is problematic with respect to fatigue, but it is also more likely to result in detectable cracks that occur during forging. Often, the forging process will beat the hard alpha into a crack that can be detected ultrasonically. The billet size was later reduced to 10-inch diameter to increase the sensitivity of ultrasonic inspection. The increased hot working required to reduce a 28-inch ingot to 10 inches instead of 16 inches increases the likelihood of any hard-alpha defect becoming a crack or void.

The 16-inch billet that eventually became the fan disk in Flight 232 was cut into eight forging blanks, each weighing about 700 pounds. The blanks were forged into a crude imitation of the final shape and shipped out for machining. After partial machining, the forging blanks were ultrasonically inspected again. At this point, one of the eight blanks was rejected because of a defect.

More Inspection

After additional forging and finish machining, the disk is ready for a final inspection before being shipped out for service. This inspection is done with a liquid penetrant. Liquid penetrant inspection works with a phenomenon known as capillary action and with the fundamental liquid properties of adhesion and cohesion.

Cohesion is the property of liquid molecules being attracted to each other. Water molecules, for example, are called dipoles, because they have an electric pole at each end. The two hydrogen atoms are asymmetrically attached to the oxygen atom resulting in asymmetric distribution of the electron charge density. Each end of a water molecule has a negative or a positive charge density. The negative end of one water molecule will attract the positive end of a nearby water molecule. This attraction leads to the property of surface tension and to water beading up into droplets.

Water molecules on the surface are attracted inward to the molecules in the interior. This attractive force minimizes the surface area by forming a sphere. Filling a glass with water to the top of the rim and carefully adding coins until the water bulges above the rim illustrates this inward attractiveness of the water molecules on the surface (Figure 5.21).

Water molecules are attracted to each other and to any other substance with a charge density. When liquid molecules are attracted to another substance, it is called adhesion. Capillary action occurs when the adhesion to the surface of a material is stronger than the cohesive forces between the liquid molecules. The process is demonstrated by dipping one end of a paper towel into a puddle of liquid and observing the liquid flow up into the paper against the force of gravity.

Adhesion versus cohesion is also observed with a liquid's ability to wet a surface. Water will bead up on a clean and waxed car hood. The water molecules are more attracted to each other than to the waxed hood (cohesive forces > adhesive forces). When the surface of the car is dirty, how-

Fig. 5.21. When coins are added to a full glass of water, the water bulges above the glass rim.

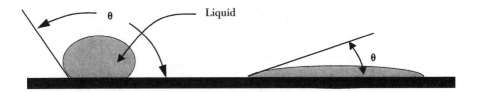

Fig. 5.22. If the contact angle θ is less than 90 degrees, the adhesive forces exceed the cohesive forces. The liquid spreads and wets the surface.

ever, the water spreads out and wets the surface; here, the adhesive forces exceed the cohesive forces. This phenomenon is described in terms of the contact angle, shown in Figure 5.22, with wetting occurring when the contact angle is less than 90 degrees.

A liquid penetrant must freely wet the surface being inspected so it can flow with capillary action into any defects that break through to the surface. The contact angle for most liquid penetrants is close to zero.

After the penetrant has had time to soak into any potential defects, normally about 20 minutes, the excess penetrant is wiped from the surface. Next, a developer is applied, usually a dry white powder, which draws the penetrant out of defects by reverse capillary action. The penetrant bleeds into the developer with colored indications that are more easily seen than a small defect or tight crack.

Recall that the General Electric designers concluded the disk had an average fatigue life of 54,000 takeoff cycles and that it was certified for 18,000 cycles. Additional safety margin against metal fatigue is provided by frequent liquid penetrant inspections after the disk is placed in service. The inspection history of the Flight 232 fan disk is given in Table 5.1.

There is no rigid schedule for disk inspection. Inspection is required every time the engine is disassembled for normal maintenance. Frequent inspections are anticipated since minor damage often occurs from foreign objects being sucked into the engine.

The disk will be permanently removed from service if there is any indication of a fatigue crack. Repairs are not allowed on this critical engine component.

Flight 232: The Manufacturing Defect and Fracture

The last fan disk inspection was done 760 cycles or flights prior to the accident. The NTSB concluded that a detectable fatigue crack about 0.5

Table 5.1. Liquid Penetrant
Inspection Schedule for Flight 232

Inspection Number	Cycles Since New
1	0
2	824
3	1,645
4	3,764
5	8,315
6	9,236
7	14,743
Accident	15,503

inches long existed at the time of the last inspection. The NTSB further stated that "the inspection parameters established in the United Airlines maintenance program . . . and the General Electric Aircraft Engines shop manual inspection procedures . . . if properly followed . . . are adequate to identify unserviceable rotating parts prior to an in-service failure."[11]

It needs to be understood that fatigue cracks are common and are factored into the requirements of design, testing, and ongoing maintenance.

Examination of Flight 232's fan disk fracture surface showed the fatigue crack initiated at a small cavity approximately 0.055 inches long on the inside diameter of the bore and 0.015 inches deep into the bore. The cavity, believed to be caused by a hard-alpha defect that fell out, was surrounded by a nitrogen-rich zone.

Unlike hard-alpha defects, which are difficult to detect with ultrasonics, a 0.055-inch-wide cavity should have been detected. The NTSB believes the hard-alpha particle fell out after the last ultrasonic inspection during subsequent manufacturing steps, thus explaining why it was not detected.

The cavity acted as a stress raiser and fatigue cracking initiated on the first takeoff and landing cycle with failure occurring after 15,503 cycles. Scanning electron microscope examination was able to find fatigue striations,[12] which mark the fatigue crack growth after each cycle, corresponding closely to the actual number of takeoff and landing cycles.

Also found on the fracture surface was a discolored zone roughly 0.476 inches long along the length of the bore and 0.180 inches deep. The NTSB concluded that this discoloration was a stain from the last liquid penetrant examination, 760 cycles before the accident.

In the final 760 cycles before the accident, the fatigue crack is believed to have grown, as confirmed by scanning electron microscopic examination and computer simulation of the fatigue crack growth, from 0.476 inches long × 0.180 inches deep to 1.24 inches long × 0.56 inches deep. At that point, the disk fractured rapidly and completely. The dimensions of the growing fatigue crack are summarized in Table 5.2.

The primary fracture surface was a radial inside-diameter-to-outside-diameter fracture separation initiating from fatigue on the inside of the bore. In the center of the fatigued region was the hard-alpha cavity. The second fracture progressed largely circumferentially, resulting in a release of approximately one-third of the fan disk.

Except for the zone on the radial fracture showing fatigue striations, the remaining fracture surfaces were typical of overstress separation.

Figure 5.23 is a National Advisory Committee for Aeronautics (the forerunner of NASA) image that shows radial cracks in a burst disk. This research was conducted in 1947 when it was not yet clear what materials were best for rotating engine components. A variety of materials were tested and rotated until they burst. Unlike the Flight 232 fracture disk with its single main fracture, this particularly brittle material had numerous crack initiation sites on the inside diameter.

The radial fractures in Figure 5.23 are consistent with the main fracture surface on the titanium fan disk. The secondary circumferential fractures occur because of differences in geometry. The geometry of the titanium fan disk differs significantly from the donut shown above, creating complex load paths and stress patterns.

Aftermath of Flight 232

The fan disk on Flight 232 was one of eight fan disk forging blanks made from the same ingot. One blank was rejected at the time of manufacture for material defects. After the Flight 232 accident, the other six were quickly

Table 5.2. Relevant Dimensions of the Defect and Fatigue Crack
on Flight 232

Size of cavity	0.055″ × 0.015″
Size of fatigue crack 760 cycles before accident	0.476″ × 0.180″
Final size of fatigue crack at time of accident	1.24″ × 0.56″

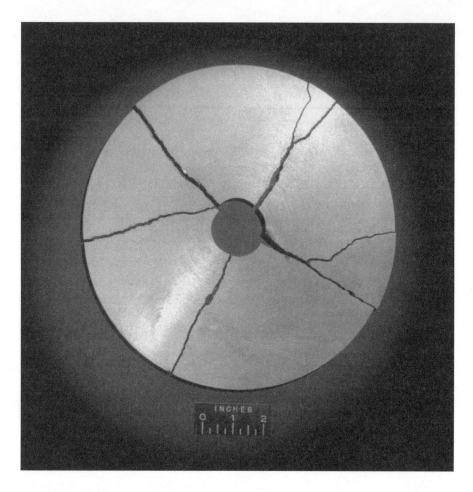

Fig. 5.23. Burst test conducted in 1947 to evaluate candidate disk materials. This brittle material had numerous fracture initiation sites.
Photo: NASA Glenn Technical Library

recalled for immediate inspection. Two were found to have hard-alpha defects and were removed from service.

The forging blank rejected 17 years earlier had accidentally been given the same serial number as the blank used to forge the disk on Flight 232. For quite some time, it was feared that the rejected blank had accidentally gone into service. This touched off a furious paper-trail investigation. Seventeen years of paper work had to be traced to find the disks. This is not a minor task. The three steps—ingot, billet, and final machining—are done by three different companies, some of which were no longer in that business. The airlines are a fourth company. Planes are often leased and resold. Of course, accurate record keeping is a very important requirement in the aircraft business for this very reason. Any anomalies in the paper trail were resolved with forensic metallurgy. For example, the ingot supplier initially thought it was not their ingot because of missing trace contaminants associated with their local conditions. Every possibility was exhaustively investigated until resolved.

In the meantime, McDonnell Douglas made changes to the no. 3 hydraulic system. Sensors were added to detect dropping fluid level in the reservoir. Added shutoff valves would automatically activate and prevent loss of all hydraulic fluid preserving partial hydraulic control of some flight surfaces.

The NTSB made numerous recommendations that triggered massive national research efforts on the ultrasonic detection of titanium defects. This research eventually evolved into changes in mandatory rules.

Hard-Alpha Research

Hard alpha has a higher melting point than titanium. This makes hard-alpha defects difficult to eliminate in the molten titanium once they form. They will, however, dissolve like a cube of sugar in hot coffee when held at elevated temperature for extended periods.

The ingot for the Flight 232 fan disk was actually remelted a second time in the same electric arc furnace in a double-melt VAR process. This was done to promote homogeneity of the microstructure and to dissolve any potential hard alpha. Because of concern for hard alpha, the process was changed in 1971 to a triple-melt VAR process. Coincidentally, the fan disk on Flight 232 was made from the last ingot produced with a double-melt process.

Shortly after the Flight 232 accident, General Electric requested and the FAA mandated additional inspection of the 52 disks made from the same raw material feedstock. This inspection was quickly expanded to all 213 disks manufactured with the superseded double-melt VAR process. General Electric announced plans to replace all of these disks within 1,500 flight cycles.

The triple-melt VAR process has improved quality, but it has not completely eliminated the hard-alpha defect problem. Other improvements, including the development of new furnaces and melt processes, have also significantly reduced hard alpha. To use a weak analogy, it's like removing infections from a hospital. There are always improvements that can be made. A hospital can develop a $10,000,000 program solely dedicated to eliminating infections. In that case, they will proudly brag about their low infection rate, but will they guarantee that there will never be another infection? Removing all contaminants from a 7,000-pound puddle of molten titanium is like removing all germs from a hospital (well, maybe not quite as hard). The engine manufacturers will brag about their progress to date (and it is worth bragging about), but they will not promise zero defects in the future.

How can they? Just like a hospital can never promise that no one will ever forget to wash their hands, engine manufacturers can never promise that a human error won't ever occur at any stage of titanium production.

The approaches to the hard-alpha problem have been to

- improve the processing to achieve cleaner titanium and minimize air leaks so that defects never occur in the first place (Defects occur because of contaminated raw stock, improperly processed electrodes, or vacuum leaks.);
- change the casting process so that the molten titanium is held at higher temperatures for longer periods of time to dissolve any defects that do form;
- improve the inspection methodology so that defects are found and removed.

The FAA and jet engine manufacturers initiated research to achieve improvements in all three areas, although hard-alpha defects are by no means the only cause of burst engines. Significant progress has been made on all fronts as evidenced by the reduction of burst engines reported elsewhere in this chapter.

In response to NTSB recommendations, a veritable alphabet soup of national committees representing joint efforts by academia, research labs, government agencies, titanium producers, and engine manufacturers were created to address the hard-alpha problem. Three of these efforts are mentioned here.

In the fall of 1989, the FAA formed the Titanium Rotating Components Review Team (TRCRT) to review design, manufacturing, inspection, and life management. In response to the recommendations of the TRCRT, the Engine Titanium Consortium (ETC) was formed in 1993 to review and improve manufacturing and in-service inspection processes for all rotating parts.

In 1990, the Jet Engine Titanium Quality Committee (JETQC) was formed to develop a database of all titanium-alloy melt-related defects found in alloy billets, bars, and forgings. The aerospace titanium food chain is limited, with only a few suppliers of titanium feedstock, melt shops that cast ingots, forge shops that make billets and forgings, and manufacturers of jet engines. It was relatively easy to arrange cooperation and keep track of all melt defects. However, it is a competitive business and companies are not anxious to call attention to their defects so reporting was done anonymously.

Often, a flurry of problems can indicate a systemic problem that may show up at other companies. Therefore, a problem with one manufacturer signals others to look for the same problem. All U.S. and European engine producers participated in the JETQC. Eventually, the reporting process became an FAA-mandated requirement.

Suppliers provide inspection reports to JETQC detailing defects found (size and type), quantity of material inspected, ultrasonic signal, and ingot location. The three-dimensional features of the defects are characterized with a scanning electron microscope, and the chemical nature of the defects is also identified, all for the JETQC report.

Annual JETQC ingot reports map the defects found, but they do not reveal who had the inclusions. Suppliers can rate their performance against the industry from these annual reports. Just collecting the data triggered more attention to the problem. A steady decrease in the number of defects has occurred since the start of data collection.

Improvements in triple-melt VAR technology have resulted in significant reduction of melt-related defects since the mid 1980s. Slowly, often through trial and error, standards were developed for cleanliness of the feedstock; required level of vacuum; allowable leak rates, power levels, melt rates, and crucible cooling; allowable interruptions in power and vacuum; cleaning of the furnace and crucible; and many other production variables.

New melting technologies were also developed. Depending on the density, the defect can float to the top, sink to the bottom, or pass through into the final ingot. The cold-hearth melting process maintains the hard alpha in the molten metal as it traverses a hearth distance; the defect is dissolved or separated into the top or bottom of the ingot where it is easily removed. (The top and bottom of the ingot are routinely discarded.) New sensors and computer simulations were developed to monitor the process and model optimized conditions.

Melting processes are now qualified with titanium nitride "seeds." Hard-alpha seeds are introduced on purpose, and acceptance of the product requires destruction of the seeds during trial ingot production runs.

Eventually, many of these requirements were compiled into an FAA advisory circular titled "Manufacturing Process of Premium Quality Titanium Alloy Rotating Engine Components." These new rules require rejection of the entire 7,000-pound ingot if a hard-alpha defect is found at any stage of production. All material from the ingot must be downgraded from "premium grade" and cannot be released for jet engine components.

One problem with studying hard alpha is its rarity. Defects are very rare with perhaps only one defect occurring per one million pounds of product. Researchers can't study what they can't find, so they had to learn how to create acceptable artificial defects.

In 1995, a power outage during melt production created a unique opportunity. A major vacuum leak resulted in an ingot with 60 "natural" hard-alpha defects. The ingot was converted into 12 billets. The billets were poked and prodded for years by researchers of the ETC. Ten of the defects were sectioned every 0.005 inches to accurately characterize their chemistry, microstructure, size, and shape. This information was used to compare ultrasonic methods and develop probability of detection models used in new design methods.

Initially, the ETC focused on the material properties of hard-alpha defects and their response to ultrasonic waves. Eventually, ETC research developed a 4× increase in defect detection with ultrasonics.

The NTSB recommended the creation of a database of worldwide engine rotary part failures to "facilitate design assessments and comparative safety analysis during certification reviews and other FAA research."[13] Also analyzed were dispersion patterns, fragment size, and energy levels of uncontained debris from burst engines. Eventually, an FAA study was completed, titled "Large Engine Uncontained Debris Analysis." Some of the information from this report, which presents a study of 65 of the best documented uncontained failures, is given later in this chapter. Following from the FAA study, updates reflecting the new information became the advisory circular titled "Design Considerations for Minimizing Hazards Caused by Uncontained Turbine Engine and Auxiliary Power Unit Rotor Failure."

New Design Rules Developed in Response to Flight 232

A significant research effort resulted in new design rules for rotating jet engine components.

Mechanical design is usually based on strength. If something breaks, make it stronger—a very simple concept. Of course, designers of anything that flies are very concerned about weight. Make it stronger, but not too heavy!

A material is tested to assess its strength, and a component is analyzed during design to assess stresses it will experience during operation. A safe margin is always provided against fracture. Extensive testing of the final component is then done for all critical aerospace structures. For example, a rotor's strength is tested by spinning it until it bursts. To guard against faulty controls or a broken shaft, the burst margin has to be at least 120% of normal rpm.

The design process is continued to verify that cyclic loads will not result in metal fatigue. For fatigue testing, the engine is run up to full throttle and then shutoff; this process is repeated until the engine fails from cyclic loading. Because of the need to test many engines for tens of thousands of cycles (a very expensive proposition), small specimens are cyclically tested to destruction. Stresses on the engine under operating conditions are calculated by computer simulation and compared to the small specimen test results.

In both design cases—burst margin and metal fatigue—centrifugal forces load the rotating fan disk as it tries to expand radially.

Fatigue data is not very repeatable and has much statistical scatter. To account for this scatter, current FAA rules use one-third of the expected life from the test results to establish a "safe life." For Flight 232, the safe life was one-third of the predicted life, or 54,000/3 = 18,000 cycles—the "certified" safe life.

Another approach is to design with statistical methods. This involves testing perhaps 10 engines and using the lowest cycle life for the rated life. However, after testing 10 engines, there is some probability that the eleventh engine, installed on a plane, will fail at a lower life. Perhaps a safer design would result if 50 or even 100 engines were tested. The expected fatigue life is now understood with higher certainty, but there is still some probability that engine number 101 will fail at a lower cycle life.

Statistical methods attempt to predict, based on the fatigue data, the probability of the next engine failing at a lower cycle life. For example, if

10 engines are fatigue tested and the fatigue lives range from 45,000 to 75,000 cycles, statistical methods might be used to calculate a 99.99% probability that all engines made with identical design and manufacture methods will have a cycle life of at least 20,000 cycles. In that case, the reliability of the engine is said to be 99.99% when used to 20,000 cycles or fewer.[14]

If 1,000 engines were tested and they all lasted longer than 30,000 cycles, a passenger should feel pretty safe about being on a plane whose engines have only 20,000 cycles, even though there is some small probability, which can be calculated with well-established principles, of one of the engines bursting. The problem is where to draw the safety line. How many cycles should planes carrying passengers be allowed to fly? How safe is safe enough?

Common sense might indicate that more test data is needed to sort out any uncertainty. However, hard-alpha defects very seldom occur. One thousand engines could be tested without finding a hard-alpha defect. How does one design for something that hardly ever happens? Testing is not the solution, because the test program will most likely end before a defect is experienced.

A damage-tolerance method, an alternate design method to safe life and strength,[15] is the answer. Damage-tolerance design considers predicted stresses (from computer modeling), fatigue crack growth data, and fracture mechanics data and methodology.

A fatigue crack grows until it reaches a critical length and then the disk flies apart. The critical length will vary with the crack location and orientation. Fracture mechanics methodology considers these variables. Crack growth rates and critical crack length are not precise values. They can be predicted with statistical methods. A one-inch-long crack might have a 20% probability of fracture (80% reliable), while a 1.5-inch crack might have only 70% reliability. The same crack at a different location with higher stresses will have a higher probability of fracture. This is standard fracture mechanics analysis used in many industries and for other parts of the airplane. Unique to titanium parts is identification of the probability of a hard-alpha defect occurring and the probability of it being detected by repeated inspections.

The permitted occurrence of hard-alpha defects per FAA's advisory circular "Damage Tolerance for High Energy Turbine Engine Rotors" is one defect in every one billion flights (a reliability of 99.9999999%). With five disks or rotors in a plane (fan disk, high and low pressure compressors, and turbine rotors), the total defect rate is five hard-alpha events per one billion flights, or 99.9999995% reliability. (Also, note that an uncontained engine failure rarely results in injury; see data later in this chapter.)

Most people have difficulty understanding very large or small numbers or very small probabilities. In a crude attempt to get into the ballpark of such small probabilities, consider the following situation.

A hard-alpha defect might occur once in every one million pounds of produced titanium. (We hope this number is less and keeps decreasing.) If a disk weighs 370 pounds, there are 370 of 1,000,000 chances for a defect to exist in that specific disk. If, like on Flight 232, the disk bursts after 15,000 flights, the probability of one flight[16] experiencing an uncontained failure is $370/1,000,000 \times 15,000 = 0.000000025$. Each flight has $1 - 0.000000025 = 0.999999975$ probability of not having a failure, or 99.9999975 reliability. This value of reliability, 0.999999975 times one billion flights, predicts 999,999,975 successful flights or 25 engine failures per one billion flights, far in excess of our goal of one failure per billion flights per disk.

The above scenario does not consider inspection. With repeated inspection, the probability decreases of a defect surviving undetected.

The actual calculation and design process is far more complicated. The total volume of the disk is divided into smaller volumes for further analysis. Each individual volume is then assessed for probability of fracture. The stresses unique to that volume are considered. For locations with increased stresses, the probability of failure is higher; for regions with less stress, the probability of fracture is lower.

The probability of defect detection, either a hard-alpha defect from initial manufacture or a growing fatigue crack that occurs during use, increases with size. The probability of detection also varies with frequency of inspection and method of inspection. For example, ultrasonics are more sensitive than liquid penetrant inspection.

All of this information is evaluated for the individual volumes of the disk. The probability of fracture for the entire disk is the sum of probabilities for all of the individual volumes. If the total probability exceeds one failure per one billion flights, the design is unacceptable. The designer can reduce the probability by

- reducing the stresses by adding more metal;
- increasing the probability of defect detection by altering the inspection method (either using a more sensitive method or increasing the frequency of inspection);

- reducing the rated life of the component. (The disk will have greater reliability if it is operated for fewer cycles.)

What is the role of probabilistic design? It is how engineers design for defects that very rarely occur. This is not to imply that defects are at any time permitted in a rotating jet engine component. On the contrary, extraordinary measures are taken to eliminate all defects during manufacturing and to continually inspect for those that may occur during use. If a defect is ever found, it is immediately removed. At no time is it acceptable to operate a fan disk with a defect. As mentioned earlier, if a defect is found at any stage of production (billet, forging blank, or final disk), all 7,000 pounds of the ingot are downgraded and are not permitted for use in jet engine components.

Probabilistic design methods consider the fact that small defects will somehow slip through the extensive quality-control process.

At no time is operation of a fan disk ever permitted with a known fatigue crack. If a fatigue crack is found during normal maintenance inspection, the disk is replaced immediately. The purpose of calculating the expected fatigue crack growth and critical fatigue crack size is not to justify operating with a fatigue crack, but to specify an inspection interval such that a crack is found before it reaches critical length and causes failure.

Jet Engine Reliability

Jet engines are quite reliable. They were a significant advance over piston engines, and they continue to be improved. As the statistics show, jet engines rarely fail, and when they do, very few of the failures result in injuries to passengers.

Most pilots can fly for an entire career without ever shutting down an engine[17] for a minor problem, let alone for a less common and more serious burst engine.

Fundamentally, jet engines are simpler with fewer moving parts than piston engines. Compared to the pounding of piston engines, jet components rotate smoothly and continuously, and they require less maintenance. The Boeing 707, the first successful commercial jet (introduced in 1958), had over 100 fewer engine-related controls, instruments, and displays in the cockpit compared to a similar piston plane.[18]

Added simplicity also means greater flight safety. For example, to increase power on a fly-around during a failed landing attempt, the pilot can adjust the throttle in a single motion. During this critical phase of flight with high pilot workload, the piston engine crew would have to make numerous engine power adjustments.

The new jet planes quickly turned into an economic advantage for airlines. The roomier, quieter, faster planes proved extremely popular from day one. The big jets carried twice as many passengers per trip and, being faster, made more trips per year. The simplistic design played out in reduced maintenance. Compared to piston engines, jets flew four times longer between overhauls and were replaced less often. The 707 cost twice as much to make, but it was four times more productive.

Extended-Range Twin-Engine Operation Performance Standards

All commercial planes have more than one engine as a redundancy safety feature. Two-engine planes, having less redundancy than planes with three or more engines, have specific requirements.

The FAA mandates rules for all two-engine planes. In case of engine failure, the plane is required to land immediately at the nearest airport. The FAA also mandates how far any two-engine plane can stray from an approved airport. The evolution of these rules, known as extended-range twin-engine operation performance standards (ETOPS), essentially tells the story of regulatory acceptance of increased jet engine reliability.

In 1953 (before jet engines), the FAA introduced the 60-minute rule for all commercial aircraft. The rule stated that an airplane shall not be farther than 60 minutes flying time from any airport. This forced an awkward, jagged path when crossing oceans.

In the late 1950s, the rule was waived for the three-engine Boeing 727, allowing it to fly transatlantic without restrictions. These rules influenced the design and development of the three-engine DC-10 and L-1011. Two-engine jets remained restricted by the 60-minute rule, but recognizing the increased reliability of jet engines, countries outside the U.S. soon allowed 90-minute flights.

The FAA acknowledged this increased reliability in two-engine jets in 1985 and granted 120-minute flights. This was quickly increased to 180 minutes in 1988. In 1998, the two-engine Boeing 777 received a 15% extension, to 207 minutes, when flying over the Pacific.

Boeing and General Electric have been pushing for an extension of ETOPS for the 777 to 330 minutes.[19] Current FAA requirements stipulate a documented engine shutdown rate of no more than two events per 100,000 hours of flight time for 180-minute ETOPS. This is a shutdown for any reason. An uncontained failure occurs thousands of times less often. In 2003, the General Electric engines on the 777 had been experiencing three events per 500,000 hours.

ETOPS is also a sophisticated maintenance program. Airlines flying under ETOPS rules are required to monitor oil consumption, rotor speeds, exhaust gas temperatures, vibrations, and other engine parameters to spot trends and service the engine before problems occur. Some airlines have discovered cost savings associated with applying ETOPS maintenance procedures to all flights.

A recent incident illustrates flight safety with one engine out. The *Wall Street Journal*[20] reported the following conversation that occurred on February 20, 2005, between air traffic controllers and the pilot of British Airways Flight 268, a four-engine 747, just seconds after taking off from Los Angeles bound for London.

> Controller: . . . you had flames coming out of the engine and it was shut down. Is that accurate?
> Pilot: We haven't shut it down. We've throttled it back and we are doing our checklist.

After a few minutes:

> Pilot: We have now shut down the no. 2 engine. We are going to consult our company and see what they require us to do.

After a further brief delay:

> Pilot: We just decided we want to set off on our flight-plan route and get as far as we can. So we'd like clearance to, ah, continue our flight plan.

The reported conversation continues with internal discussion among three air traffic controllers.

> Controller 1: Remember that [British Airways 747] I told you about?
> Controller 2: Yeah.

Controller 1: He's engine out. He's going to continue to his destination or as far as he can get.
Controller 2: OK, I have no flight plan on him.

Assuming that the plane would land immediately, air traffic control had deleted the flight plan in their computer. They had to construct another one.

Controller 3 [who saw the engine flames]: Is he going?
Controller 1: He's going.
Controller 3: If you saw what we saw out the window, you'd be amazed at that.

According to the British Air Accident Investigation Branch (their NTSB), the underpowered plane proceeded across the United States at a slower speed and lower altitude than usual without further incident. The flight crew decided to cross the Atlantic. Air drag is higher at lower altitudes. Also, unfavorable winds caused more fuel usage than normal when crossing the ocean. After a minor concern about accessing fuel in one of the four wing fuel tanks, the flight crew became worried about running out of fuel, declared an emergency, and landed just short of London in Manchester. British investigators later determined that there had been enough fuel to safely reach London.

The *Wall Street Journal* reported that if two engines on the same side failed "[the plane] would have more difficulty flying . . . forcing extensive rudder use to keep flying straight."[21] The aircraft would have to fly lower and consume more fuel, making it more difficult to safely reach an alternate airport.

After Flight 268 landed, the U.S. and U.K. opened investigations. The U.K. investigator determined that British Airways had flown 747s with one engine out on 15 occasions since April 2001 (in one case for 11 hours from Singapore to London) and concluded that the practice was safe.

The incident with Flight 268 set off a feud between U.K. and U.S. regulators over jurisdiction. A fight was avoided with a compromise. The FAA conceded that international law gave Britain's Civil Aviation Authority (their FAA) oversight of British Airways, and the airline agreed to change its engine-out procedures, at least when flying in U.S. air space.

Engine Shutdowns

Engines can effectively shut themselves down if the mechanical damage is severe enough. Engines may also be shut down by the pilot for a variety of benign reasons such as low oil pressure or excess vibrations, temperature, or pressure. The NTSB web site[22], quoted below, explains the difference between an innocent, but potentially frightening, malfunction and a more serious uncontained failure considered a "flight risk" and requiring mandated investigation.

> Engine malfunctions or failures occasionally occur that require an engine to be shut down in flight. Since multi-engine airplanes are designed to fly with one engine inoperative and flight crews are trained to fly with one engine inoperative, the in-flight shutdown of an engine typically does not constitute a safety of flight issue. In fact, these events are generally not reportable to the NTSB. Following an engine shutdown, a precautionary landing is performed with airport fire and rescue equipment positioned near the runway. Once the airplane lands, fire department personnel assist with inspecting the airplane to ensure it is safe before it taxis to the gate.
>
> Most in-flight shutdowns are benign and likely to go unnoticed by passengers. For example, it may be prudent for the flight crew to shut down an engine and perform a precautionary landing in the event of a low oil pressure or high oil temperature warning in the cockpit. However, passengers may become quite alarmed by other engine events such as a compressor surge—a malfunction that is typified by loud bangs and even flames from the engine's inlet and tailpipe. While this situation can be alarming, the condition is momentary and not dangerous.
>
> Other events such as a fuel control fault can result in excess fuel in the engine's combustor. This additional fuel can result in flames extending from the engine's exhaust pipe. As alarming as this would appear, at no time is the engine itself actually on fire. Also, the failure of certain components in the engine may result in a release of oil that can cause an odor or oily mist in the cabin. Despite these observations, such occurrences do not necessarily indicate an unsafe condition that must be investigated by the National Transportation Safety Board.
>
> Two terms are helpful in describing the nature of engine failures. A "contained" engine failure is one in which components might separate inside the engine but either remain within the engine's cases or exit the engine through

the tail pipe. This is a design feature of all engines and generally should not pose an immediate flight risk. An "uncontained" engine failure can be more serious because pieces from the engine exit the engine at high speeds in other directions, posing potential danger to the aircraft structure and persons within the plane. The Board will likely investigate any uncontained engine failure involving a transport category aircraft.

Compressor Surge

The compressor blades force air, under pressure, into the pressurized combustion chamber. Blades attached to the spinning compressor fan disk push air into the engine. Just like wings can stall with insufficient lift, the compressor blades can fail to provide sufficient air compression to maintain flow in the correct direction.

With insufficient compression (or pressure), the airflow in the compressor can become unstable and reverse violently in a loud bang or pop known as a compressor surge. The high-pressure air farther back in the engine escapes forward through the compressor and out the engine inlet.

A variety of things can disrupt the airflow through a jet engine including engine deterioration, ingestion of excessive rain or other foreign objects, internal failure such as a broken blade, and even a crosswind over the engine's inlet.

Once the air from within the engine escapes, the instability often self-corrects, and the compression process may reestablish itself. A single surge and recovery will occur quite rapidly, usually within fractions of a second.

Compressor surge has been mistaken for blown tires or a bomb in the airplane. The flight crew may be quite startled by the bang, and in many cases, this has led to a rejected takeoff above V1 (pilot jargon for safe takeoff speed). Usually there is insufficient runway to safely stop if a takeoff is rejected above V1. High-speed rejected takeoffs have sometimes resulted in injuries, loss of the airplane, and even passenger fatalities. One such accident is described in Chapter 2.

Foreign Object Damage

One problem unique to jet engines is the amount of air sucked in by the fan and compressor blades. New engine certification tests have to demonstrate safe operation after ingesting a variety of items including gravel, sand, water, snow, ice, and small birds.

The bird ingestion requirements are quite specific. Tests are done to ensure that an engine can operate safely after ingesting one small bird (e.g., 2–4-ounce starling) for each 50 square inches of inlet area, up to a total of 16 birds. (Recent tests have used artificial birds.) In a separate test, safe engine operation must be verified after ingestion of one medium bird (1–2-pound gull, duck, or pigeon) for each 300 square inches of intake, up to a maximum of 10 birds.

Safety is also verified by testing the ingestion of much larger objects such as a fan blade, 4-pound bird (geese, buzzards, larger gulls and ducks), tire tread, or hand tool. One test requires a fan blade to be released with an explosive charge during full rpm. Part of the test involves verification of safe containment of the released blade. Ejection out the back of the engine is also acceptable.

The engine is not required to survive a blade release or ingesting a 4-pound bird or a hand tool. To pass the test, the engine must not explode, disintegrate, or sustain an uncontrollable fire. Titanium ignites at a lower temperature than it melts at, so any titanium rubs will ignite a high-intensity fire. Fuel leaks are an additional source of engine fire. Of course, fire extinguishers[23] are mounted inside each engine.

The new General Electric 128-inch-diameter GE90 engine (a Boeing 737 fuselage would almost fit inside), designed for the Boeing 777, successfully ingested four 2.5-pound birds during certification in 1995. There was no loss in thrust during the required 20 minutes of operation following the ingestion. All fan blades were in excellent condition after the test and have continued to run in other engine tests.

Remarkably, the GE90, under test conditions, also ingested an 8-pound bird. In this instance, the engine successfully met the less stringent requirement of not being a safety hazard during shutdown.

Uncontained Engine Failures: The Statistics

It is initially disturbing to discover the frequency of uncontained engine failures. Somewhat reassuring is how seldom injuries result. More comforting still is the significant reduction of uncontained failures in recent years.

An uncontained engine failure usually has the following three characteristics:

1. A very loud bang (A major compressor or turbine component has broken and crashed through the engine casing.)
2. A violent shudder (The unbalanced engine is vibrating.)
3. No noise or vibration after the pilot shuts off the engine

If flying debris did not injure any passengers or crew or critical airplane equipment and the pilot appears to have control of the plane, there is most likely no further safety hazard.

During the period 1962–1989, there were 676 uncontained jet engine failures. Only 13.7% of those resulted in significant aircraft damage, and they did not prevent the plane from landing safely. Just 2% caused serious injury, fatalities, hull loss destruction, or a crash landing. An analysis of the data shows that most engine bursts occur during run-up to full power with 77% occurring during takeoff or the climb to cruising altitude.

During this 28-year period there were over one billion operating hours on commercial transport planes. The uncontained failures were caused by bird ingestion or other foreign object damage, corrosion or erosion, manufacturing or material defects, mechanical defects, and human error (maintenance, inspection, and operational error).

It is surprising how often uncontained failures have occurred, but it is even more surprising how seldom they actually result in serious damage or injury. This can be explained with simple geometry. Assume that uncontained engine fragments can fly off in any direction parallel to the plane of the page, over a full 360-degree arc. The arc representing the damage zone or the zone that can actually injure a passenger or damage critical aircraft equipment is much smaller. Therefore, the damage zone is a much smaller percentage of all possible trajectories. Even if the engine fragments are "aimed" at the fuselage and passengers, there is still a good chance that everything critical (including passengers' heads) will be missed. Most of the damage zone is occupied by noncritical sheet metal (Figure 5.24).

Since all engines rotate in the same direction, the fragments that impact the plane from engines mounted on the left wing come from a tangent point near the top of the engine. The fragments directed towards the plane from the engine mounted on the right wing will come from a tangent point near the bottom. Fragments that break off elsewhere will miss the fuselage altogether.

Studies show a well-defined fragment pattern of flying debris. The "fragment spread angle" (Figure 5.25) is defined to be ±3 degrees for one-third of a disk fragment (the heaviest and potentially most damaging fragment), ±5 degrees for intermediate fragments, and ±15 degrees for small fragments. Heavier segments generally stay within the plane of rotation, while the smaller pieces get deflected to larger trajectory angles.

The angles, shown in Figure 5.25, occur from the centerline of the engine and run along the length of the engine corresponding to the location

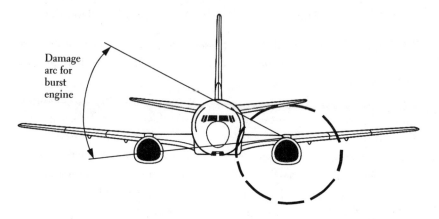

Damage
arc for
burst
engine

Fig. 5.24. The percentage of fragment trajectories that can damage the plane or passengers is reasonably small.

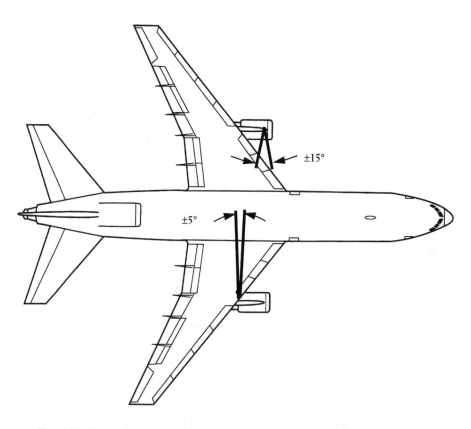

±15°

±5°

Fig. 5.25. Design fragment spread angles for heavy components (±5°) and lighter components (±15°)

of fan disks, compressor spools, and turbine disks. (Flight 232's damage patterns were consistent with these angles.) Engineers use these numbers to assess the damage potential from the kinetic energy of flying fragments. The most critical components—those that if damaged would most jeopardize flight safety—must be relocated or shielded accordingly.

Flight 232 was a DC-10, which has its tail engine as far back as possible to minimize the possibility of critical damage. The damage to Flight 232 was incredibly unlucky, with all possible trajectories just barely lining up simultaneously with all three hydraulic systems.

Sixty-five uncontained failures of "large" engines from particularly well-documented accidents were studied to identify the size and speed of typical fragments. The estimated velocities are based on assuming 25% loss of energy after passing through the engine containment barrier. This fits the experimental evidence of larger fragments losing more velocity than smaller fragments as they pass through a barrier. Typical fragment sizes and estimated velocities for the uncontained failures studied are shown in Table 5.3.

The statistics are studied to provide continual improvements. Safety can be improved with relatively minor design changes. For example, critical components can be relocated (behind bulkheads, noncritical ducting, or

Table 5.3. Fragment Weights and Estimated Velocities

	25th Percentile	50th Percentile	100th Percentile
Fan Blade, High Bypass Ratio Engines, Blade Tip Diameter = 90", 3,600 rpm			
Fragment weight (lb)	1.8	4.5	9.0
Fragment velocity (ft/sec)	1,288	1,162	911
Fan Disk, High Bypass Ratio Engines, Disk Diameter = 30", 3,600 rpm			
Fragment weight (lb)	27.8	45.0	84.5
Fragment velocity (ft/sec)	390	303	276
Compressor Blade, Diameter = 20", 10,000 rpm			
Fragment weight (lb)	0.13	0.19	0.25
Fragment velocity (ft/sec)	960	1,004	1,047
Turbine Disk, Disk Diameter = 20", 10,000 rpm			
Fragment weight (lb)	19.0	70.0	81.6
Fragment velocity (ft/sec)	736	571	520
High Pressure Turbine Blade, Disk Diameter = 27", 10,000 rpm			
Fragment weight (lb)	0.05	0.13	0.25
Fragment velocity (ft/sec)	1,475	1,440	1,353

other structural reinforcements of the fuselage) to minimize hazards to them from engine debris. If a part cannot be moved, it has to be shielded to withstand the kinetic energy of any potential fragment.

BY APPLYING SOME OF THE METHODS briefly outlined in this chapter, significant progress has been made in the reduction of uncontained engine failures. This progress is demonstrated by the most recent NTSB statistics on uncontained failures. In the period from 1990 to the end of 2005, there were 64 events, 25 of which did serious damage to the plane but did not prevent safe landing. Only 3 cases resulted in fatalities. These reductions in uncontained engine failures occurred despite a six-fold increase in air traffic since 1970, including a doubling since 1990.

6...

Metal Fatigue: Bending 777s and Paper Clips

Most people would be alarmed to discover that all aircraft will eventually fall apart, perhaps catastrophically (as occurred in examples in Chapters 4 and 5 where fuselages and engines exploded violently), from metal fatigue.

This chapter will explain metal fatigue as a fundamental mechanical property, explain how fatigue testing occurs, including full-scale testing of airplanes, and trace the history of aircraft structural design methods and rules. The basic concepts will be explained in terms of atomic crystal structure and demonstrated by bending a paper clip.

Metal Fatigue Defined

Metal fatigue is structural damage that occurs when loading is repeated or "cycled" at relatively low loads. If a component can support 100 units of load before failure, it might start accumulating fatigue damage at 20 units of cyclic load or less. If the loading is repeated, failure will eventually occur.

Although first described in the nineteenth century as the metal becoming "tired" (hence the name), metal fatigue differs from physiological fatigue. The damage is progressive and permanent. There is no recovery after periods of rest. Metal fatigue has three stages:

1. Crack initiation occurs on an atomic scale, as explained later in this chapter.
2. After initiation, a crack grows with every load cycle. For aircraft, the major load cycle is takeoff and landing. The fuselage is pressurized, the engine spins up to speed, the wings receive their full lift, and the landing gear goes through one major load cycle. Crack growth was observed in a "desktop" experiment in which a strip of aluminum was repeatedly bent (see Chapter 4).

3. Eventually, the fatigue crack grows to some critical length, at which time the component can rapidly break in a violent manner.

Engineers go to great pains to understand and predict how fast a fatigue crack is growing and when it will reach a critical stage. This is done to ensure that the structure is safe even if it contains fatigue cracks. Added safety is provided by frequent inspections at intervals coordinated with the predicted rate of crack growth. The fatigue crack is found and immediately repaired long before it affects flight safety.

Fatigue Data for a Paper Clip

Metals are very strong and not easily stretched without special equipment. The wire in a paper clip is certainly not easily stretched with "finger forces." A paper clip is, however, easily bent back and forth and can be used to demonstrate many of the basic features of metal fatigue. There is nothing special about the following test setup except that it easily creates fatigue data to illustrate basic concepts.

HOW IS FATIGUE DATA CREATED?

A method is needed so that small, inexpensive, easy-to-produce test samples can be used to predict the fatigue performance of a complex structure such as an airplane.

First, small test specimens (commonly 4" long and 0.3" in diameter) are tested to characterize the fatigue performance of the material. Loads are applied repeatedly to a test specimen until it fractures, and the number of cycles to failure is noted. The weight divided by the cross-sectional area of the rod results in mechanical stress with units of pressure or pounds per square inch. (A 50,000-lb weight hanging from a rod with a 2-in × 1-in cross section will have a mechanical stress of 50,000 lb divided by 2 in^2, or 25,000 psi of mechanical stress.)

The test is repeated with a new sample and a different load. Testing is continued with enough specimens until a curve of cyclic stress versus cycles to failure is defined to span the range of interest to the engineers. Hundreds, if not thousands, of small specimens may be tested.

continued

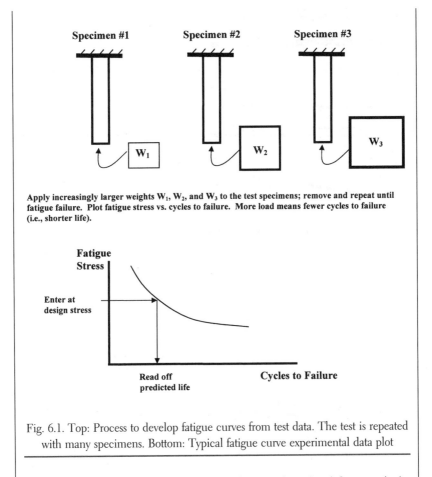

Specimen #1   **Specimen #2**   **Specimen #3**

W_1 W_2 W_3

Apply increasingly larger weights W_1, W_2, and W_3 to the test specimens; remove and repeat until fatigue failure. Plot fatigue stress vs. cycles to failure. More load means fewer cycles to failure (i.e., shorter life).

Fatigue Stress

Enter at design stress

Read off predicted life

Cycles to Failure

Fig. 6.1. Top: Process to develop fatigue curves from test data. The test is repeated with many specimens. Bottom: Typical fatigue curve experimental data plot

As increasingly complex components are made, manufacturing defects can be introduced that degrade fatigue life. Fatigue testing must continue throughout the design and development of the aircraft at the component level. Component testing is more expensive, so fewer fatigue tests occur. Eventually, the process ends with fatigue testing of a complete aircraft. Fatigue testing is also very important in many other industries, such as the automotive and railroad industries.

Use a "jumbo" paper clip, approximately 1.8 inches long, and unbend it as shown in Figure 6.2. Insert the small end of the paper clip between the two hex nuts on a bolt and tighten (Figure 6.3). Cycle the paper clip between the bottom of the top nut and the bottom of the bolt head (Figure 6.4) until the paper clip fractures. The stresses in the paper clip are not easily measured.

Step 1.

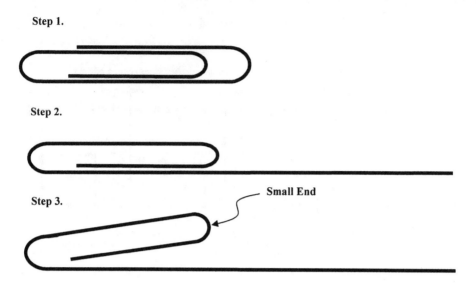

Step 2.

Step 3.

Small End

Fig. 6.2. Unbend paper clip for insertion into "test fixture" (see Figure 6.3).

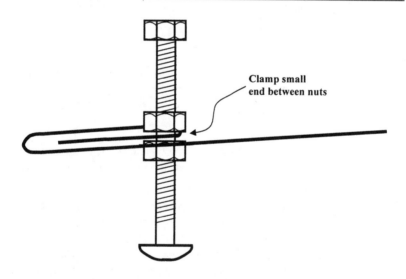

Clamp small
end between nuts

Fig. 6.3. Insert unbent paper clip test specimen into test fixture.

(They can be measured, just not easily.) The wire rotation is more convenient to measure and can be correlated with mechanical stress.

The fatigue crack initiates first and is followed by crack growth. Once
the wire starts to crack, it takes noticeably less force to bend.

Record the number of cycles to failure and the angle swung for each
cycle. Repeat the experiment four or five times using a new paper clip each

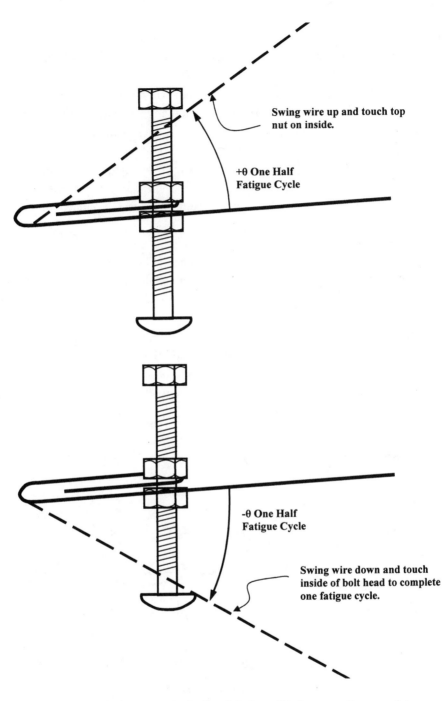

Swing wire up and touch top nut on inside.

$+\theta$ **One Half Fatigue Cycle**

$-\theta$ **One Half Fatigue Cycle**

Swing wire down and touch inside of bolt head to complete one fatigue cycle.

Fig. 6.4. Cycle the paper clip back and forth until it fractures. One complete fatigue cycle is $+\theta$ followed by $-\theta$, or simply count how many times the wire is touched to the bolt head.

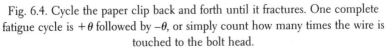

time. Repeat the experiment with a new rotation angle by changing the location of the nuts.

A small rotation angle, perhaps ±2 degrees, will result in an average life of 20,000 cycles, and an even smaller rotation angle will allow 50,000 cycles—a number more typical of aircraft design lives. Sample data for the paper clip experiment is shown in Figure 6.5. Note the statistical scatter for apparently identical tests and the well-known relationship between increased rotation (mechanical stress) and decreased life. Also note the similarity in shape with Figure 6.7, which shows fatigue data developed for the DC-10.

The paper clip data are perfectly valid fatigue data suitable for design purposes—if, for some reason, someone wanted to design a device that depended on bending a paper clip back and forth. The experiment also mimics full-scale fatigue testing done on complete airplanes. To carry the analogy farther, the paper clip should only be cycled from the horizontal or zero-degree position to the vertical or positive rotation. This would simulate wing loading starting from a no-load condition and then lifting up $+\theta$ every flight. You might think that with rotation reduced by one-half, the paper clip's life would increase by two. Actually, its life increases significantly more. Try it and see for yourself.

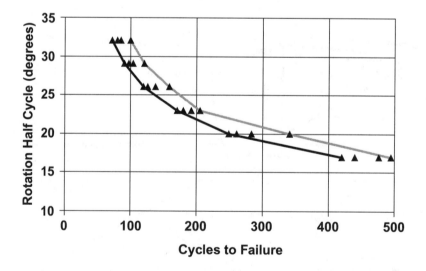

Fig. 6.5. Typical paper clip fatigue data. Note similar shape to DC-10 test fatigue data in Figure 6.7.

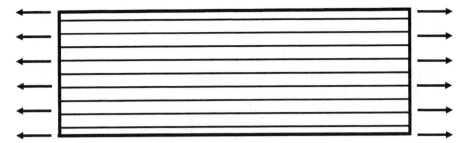

Imaginary lines of force flow unobstructed through a structure.

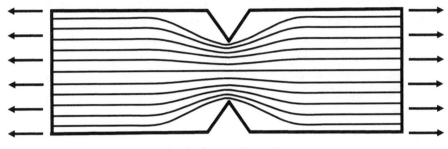

The notch concentrates the imaginary lines of force.

Fig. 6.6. A notch pinches or concentrates the imaginary lines of force flowing through the structure. The effect results in "stress concentration," or higher stresses in the notch.

There are other differences between the paper clip and the airplane. The paper clip will grow a single fatigue crack and rapidly break. The aircraft during normal use (or full-scale fatigue testing) will slowly grow multiple minor fatigue cracks and not break. The airplane, by design, is said to be a redundant structure.

Although there is a rate effect (very fast cyclic motion will fail differently than slow motion), rest periods have no effect. The plane (or the paper clip) will fail after the same number of cycles whether taking off 10 times per day or 10 times per year (perhaps another experiment for the paper clip test rig).

Stress Concentration

Fatigue is greatly affected by every little scratch, nick, and machining mark. Relatively minor differences in manufacturing process that result in visually undetectable differences can result in fatigue life reduction by a factor of 20. To understand why, one must understand the concept of stress concentration (Figure 6.6).

Notches result in "stress concentration," or an increase of stress in the metal of up to several hundred percent. One way to visualize this increased stress is to consider imaginary force lines flowing through a structure like a slow-moving stream of water (except the "flow" is in two directions as if you were trying to pull the structure apart). If unobstructed, the lines will flow uniformly on parallel paths and remain straight. If they must flow around a notch, the lines are pinched or concentrated.

An obstruction in a river will see a similar effect of concentrated erosion. The notch disrupts the flow of stress and imaginary lines of force just like an obstruction in a river disrupts the flow of water.

Given enough stress and load cycles, a fatigue crack can initiate with or without a notch. It is, however, a well-known and experimentally verifiable fact that notches increase stress and result in significantly fewer load cycles to failure (less life). Sharper notches result in greater stresses and greater reduction of cycle life.

A Case Study of Design and Testing: The McDonnell Douglas DC-10

Because it takes a few years to design an airplane, build a prototype, and complete the flight tests required for FAA certification, the detailed design and preliminary fatigue testing proceed in parallel. In fact, full-scale fatigue testing, which took 19 months to complete for the DC-10, actually ended almost a year after commercial service began. Current rules require the number of cycles on the test rig to exceed the cycles on a commercial plane by a factor of two. In other words, if the test rig has completed 2,000 cycles, the commercial plane is permitted to fly up to 1,000 cycles. As the test rig completes more cycles, the commercial plane can too.

The accelerated schedule is driven by economics. The airplane manufacturers cannot earn income on the billions of dollars spent on design, development, and testing for a new plane until delivery starts. Of course, safety is and remains the number-one concern. Delivery followed by later testing has to be justified by sound engineering practice. The relatively minor fatigue problems uncovered during full-scale testing will take years to occur in the commercial fleet and are easily fixed with minor design changes.

Full-scale fatigue testing after start of delivery proved fatal for the de Havilland Comet (see Chapter 4). Learning from the Comet's mistakes and

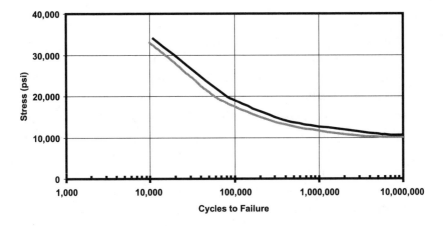

Fig. 6.7. Sample fatigue data for DC-10 fuselage joint design. Adapted from Stone, AIAA Paper No. 73-803

having the benefit of better fatigue data (and better design methods), full-scale testing after the start of commercial flight can now be justified. The procedure is supported by thousands of small-scale specimen tests and larger component testing during the design phase. McDonnell Douglas engineers were also able to build on previous experience and use about 2,000 fatigue test results from the DC-8 and DC-9 projects. These test results were used to establish the initial configuration of the DC-10.

An additional 1,700 small-scale specimen tests were completed to determine the material, fastener design, and manufacturing processes. The testing included about 300 individual tests for both wing and fuselage joint designs. The fuselage joints are particularly vulnerable (see Aloha Flight 243 accident in Chapter 4).

The fuselage is made of overlapping sheets of aluminum riveted together to form a "lap joint." The variables that affect the fatigue life of the lap joint are fuselage skin thickness; joint reinforcement (thin strips of added sheet metal); rivet type, size, and spacing; and assembly methods. Sample fatigue data used during development of the DC-10 fuselage lap joint design are shown in Figure 6.7.

One way to interpret the curve in Figure 6.7 is to enter the vertical axis at the stress resulting from pressurization of the cabin, 14,350 psi. The fatigue life corresponding to 14,000 psi stress is a little over 200,000 cycles. Note the similarity in shape with the paper clip fatigue data (see Figure 6.5).

After small-scale tests for the DC-10, larger panels of the fuselage were constructed and fatigue tested, as were window panels and other joint details.

Fail-safe testing of large sections of fuselage (168" × 104" panels with curvature equal to the radius of the plane) was conducted to verify that the fuselage was safe, even with a 40-inch-long fatigue crack straddling a cracked reinforcing rib. In one such test, the panel fractured when the fatigue crack grew to a 45.56 inches in length with a simulated cabin pressure 1.53 times normal. This type of failure could cause the plane to catastrophically break up at 40,000 feet. In order for that to happen, however, the following must occur.

The fatigue crack must grow undetected for years (after numerous visual and teardown inspections) to a length of 45 inches, the pressure control systems must fail (including the pressure relief system designed to automatically limit maximum cabin pressure), and numerous alarms must also fail. All these systems must fail simultaneously. Additionally, the flight crew must somehow miss all equipment failures and fail to respond as trained.

Occasionally, a large fuselage fatigue crack is found before it endangers passengers. A fatigue crack in the fuselage may make news in the aviation community; but nobody else seems terribly interested in a safe and uneventful landing.

After completion of hundreds of small-scale specimen tests and numerous larger component tests, the full-scale fatigue test of an entire DC-10 is almost anticlimatic. In a test fixture (like the paper clip), an entire DC-10 plane was cycled 84,000 times, or twice its design life.

Full-scale fatigue testing is done to evaluate fatigue strength under the combined loads of cabin pressurization, flight loads (lift), and ground loads (landing gears). Other objectives involve demonstrating fail-safe characteristics of the aircraft, obtaining crack growth data for damage-tolerance design methods (described later), and developing inspection and repair procedures.

Minor fatigue cracks did occur during testing of the DC-10. For example, two cracks occurred in the corners of a passenger door after 54,000 and 57,000 simulated flights. These cracks propagated slowly, growing just 3.75 inches and 5 inches after 29,000 and 27,000 additional simulated flights, well within the structural safety limits of a 40-inch-long crack.

All fatigue cracks found during full-scale testing must be resolved. The choices are to make structural design changes immediately, to repair the entire fleet as needed at some future date, or to require additional inspection. Usually, the decision is made based on an economic analysis. All choices are equally acceptable with respect to flight safety.

Aircraft Structure: Historical Perspective

The first riveted aluminum fuselage was introduced in 1930 by Northrup Company. The real trendsetter, though, was the highly successful Douglas DC-3, first manufactured in 1935.[1]

Except for changes in rib design, details of attachments, and other minor tinkering, the DC-3's riveted aluminum fuselage with reinforcing ribs has remained the standard configuration for large commercial planes up to the present. The new Boeing 787, the first plane to break the mold, will have a composite fuselage consisting of plastics reinforced with carbon and glass fibers. The latest Airbus is aluminum.

Until the 1950s and the catastrophic discovery of metal fatigue, the design philosophy for structural integrity changed little from the time of the Wright brothers.

The flight loads were identified during the design process. Loading conditions included various combinations of wing flaps up/down, rolling left/right or diving, full rudder left/right, engine out, and so on. Because of increasing computerization, design analysis keeps becoming more sophisticated. The number of design conditions has escalated from fewer than 100 on early Boeing models to thousands on more recent designs.

The structural integrity of a plane design is proven by a one-time static loading at 1.5 times the design loads. This method of static testing is still used to certify new wing designs, fuselage pressure containment, landing gear, and other structural components. For example, during mandated certification testing, the wings on a Boeing 777 were bent over 24 feet before breaking off. The test was done to certify wing strength equal to at least 1.5 times the flight loads experienced during recovering from an emergency dive maneuver.

Have you ever noticed a plane in minor turbulence flapping its wings like a bird? It's not a problem. Interestingly enough (and somewhat reassuringly), the wings have never broken off a modern commercial jet. The emergency dive design condition and 24-foot wing bend test (for the 777) simulate a dominant load that makes a few feet of cyclic wing tip deflection insignificant. This differs from the fuselage, which is designed for pressurization. At every takeoff, the fuselage is pressurized up to the design pressure load. The margin of over-design of the fuselage is much lower than for the wings. Consequently, undangerous fatigue cracking in the fuselage occurs occasionally (catastrophic fuselage fatigue cracks remain extremely rare).

Although engineers were beginning to think about metal fatigue in the early 1900s, the problem was not a major design and test focus until the Comet disasters in the early 1950s. In a sense, planes are still designed and tested much as they were nearly 100 years ago. When new design methods appear, the old methods don't disappear. Why should they? The old procedures almost always work and are always easier (and cheaper) than newer, more complex technology. New rules simply add to and supplement old rules. The history of aircraft design unfolds during the design of a new plane. The "first pass" design is based on the easiest methods, essentially rules of thumb passed down and refined over the years. Layers of newer technology are applied as the design is refined. Of course, the first crude design is probably as safe as a 1930s vintage plane. A few years (and a few billions of dollars) later, a twenty-first century design emerges.

Full-Scale Fatigue Tests on Planes

The first full-scale fatigue test was done on the de Havilland Comet, unfortunately as part of the crash investigation after two in-flight breakups due to metal fatigue. Being able to take advantage of the Comet's unfortunate experience, Boeing fully tested the 707 before commercial use. The de Havilland Comet and Boeing 707 were fatigue tested in a water tank for safety, because of the concern that an explosive decompression could occur as stored energy in the compressed gas is rapidly released (see Chapter 4).

As more experience and knowledge about metal fatigue were gained, air testing, which is much faster and cheaper (but potentially dangerous), was substituted.

An air test of the Boeing 777 in 1996–97 required 62 pipes to pump 46,000 cubic feet of air into the fuselage in 15 seconds. Over 100 computer-controlled hydraulic actuators were used to simulate loading on the wings and landing gears. A complete cycle simulated the plane's weight gradually lifting off the landing gear while the wing loading ramped up to the weight of the plane. Simultaneously, the cabin was pressurized. The process was reversed during the landing simulation.

Table 6.1 gives design service objectives and fatigue cycles performed during full-scale tests. Boeing has also obtained old, decommissioned planes with many flight cycles and added simulated cycles, as noted in Table 6.1.

The "design service objective" is a euphemism for design life. The design life is based on the marketing needs of airlines. For shorter, more frequent flights, planes with higher cycle lives such as the Boeing 737, are needed. For longer, less frequent flights (e.g., flying over oceans), planes with lower cycle lives, such as the Boeing 747, are appropriate.

Aircraft manufacturers do not like to use the phrase "design life," because it implies that at some point the design life is used up, and the plane is no longer safe to fly. Planes are considered to be safe forever if properly maintained. (Consider the DC-3, still flying after 60–70 years.) If a fatigue crack is removed and replaced with new metal, the fatigue life for the new metal starts at the beginning. Some people will even argue that older planes are safer because all the problem areas are well known, properly inspected, and carefully maintained. Maintenance personnel know where to look for the problems.

Table 6.1. Design Service Objectives (Design Life)
and Number of Fatigue Tests Conducted

Airplane	Design Service Objective (takeoffs)	Fatigue Test Cycles
707	20,000	50,000
737	75,000	150,000
		129,000*
747	20,000	20,000
		40,000†
		60,000‡
757	50,000	100,000
767	50,000	100,000
777	44,000	120,000§
DC-10	42,000	84,000

*In-service aircraft with 59,000 actual flight cycles plus 70,000 pressure cycles
†In-service aircraft with 20,000 actual flight cycles plus 20,000 pressure cycles
‡Pressure cycles on sections of a fuselage
§Aircraft also subjected to 20,000 pressure cycles

New rules kick in for planes flown past their original design service objectives. Essentially, they involve more of the same things already described in this chapter: more analysis of repairs, reanalysis reflecting the current number of cycle lives on the structure, and more inspection and maintenance. (Corrosion also becomes an issue with age and requires additional inspection and special maintenance requirements.) With rare exception, older planes are not considered unsafe to fly, but they are potentially too costly to maintain. A plane operated past its design service life might cost two to three times more than a modern plane to maintain.

The chief of Boeing's 747 fleet support said that there is no life limit on a 747, although older planes need additional costly maintenance. He further stated that the man hours needed to maintain a 25-year-old plane might be double or triple the hours needed for a 15-year-old plane.

Mandated Design Rules

New design rules typically lag behind disasters by a number of years. Sometimes, it takes a year or so to identify any potential research areas to address the problem. It then takes a few years to perform the research. Upon completion of the research (or completion of enough to draw useful conclusions), Boeing, Airbus, the FAA, the NTSB, and other interested parties negotiate rules that must be applied universally. Typically, the mandated rules represent a compromise. Meanwhile, each manufacturer usually responds to the new information resulting from a crash investigation much faster than the process described above and implements more stringent design and/or manufacturing procedures unique to their proprietary practices. The design engineers lead the regulators. For example, full-scale fatigue testing became routine after the Comet disasters in the 1950s, but it was not actually mandated until 1998.

Design for Metal Fatigue

There are three design methods used for metal fatigue and required by regulators. Listed in terms of sophistication and historical development, they are

1. safe life design;
2. fail-safe design;
3. damage-tolerant design.

Safe life design. Certain critical parts, such as landing gear and rotating engine parts, are replaced after a fixed number of cycles regardless of any evidence of fatigue. If a critical engine or landing gear component fails, there is no backup or redundant structure. A single fatigue crack in an engine rotor or landing gear will cause complete failure of that component and put the aircraft at risk of a more serious accident.

Landing gear and rotating engine parts are certified by the FAA to operate for a number of cycles (below the expected fatigue life) corresponding to a "safe life."

Safe life involves predicting component life with a combination of testing and design stress analysis. Computer modeling is used to predict the stresses in complex components. The calculated stresses are then used to predict component life based on experimental data of smaller test samples.

There are a few problems with the process. As demonstrated with the paper clips, the experimental data are not very repeatable. As mentioned, fatigue is sensitive to every manufacturing scratch and machining burr. Test specimens carefully polished to eliminate all manufacturing imperfections will have less statistical scatter but will still fail to give complete repeatability. Imperfections and random "stuff" also exist on a microscopic level inside the metallic crystals.

Engineers deal with the problem of repeatability in two ways. First, there is a need to repeat the tests often and treat the results statistically. This leads to predictions of percent reliability, a bit like predicting rain with 60% confidence. Of course, much better reliability is required. Second, additional safety is provided with frequent inspections for metal fatigue. The first of many inspections occurs long before any predicted fatigue failure.

A need remains for limited full-scale testing; even complete airframes and engines are tested to account for all manufacturing tolerances, fit-up errors, tool marks, etc. However, it rapidly becomes very expensive to test multiple $200-million-dollar airplanes to destruction. Engineers compromise by testing hundreds of small specimens, scores of panel sections, maybe a few fuselage sections, and one or two complete airplanes.

Although fatigue is very sensitive to manufacturing imperfections, even this aspect is successfully managed with 50 years of experience, hundreds of millions of dollars of test data, and extreme attention to detail. Airplane manufacturers really do know what they're doing, and they do it well.

Fatigue is controlled by good craftmanship, design practice, manufacturing, and maintenance.

The fan disk of a jet engine can be used as an example of how a component's safe life is determined. The designers estimated that each fan disk has an "average" fatigue life of 54,000 cycles. The FAA certified the safe life at $54,000/3 = 18,000$ cycles. Therefore, the component is replaced after 18,000 cycles, regardless of any evidence of fatigue. For added safety, the fan disk also receives in-service inspections. On infrequent occasions, the safe life does not provide an adequate margin. As described in Chapter 5, a fan disk on United Flight 232 disintegrated before it reached its certified safe life. When the fan disk cracked, no additional structure remained to support the centrifugal forces of the rotating engine part. Despite seven inspections for metal fatigue during its service life, the fan disk failed because an undetected manufacturing defect unexpectedly reduced the fatigue life.

Fail-safe design. Fail-safe design practice, first adopted in the 1950s, was not mandated by the FAA until 1964. This approach requires a minimum residual strength for safe flight to be maintained even after complete or partial fracture of a primary structural component.

The structure is designed either to "fail safely," because other components are available to carry the load, or to have adequate strength even with a fatigue crack present. "Fail-safe" is the preferred design procedure and is always used if possible.

An example of fail-safe design is the fuselage. The fuselage is designed to safely operate with a 40-inch-long fatigue crack. As described in Chapter 4, when a fatigue crack reaches a reinforcing tear strap, the crack will turn the corner and create a big flap. The flap opens, allowing the pressure in the fuselage to safely leak out, thereby removing the driving force and mechanical stress trying to propagate the crack. This allows a growing crack in the fuselage to fail safely.

Fail-safe design of structural components uses a lower reliability than safe life design of critical components. Boeing's design practice gives all structures a 20-year life (number of cycles varies with flight cycle time) with 95% reliability. More critical components are designed for a 30-year life. The expectation is that the structure will remain free of fatigue cracks for 20 years and require only minor repairs in years 20 through 30.

The 95% reliability is not meant to imply that 5% of the fleet is expected to fail. After 20 years, 5% of planes might begin to grow fatigue cracks that

will be detected during normal inspections. Also, these components are redundant with other structures; a single fatigue crack will not jeopardize
flight safety.

Damage-tolerant design. One or two crashes have resulted when an aircraft structure did not fail safely. This situation occurred when multiple fatigue cracks grew in remote, difficult-to-inspect locations of the plane. It
was recognized that inspection intervals must be set to ensure that fatigue
cracks are found before becoming dangerous.

Fail-safe design was modified to incorporate the science of "fracture
mechanics," or the study of crack growth rates and determination of exactly when a crack reaches a dangerous length. Although aircraft manufacturers were using these new methods in the late 1960s, the new
rules, referred to as "damage tolerance" rules, were not mandated until 1978.

Under the old rules, structural safety was sought through redundancy.
Safety was maintained even if component failures occurred. Under the
new rules, fatigue cracks are assumed to exist and grow, but they are
managed by ongoing inspection and repair. Improvements in inspection
technology and increased ability to reliably find fatigue cracks are integral to this new approach. With increased confidence in the latest damage-tolerance design methods and inspection technology, the design
service life for the Boeing 777 was increased from the traditional 20 years
to 30 years.

Inspection intervals have historically been based on service history. New
intervals have been adjusted by analysis and testing to verify that cracks are
found before reaching a critical length. Since planes seldom crashed under the old rules, only relatively minor (but important) adjustments to inspection schedules were needed.

An example of a damage-tolerance design method is the analysis and
testing done to verify the safety of a 40-inch crack in the fuselage. A 40-
inch-long defect could occur from a burst engine, terrorist bombing, or
growing fatigue crack (hence the name damage tolerance). Even with such
damage, a plane is expected to be able to land safely. Fuselage design details are described in Chapter 4. Damage-tolerance design methods for engine components, with significant increases in reliability, are described in
Chapter 5.

What happens if a gun is fired inside the cabin? Nothing. A bullet hole
in the fuselage is insignificant compared to a 40-inch-long fatigue crack. A

window hit by a bullet might not fare as well, but even a shattered window does not threaten flight safety. The windblast out the window can be a problem, of course, just like in the movies. Blown out window accidents are described in Chapter 4.

The Latest FAA Rules for Fatigue Design and Testing

The rules for fatigue design and testing keep evolving as new problems occur. Aloha Flight 243,[2] described in Chapter 4, surprised the industry and partially defeated the damage-tolerance design methods. Essentially, a series of very small fatigue cracks grew out of a line of consecutive rivet holes. The remaining strength of the fuselage, with a series of small cracks, was considerably less than previous analysis predicted. Also, small cracks can link up to form a dangerous long crack. The phenomenon is referred to as widespread fatigue damage. Widespread fatigue damage is most likely to occur in riveted fuselages with thousands of identically loaded rivets. Since 1998, the FAA has required full-scale fatigue tests to demonstrate that widespread fatigue damage does not occur during an airplane's design service objective.

Fatigue testing of the Boeing 777 began in January 1995. The 120,000 simulated cycles, occurring every four minutes around the clock with stops only for routine inspections, were not completed until March 1997. Meanwhile, the first commercial flight occurred on June 7, 1995, just six months into the two-year fatigue test.

Also in response to the widespread fatigue damage problem, intense inspections of high-cycle planes have been conducted, as well as tear-down inspections of high-cycle out-of-service planes. After the Flight 243 accident, the whole subject of crack growth and critical crack growth had to be restudied with testing and analysis to evaluate the effect of multiple small cracks.

Aircraft are also rated for a fixed number of flight hours. This relates to cyclic loading from random wind gusts. Wing fatigue damage is caused by the ground-to-air-to-ground cycle, by pilot-induced maneuvers, and by turbulence in the air. Flight hours are more important for analysis of wing fatigue than of fuselage fatigue. By now, wind gusts are well understood. New wind gust design data have been developed for many aircraft models by flying instrumented planes for thousands of hours and recording gust data for statistical processing.

Aircraft Maintenance

Maintenance service checks are referred to as A, B, C, and D checks. The actual inspection schedules vary with different aircraft designs and from airline to airline, depending on how often the plane is flown. The schedule might be specified in terms of calendar time, flight cycles, or flight hours.

The A check is the aircraft version of a service check of windshield wipers, brakes, tires and fluid levels (oil, hydraulic, etc.). The B check might sample fluids (stray contaminants indicate potential problems), replace filters, and lubricate parts of the plane. Unless there is a service bulletin requiring a specific inspection for fatigue, only a brief visual examination occurs for metal fatigue during A and B checks. An A check might occur every month or so and a B check every three months. Both A and B checks are usually done overnight at an airport gate.

The C and D checks are more in-depth inspections for fatigue cracks and other routine maintenance. In a C check, sections of the plane might be stripped to bare metal for fatigue inspection. The entire plane is stripped to the bare shell during a D check. These are extensive overhauls done at maintenance centers.

An airplane is delivered with a service manual written by the manufacturer with FAA participation. The manual specifies required maintenance, inspection, and part replacement schedules. As the airlines gain experience, they begin to develop their own nuances and schedules.

The C and D check schedules get particularly scrambled by a variety of factors. An airline may decide to extend their C check effort to minimize downtime during the next D check. As problems become known, the FAA will issue Airworthiness Directives requiring additional inspection and repairs. Airlines may expand the mandated inspection and alter their C and D schedules.

The FAA issued an Airworthiness Directive for the 747's forward fuselage section. A fatigue cracking problem became known to the industry in 1986, 16 years after the plane was introduced. Inspection and repair is now mandated on the fuselage reinforcing ribs at 10,000, 13,000, 16,000, and 19,000 cycles. Total replacement of parts is required by 20,000 cycles.

Inspections and maintenance are expensive. "Used" planes advertised for sale will state when the next major inspection is due. One such In-

Table 6.2. Example Inspection Schedule
for a Boeing 737-200

A check	Every 250 hours of flight time
B check	1,250
C check	2,500
D check	15,000

ternet ad for a Boeing 737-200 gave the inspection schedule shown in Table 6.2.

The baseline inspections for a Boeing 747-400 are

A check every 600 flight hours;

B check not used;

C check every 5,000 flight hours or 15 months, whichever occurs first;

D check every six years (no flight hour restrictions).

The chief of Boeing's 747 fleet support states that the A check takes one day, the C check one week, and the D check one month or longer.

Lufthansa gives a few details on a D check of a new 747 after six years of service. About 230 specialists work around the clock, taking approximately 60,000–70,000 man hours to complete the task in six weeks. Keeping track of the Boeing 747 service manuals—over 41,000 pages including 5,000 pages of checklists—is itself a lesson in maintenance management. Although everything is checked (engines, electronics, hydraulics, brakes, etc.), the main purpose of the D check is to strip all exterior paint and remove all interior furnishings to expose the bare metal for fatigue inspection.

All three million of the plane's fasteners are visually inspected and up to 10,000 individual rivet holes are inspected for fatigue with a process known as eddy current[3] testing, in which high-frequency electrical current is imposed on the part. This method will find cracks as small as a few hundredths of an inch long. (Small cracks radiating out of a series of consecutive rivet holes are a concern.) Additionally, the window and door cutouts are x-rayed for cracks, similar to the way a doctor looks for a fractured bone.

Lufthansa also reports extensive structural inspections and system checks at 8- to 18-month intervals and again every 48 months.

Airlines are permitted to make limited repairs to fatigue cracks per the manufacturer's service manual. The basic idea involves cutting out a crack, grinding out all scratches and machining burrs, restoring structural integrity, and maintaining the pressure seal with a patch.

Depending on the length and depth of a crack, increasing safety restrictions are imposed. The service manual might require a permanent repair to be done within a few flights, or the plane may be required to immediately fly to a repair facility. Until repaired, the maximum cabin pressure might be limited or even reduced to zero. If the fatigue crack exceeds certain parameters, the plane cannot be flown without engineering evaluation by the manufacturer.

Mechanics Are Regulated, Too

Mechanics are highly trained professionals who take their work very seriously. They are licensed, have random drug tests, and go through yearly training, just like pilots. Everything done to an airplane can be traced to a specific mechanic and parts supplier. Any time a repair is made, even the replacement of a single screw, the mechanic has to sign a logbook. A supervisor has to check the work and also sign. There is direct accountability, and fines and other penalties occur when appropriate.

An incorrect screw (either a wrong part or improperly installed) can be the source of mechanical failure, including metal fatigue. For this reason, a single screw does in fact receive a complete paper trail, just like any other part.

Metal Fatigue and Atomic Structure

There is a simple explanation for metal fatigue using atomic structure and crystals. In practice, atomic theory fails to accurately predict metal fatigue because of sensitivity to minute manufacturing imperfections.

Atomic Structure and Crystals

A metal bar is amazingly strong. A one-inch-diameter bar of ordinary steel or high-strength aluminum can support about 50,000 pounds, or the weight of about 15 automobiles.

A metal bar loaded in tension will elongate (Figure 6.8). When the load is removed, the bar springs back to its original length. With reversed loading (compression), the bar will compress and spring back upon load removal.

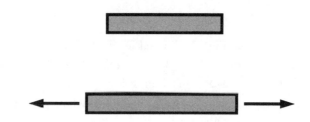

Fig. 6.8. A metal bar elongates when loaded in tension. Elongation shown is greatly exaggerated.

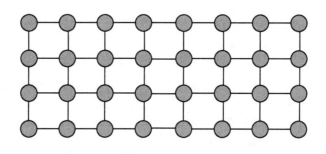

Fig. 6.9. Two-dimensional slice of a three-dimensional metallic crystal. The dots are atoms, and the lines are atomic bonds.

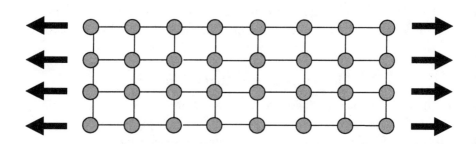

Fig. 6.10. Applying a load to the crystal stretches the atomic bonds like tiny springs.

The springiness of a metal bar can be explained with atomic theory. Metallic atoms are tightly packed in a crystalline structure, a regular arrangement of atoms that repeats itself. In Figure 6.9, each dot represents an atom, and each line represents an atomic bond between the atoms. When the crystal is stretched with a force, the atomic bonds stretch like

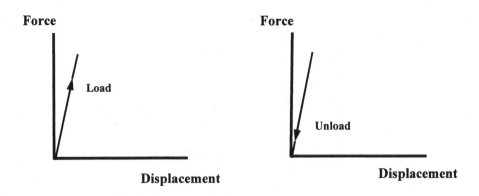

Fig. 6.11. Force versus displacement plots of elastic loading and unloading for a metal bar. During elastic deformation the bar starts with zero force and displacement. Both values return to zero with load removal.

tiny springs (Figure 6.10). The atomic bonds spring back to their original configuration when the force is removed.

The atomic bonds are very "stiff," with relatively little deformation occurring. An aluminum bar ten inches long and one square inch in cross section will stretch only 0.047 inches when loaded with 50,000 pounds. A similar steel bar will stretch about one-third as much.

If the same aluminum or steel bar were loaded in compression, it would compress by the same amount as it stretched. In either case, the bar will spring back to its original ten-inch length when the tensile or compressive load is removed. This type of loading is called "elastic;" the bar is said to spring back elastically upon load removal.

A graph of force versus displacement shows a straight line during loading (Figure 6.11).[4] The path completely retraces itself during unloading in the opposite direction, resulting in zero force and displacement with total load removal.

If the rod is loaded beyond some critical value, it does not return to zero displacement with load removal. The rod is stretched to some permanent new length. A plot of force versus displacement for this new condition is shown in Figure 6.12. Permanent elongation or "plastic" deformation has occurred.

For aircraft design, the plane's metal must remain elastic for all design loads. A new plane is certified to operate safely by verifying with testing that the structure does not fail at 150% of the design loads. Plastic deformation will occur during this overloading condition. The plane is expected

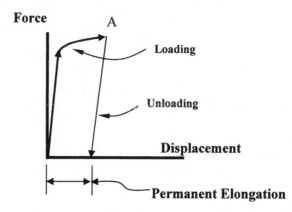

Fig. 6.12. Plot of force versus displacement for loading and unloading of a permanently stretched metal bar.

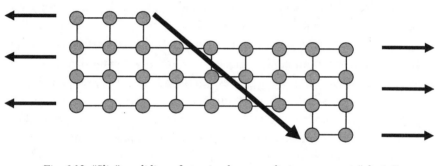

Fig. 6.13. "Slip" or sliding of atomic planes results in permanent "plastic" deformation.

to fracture (or buckle) shortly after meeting the 150% overload requirement.

When the plane's structure permanently deforms, a new mechanism is taking place other than stretching of atomic bonds. This new mechanism is called "slip." Planes of atoms are sliding on each other, as shown in Figure 6.13.

The force required to break all the atomic bonds can be estimated. The number of atoms in a block of metal can be accurately determined using atomic theory. When the block of metal is heated to a molten state, the amount of heat can be carefully measured. From this, the energy required to break all atomic bonds can be estimated.

With this knowledge, the forces required to simultaneously break all the bonds and slide a plane of atoms, as shown in Figure 6.13, can be calculated. Unfortunately, this calculation estimates a strength up to a thousand times greater than that indicated by the experimental data.

Crystalline Imperfections: Dislocations

Crystalline imperfections known as dislocations[5] were hypothesized to explain how atomic bonds could slip one at a time. During solidification, the random nature of crystal formation leads to imperfections or disruptions in the perfect crystalline pattern. Dislocations can take many forms; the easiest to draw on a two-dimensional sheet of paper is called an edge dislocation. An edge dislocation is an extra half-plane of atoms in a crystal.

When crystals with edge dislocations are loaded in sideways shear, only one atomic bond at a time needs to be broken for a plane of atoms to slide off the edge of the crystal (Figure 6.14). This greatly reduces the force needed to obtain permanent plastic deformation.

The process is often compared to moving a large rug. Pushing a hump through a rug (Figure 6.15) is far easier than pulling on one end and trying to overcome the friction of the entire rug.

Shear stresses always occur, no matter how a structure is loaded. The shear stresses in a rod pulled in tension can be visualized by considering a glued angle joint in the rod, as shown in Figure 6.16. The sliding stresses that break the glue joint are shear stresses.

A dislocation sliding or slipping of a crystal under shear stresses caused by a tensile pull is shown in Figure 6.17. In Figure 6.17, a dislocation to the right of the slip plane slips down and off the crystal. One must imag-

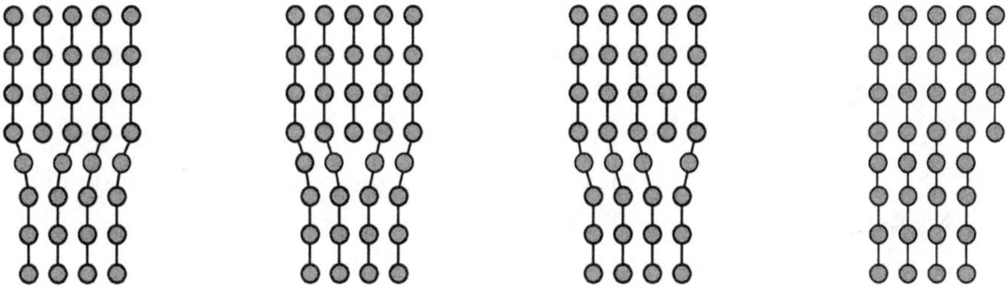

Fig. 6.14. An edge dislocation loaded sideways in shear is shown moving through a crystal. Dislocations make slip easier by requiring atomic bonds to break one at a time.

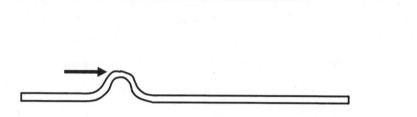

Fig. 6.15. A dislocation is often compared to a hump in a rug. It is easier to move the rug by pushing on the hump than by pulling on the end of the rug.

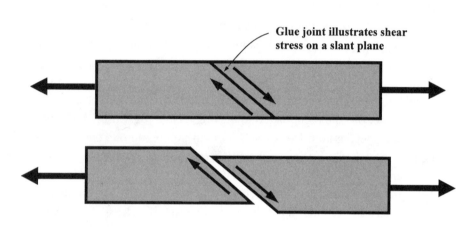

Fig. 6.16. Shear stress can be imagined as the sliding that occurs in a slant-angle glue joint pulled in tension.

ine a second dislocation to the left of the slip plane that slips up and off the crystal with the same shear stress. The two dislocations slipping off the crystal create the step shown in Figure 6.18.

A dislocation is like a tiny hole in the crystal. As dislocations accumulate at the same location, the tiny hole grows and strains the crystal. This strain resists additional dislocation movement into the growing hole. The net effect is that the accumulation of slip in one plane strengthens the crystal as the dislocations "pile up" and interfere with each other. This shifts slip from a strengthened slip plane to an as-yet unstrengthened slip plane. When many planes are strengthened, the end result is a stronger metal. All metals strengthen in this manner, which is known

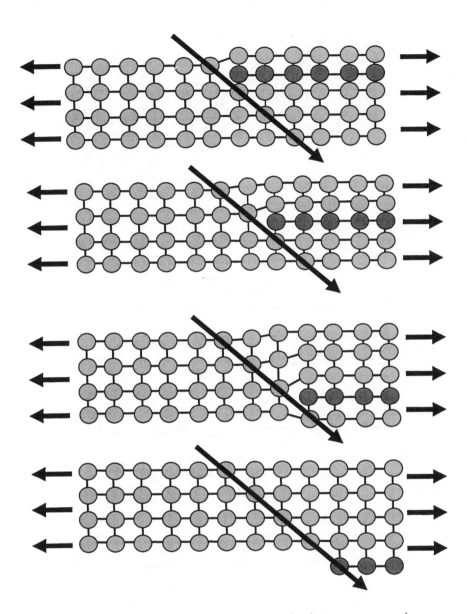

Fig. 6.17. When a crystal is loaded in tension, the shear stresses cause the dislocation to "slip" along a slip plane.

as work hardening. The effect is very noticeable in copper tubing. Bend an 8-inch length of 0.25-inch copper tubing in one direction and then flip it around and bend it in the opposite direction. The first bend is very easy, but the second is noticeably difficult. The copper can be "annealed" with a torch, in which case the copper atoms diffuse around un-

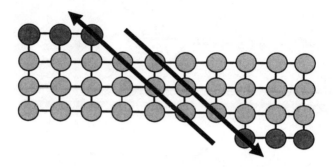

Fig. 6.18. Dislocations to the right and left of the slip plane slide off the crystal and create a permanent crystalline offset deformation, an increment of plastic deformation.

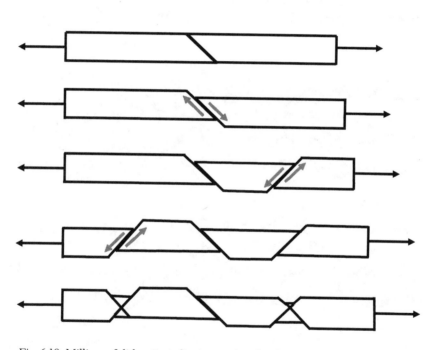

Fig. 6.19. Millions of dislocations slipping result in bar becoming longer and thinner.

til they undo most of the accumulated dislocations. Both effects are well known to plumbers.

Slip occurs preferentially at ±45 degrees, the orientation of the highest shear stress (at least during a tensile pull). Repeated slipping at ±45 degrees (on multiple slip planes) causes the bar to become uniformly longer and thinner, as shown in Figure 6.19. Every time an atomic plane slides

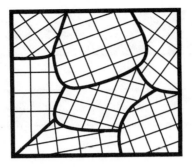

Fig. 6.20. Each square represents an atom inside a single grain. The diagram shows seven grains.

one atomic spacing, the crystal becomes one atomic spacing longer. It also gets thinner locally at the slip plane by one atomic spacing.

A paper clip might have over 1,000,000,000,000,000,000,000 atoms that form millions and millions of tiny crystals. These crystals have random orientations that tend to smooth out and make even more uniform the slip that is occurring.

When a molten metal begins to solidify and crystallize, the crystals form at random sites with random orientations. When a crystal with one orientation meets a crystal with a different orientation, a grain boundary is formed. One must imagine millions of randomly oriented tiny crystals called grains outlined by grain boundaries (Figure 6.20).

Each crystal will have a preferential direction of slip, but the net slip effect is greatest at ±45 degrees, the angle of highest shear stress.

The net effect of many, many atomic planes sliding at ±45 degree planes is that the bar becomes longer and uniformly thinner.[6] Many metals are quite ductile, meaning that they can be easily elongated or bent (this differs from a brittle metal that elongates very little). Depending on their alloying and processing, they may in fact elongate with approximately the scale shown in Figure 6.21.

Elastic and Plastic Deformation with a Paper Clip

Straightening a paper clip and pulling on it will illustrate all the concepts of elastic and plastic deformation. However, the elastic elongation is very small and difficult to measure without special instrumentation. (A 60-pound pull will elastically elongate a straightened 6-inch paper clip by about 0.010 inches.) It is much easier to illustrate elastic and plastic deformation by

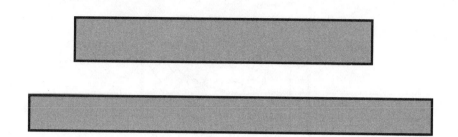

Fig. 6.21. With slip occurring in millions of crystals, the macro effect is that the rod gets longer and thinner. The deformation shown is approximately true to scale for many metals.

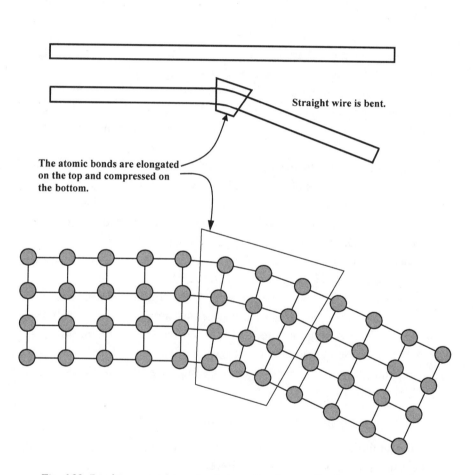

Straight wire is bent.

The atomic bonds are elongated on the top and compressed on the bottom.

Fig. 6.22. Bending a metal is easier than performing tensile elongation. Bending results in stretching of atomic bonds on the top edge and compression on the bottom edge.

bending a paper clip, because leverage can be used to trigger slip with lower forces.

First it must be understood that bending results in elongation of atomic bonds on one side and compression on the other, as shown in Figure 6.22. When the load is removed, the stretched (and compressed) atomic bonds will spring back to their original position. Bend the paper clip at a small angle and let go. The wire elastically springs back. Bend it beyond some critical value and slip occurs. The wire stays permanently bent. If the wire is bent at a large angle, say 90 degrees, it will still spring back a small amount, maybe 5 degrees or so. This permanent large bend results from the sliding of planes of atoms to new permanent locations. However, there are still stretched (and compressed) atomic bonds waiting to spring back slightly after load removal.

In one final experiment that demonstrates sliding of atomic planes, the sliding generates heat from internal friction. If a paper clip is rapidly bent back and forth and quickly placed against delicate skin (dry lips will do), heat generation is easily felt. Grabbing the end near the large radius and rotating it rapidly from 0 to 45 degrees about 15 times seems to do the trick. (Use a large jumbo clip. The rotation should be in the plane of the clip to create bending, not in torsion as described in the previous experiment.)

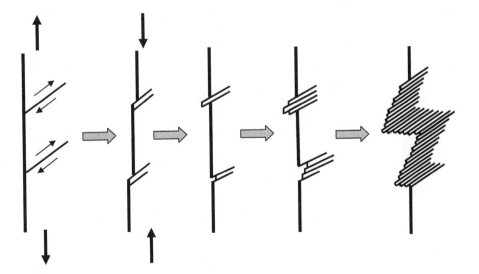

Fig. 6.23. Repeated back and forth cyclic loading accumulates slip damage and initiates a fatigue crack. *Source*: Stephens et al., *Metal Fatigue in Engineering*

Repeated Slip Initiates Metal Fatigue

Repeated slip—repeated back and forth sliding of atoms—results in metal fatigue. As shown in Figure 6.23, an upward slip occurs first. When the load is reversed, a downward slip occurs. Repeated push-pull cycling results in accumulated upward and downward slip. These repeated, accumulated slips are known as extrusions and intrusions and can sometimes be seen with an electron microscope. The delicate structure is easily damaged during cycling or subsequent handling.

Once a fatigue crack has grown a certain length, it starts to develop its own stress field and stress concentration.

Fatigue is a matter of continuous life and death for all aircraft. The problem has more or less been solved by extensive study, analysis, and testing. Fatigue cracks still occur (they are expected to occur), but procedures are in place to safely manage their occurrence. Fatigue cracks can still become dangerous due to human error, however, so the inspectors have to find the fatigue cracks, and a proper repair must be made.

7...

Combustion: Fire and Explosion

Swissair Flight 111, a McDonnell Douglas MD-11,[1] departed New York's JFK Airport at 8:18 p.m. on September 2, 1998, bound for Geneva, Switzerland. The following story unfolded in the newspapers immediately after the disaster.

Almost an hour later (9:14 p.m.) and 320 miles northeast of Boston, at 33,000 feet, the pilots radioed to report smoke in the cabin from an unknown source. The pilots proposed going back to Boston and started to turn when the Canadian controller suggested Halifax, just 70 miles away, instead. For the next 11 minutes, the plane gradually descended to 9,600 feet.

Flying at 8,000 feet off the southeastern shore of Nova Scotia and just 40 miles from Halifax International Airport (7–10 flying minutes), the plane suddenly plunged into the sea at 9:31 p.m. The crash site was five miles off Peggys Cove. Local residents reported hearing a loud noise.

Peggys Cove is one of Canada's most popular tourist destinations. The *New York Times* reported that "Peggy's Cove . . . was still known mainly as the most visited, most admired old fishing village on Nova Scotia's craggy east coast—necklaced by great white rocks, serenaded by crashing surf and crowned by a red-tipped lighthouse worth driving hours to see . . . and a place where daring rescues at sea are a proud tradition."[2]

Many local fishermen and lobstermen, familiar with the rocky coastline, immediately took to the sea hoping to rescue survivors. Within 40 minutes, the smaller boats were joined by Canadian Navy and Coast Guard frigates, cutters, and helicopters. However, heavy weather and 10-foot swells forced a suspension of all rescue activity till dawn. The next day, over 1,500 navy and coast guard personnel and private citizens were involved in the search and rescue operation. No survivors were found.

An aircraft recovery effort goes through stages. Initially, there is a rush to find survivors. The search for survivors eventually becomes a search and recovery of victims. Only later does locating the black boxes become a priority. As part of the wreckage recovery, the flight data recorder and cockpit voice recorder are always the first items sought. They are often the easiest to find because of their "pinging" radio transmitters activated by water immersion.

The official report issued by the Transportation Safety Board of Canada adds more detail to the timeline.

At 9:10 p.m., the first whiff of smoke was almost indiscernible. The pilots decided the amount was small, momentary, and no longer detectable. Two-and-a-half minutes later, the smoke returned, and within a minute the pilots reached for their oxygen masks and made plans for a diversionary landing. The flight crew radioed "Pan, Pan, Pan," indicating a serious but not life threatening situation.[3]

At 9:17 p.m., the pilot briefed the head attendant about an emergency landing at Halifax in 20–30 minutes.

By 9:21 p.m., the flight crew decided to turn south, away from the airport, to dump fuel over the ocean.

Later, the Canadian investigators concluded that even without the diversion to dump fuel, there was not enough time to safely land the plane. Although the plane was relatively close to the airport, it takes time to come down from altitude. In hindsight, the flight crew should have decided to dump fuel over land or not at all. However, the pilots were dealing with a small, hidden fire causing unpredictable electrical damage, and they didn't think time was so short.

Landing procedures assume that the landing weight is considerably less than the takeoff weight. A significant percentage of the takeoff weight is fuel (33% in the case of Flight 111). The weights for Flight 111 are given in Table 7.1.

Table 7.1. MD-11 and Flight 111 Weights

Category	Weight (lb)
MD-11 empty	290,586
MD-11 certified maximum takeoff	630,500
MD-11 certified maximum landing	531,534
Flight 111 takeoff	439,998
Flight 111 fuel	143,961

In an emergency, there are no structural limits; the aircraft can safely land at any weight. However, excess weight will affect the plane's ability to stop in a safe distance. A common crash scenario is running off the end of the runway (see Chapter 2).

At 9:23 p.m., after the flight crew had gone through the Smoke/Fumes Unknown Origin Checklist, they turned off the cabin power switch. (Because intermittent electrical shorts can cause smoldering and smoke before ignition and fire, the procedure is to shut off all unnecessary electrical systems.) The cabin power switch shuts off the recirculation fan and changes the airflow patterns in the cockpit ceiling. Instead of being drawn away from the cockpit, smoke will tend to move towards the cockpit.

At 9:24 p.m., the autopilot disconnected because of electrical damage caused by the fire. For the next one-and-a-half minutes, the flight data recorder was reporting a series of electrical failures, which initiated various caution lights and alerts on the control panel. Forty seconds later, the captain declared an emergency and shortly after told air traffic control that they had to land immediately.

The cockpit voice recorder confirmed that the pilots commented on something dripping from the ceiling at 9:25 p.m. The investigators later found molten plastic on the floor.

A moment later, the flight data recorder stopped recording and the control panel went dark. The crew had to fly the plane with small backup dials in a smoke-filled cockpit. (The primary flight controls are hydraulic and do not require electrical power to operate.) Finally, at 9:31 p.m., the aircraft struck the water. While staring out the window into the dark sky over an ocean with no visual references, the pilot accidentally rolled the plane over into the ocean. The aircraft broke up severely on impact.

Flight 111: Recovery, Reconstruction, and Fire Damage

Starting with 15 months of recovery and sorting, the investigation lasted four-and-a-half years. Ninety-eight percent of the plane was retrieved. Much of this effort involved the examination and placement of thousands of individual pieces of wreckage onto a reconstruction mockup (Figures 7.1 and 7.2). About 33 feet of the cockpit area was reconstructed.

To aid in the reconstruction, Boeing provided three-dimensional CAD (computer aided design) drawings that could be zoomed and viewed from any angle. Other software was used to create a virtual reality tour with

Fig. 7.1. Swissair Flight 111. Port side view of the full-scale reconstruction jig.
Photo: Transportation Safety Board of Canada; used by permission.
Source: SR Flight 111—Wreckage recovery and wreckage reconstruction photos
www.tsb.gc.ca/en/media/photo_gallery.asp, Transportation Safety Board of Canada,
2003 and 2005. Reproduced with the permission of the Minister of Public Works
and Government Services Canada, 2006.

panoramic views of the affected areas. Minute examinations took place of the panels, fabrics, wires, and so on.

Since the fire was immediately doused on impact, there was no post-crash fire to confuse the investigators. The fire damage was relatively modest and did not, for example, burn through to the outside of the fuselage. Unfortunately, the fire was in a particularly sensitive location (the cockpit) where most of the control wiring was located.

It quickly became apparent that the fire damage was centered in the cockpit ceiling and had worked its way down. No burn damage occurred in the back of the plane.

Heat damage and soot patterns were studied extensively. The aircraft skin was painted on the outside with white paint and on the inside with green primer. The fire did not burn through the thin fuselage skin, nor was

Fig. 7.2. Swissair Flight 111. Inside the reconstruction jig and looking forward into the cockpit door. *Photo:* Transportation Safety Board of Canada; used by permission. *Source:* As for Fig. 7.1

the exterior paint discolored. There were varying degrees of discoloration on the inside primer.

To assess the intensity of heat damage caused by the fire, pieces of comparable materials were intentionally exposed to heat at various temperatures for specified time durations. The interior primer paint on the aluminum was found to incrementally change color from light green to dark brown when exposed to increasing temperatures, and the paint disappeared altogether when baked at 900 to 1,100°F for 10 minutes. Soot patterns were also mapped extensively.

The most severe heat damage was identified by the presence of minute amounts of resolidified aluminum metal (i.e., metal that had once been molten or near-molten) in the cockpit. Depending on the alloy, aluminum melts at 1,175 to 1,250°F.

The pre-fire airflow patterns were studied (including two full-scale tests) to determine potential locations from which smoke could enter the cockpit, but not the cabin, as reported by the flight crew. The list of potential sites was further narrowed by the obvious requirements of an ignition source

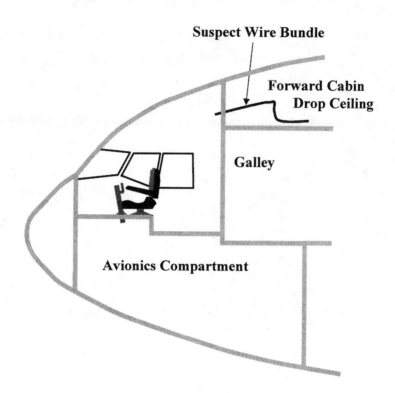

Suspect Wire Bundle

Forward Cabin
Drop Ceiling

Galley

Avionics Compartment

Fig. 7.3. Swissair Flight 111 fire initiated in forward cabin drop ceiling. Adapted from Transportation Safety Board of Canada, Report No. A98H0003

in the presence of combustible material. A fire growing from each location had to explain and match up with the recorded sequence of system failures. One location, in the top of the cockpit rear wall, seemed to meet all the requirements.

In January 2002, computer modeling of the fire was conducted by the University of Greenwich. The computer model incorporated a three-dimensional representation of the interior construction details to correctly simulate airflow, smoke, and the spreading fire. Also incorporated were fire burn test data (how fast various materials burn and how much heat they release), air flow physics, heat transfer rates, and smoke paths.

The fire ignition site was narrowed down to a two-foot square area in the forward cabin drop ceiling (Figure 7.3) by evaluating the following evidence. The investigators examined soot patterns, heat damage, the sequence of system failures (correlated with known wire locations), air and smoke patterns, and the presence of electrical wires as potential ignition sources near flammable insulation blanket material. Because of the complete absence of any equipment that could overheat and initiate a fire, the only plausible explanation was a short circuit and arcing event.

Most of the plane's 150 miles of wiring was recovered and examined. The investigators found 14 wires with arcing damage. Arcing of copper wires is easily identified by the presence of resolidified copper. Since the temperatures from this relatively modest fire did not approach the melting point of copper (1,981°F), any melted copper present on Flight 111 had to be from arcing. Also, an arcing indication is usually limited to a relatively small spot on the wire.

It was three years into the investigation before the wire believed most likely to have initiated the fire was found. Arcing can cause fire, but fire can also result in arcing, so it is impossible to state with 100% certainty which happened first. However, only one arced wire was found from the identified fire ignition region.

An electrical arc is believed to have ignited the plastic cover on the insulation blankets. Insulation blankets are installed inside the fuselage skin to maintain comfortable temperatures and reduce noise entering the cabin. There are several configurations of insulating blanket. One of the more popular is a batting of fiberglass. Fiberglass is essentially melted glass spun into very fine fibers (approximately 3–5 microns in diameter) that are fluffed and tangled to form a blanket.

Glass itself is a relatively good conductor of heat. Fiberglass insulates by trapping air and preventing convection currents. The insulating properties are reduced if the blanket is matted from moisture.

On a typical flight, gallons of water are emitted by passengers breathing. Most of this water remains as vapor and is vented through the normal ventilation system. However, since it is –60°F at 40,000 feet, any moisture that makes its way to the fuselage skin quickly condenses. All airframes have carefully designed collection and drain points for sidewall condensation and spills from lavatories and galleys.

The fiberglass insulating blanket is covered with a thin moisture barrier. One of the most common is polyethylene terephthalate (PET), also known by the more common DuPont trade name, "Mylar." The fiberglass is not flammable, but the Mylar is. After ignition, other items will support combustion (adhesives, tapes, plastic hangers, etc.).

The flammability test used to certify the Mylar-covered insulation blankets was the "vertical Bunsen burner test." This test involves exposing the bottom edge of a 2-inch × 12-inch strip of insulation vertically over a Bunsen burner. The flame is applied for 12 seconds and removed. To pass the test, the material must self-extinguish within 15 seconds, the burn length must not exceed 8 inches, and drippings must not flame for longer than 5 seconds. The Mylar-covered insulation blanket passed this test by immediately shriveling up and shrinking away from the burner. In this way, it avoided ignition during the test. The test was later judged to be inappropriate for the conditions in the plane and was eventually replaced.

Tests specific to the conditions of Flight 111 were developed and conducted to determine if ignition of the insulation blankets could take place.

Short Circuit

A "short circuit" is formed when an electrical current finds a path other than through the intended wire.

The short circuit provides a path with reduced resistance to electron flow. This results in increased electron flow. The relationship between voltage (the potential or "pressure" trying to drive the electrons), current (a "flow" of electrons), and resistance to the electron current flow is related by Ohm's Law:

$$\text{voltage} = \text{current} \times \text{resistance, or}$$
$$\text{current} = \text{voltage/resistance.}$$

A resistor is anything that resists the flow of electrons. Usually, a resistor is a component using the flow of electrons to dissipate energy (e.g., light bulb, motor, heater, etc.).

Increased current in a short circuit is expected to trip the circuit breaker and stop all current, thereby protecting the circuit (Figure 7.4). A short circuit is a mistake, a breakdown in the normal connections of the electrical equipment. For Swissair Flight 111, the short circuit was probably two bare wires touching. Bare metal becomes exposed when the wire insulation is damaged by a variety of means. Cracks can occur in the insulation from aging, vibrations and chaffing, and damage from maintenance crews.

If the short-circuited wires are separated by a small gap, the increased current can jump the gap like a tiny bolt of lightning. This is known as arcing. Arcing is a high-temperature electron discharge across a gap. In

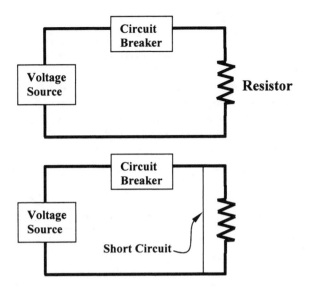

Fig. 7.4. A typical circuit with and without a short circuit. The short circuit provides an alternate current path that bypasses the normal circuit resistors. The circuit breaker is supposed to trip, or open, and stop all current flow, thereby protecting the circuit.

spite of the enormous temperatures involved (temperature at the center of an arc can be over 9,000°F), it is not easy for a tiny spark to actually ignite anything. The arc does not contain large amounts of energy.

Energy has many forms. One form, heat energy, is the ability to heat up a mass. A large spark may have the same arc temperature as a small spark but contain considerably more energy. A spark with twice the energy will have the ability to heat up twice the mass to the critical ignition temperature. The small amounts of energy in a spark make ignition difficult, but not impossible.

Tests were conducted to determine if ignition of the insulation blankets would occur when exposed to arcing from wires carrying currents and voltages similar to those in the MD-11 electrical systems: 115 volts AC and 28 volts DC. Ignition during the tests was sporadic, sometimes occurring after the first arc strike and sometimes after numerous strikes. Typically, the circuit breakers were tripped, but not always.

Arcing can also occur when the bare wires are separated by up to 0.25 inches and contact is assisted by metal shavings left over from maintenance crews or by moisture from condensation dripping on damaged wires.

CIRCUIT BREAKERS

A circuit breaker is designed to heat up and trip with excess current, stop all electrical flow, and thereby protect the circuit. Most aerospace circuit breakers consist of a bimetallic element and two electrical contacts. Each metal in the bimetallic element has a different coefficient of thermal expansion. This means that upon heating, the two metals expand differently, bending the element and thus breaking electrical contact, as shown below.

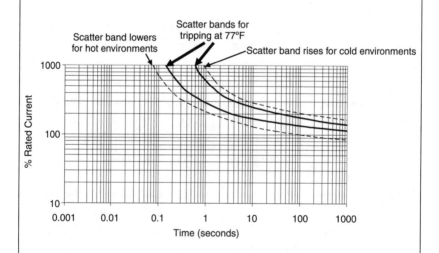

Fig. 7.5. Bimetallic element in circuit breaker. Upon heating, bottom element expands more. The element bends and breaks electrical contact inside the circuit breaker.

The level of heat required to trip the circuit breaker not only depends on voltage and current, but also on the duration of the event. A very high current flow for a brief period of time may not trip the breaker, as shown by a typical aerospace circuit breaker trip diagram below.

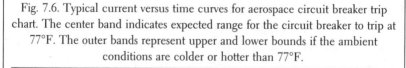

Fig. 7.6. Typical current versus time curves for aerospace circuit breaker trip chart. The center band indicates expected range for the circuit breaker to trip at 77°F. The outer bands represent upper and lower bounds if the ambient conditions are colder or hotter than 77°F.

continued

The diagram is read as follows: A current at 100% of the circuit breaker's design current will never trip the circuit breaker. Currents and times to the left of the scatter band will not trip, but values to the right will. For example, a current at 1,000% of the design value, but for only 0.1 seconds, will probably not trip the breaker. This combination of current and time is to the left of the scatter band for activation at 77°F, indicating that tripping will not occur. These same values are also to the right of the lowest curve indicating that tripping will occur if the ambient temperature is high.

Two of the cables supplying power to the in-flight entertainment network on Swissair Flight 111 had arcing events that did not trip the associated circuit breakers. The breakers most likely did not trip because the electrical characteristics of the arcs were outside the current versus time curve (see Figure 7.6) required to trigger the circuit breaker.

The Swissair Flight 111 investigators concluded that the thermal acoustic insulation blankets used in the fuselage were flammable. An arced wire was found in the in-flight entertainment power supply on the right side of the cockpit overhead controls and was most likely the ignition source. The circuit breakers will not trigger for all arcing events.

The in-flight entertainment system, believed to be the source of the electrical arcing, was decertified. Additional electrical changes were made to similar systems. Less flammable insulation blankets were mandated for DC-9, MD-88, and MD-90 planes. Other Boeing designs did not use the Mylar-covered insulation blankets. Also, new flammability test criteria were eventually developed and required for thermal insulation blankets.

Other Sources and Examples of In-Flight Fires

Hidden electrical fires are particularly challenging because of access difficulties. It is difficult to fight the fire if it can't be located. Between 1979 and 2004, the NTSB documented four hidden fires behind sidewall or ceiling panels with electric sources. In these four cases, the fires took 15–20 minutes between initial awareness and fatal in-flight disaster.

There are various sources of in-flight fires. Overheated equipment (especially motors), arcing electrical wiring, and lightning strikes have all been known to cause fires. Smoking in the restrooms has been a historical problem. Bleed air from the engines, used to pressurize the cabin, can also overheat and cause fires.

In another tragic and very unusual example of in-flight fire, Saudi Arabian Airlines Flight SV163, a Lockheed L-1011, left Riyadh, Saudi Arabia, on August 19, 1980 (Table 7.2). A few minutes into the flight there were indications of a fire in the cargo hold. The pilot decided to turn back to Riyadh, just 60 miles away. The fire spread into the cabin and panic ensued among the passengers. After a perfect landing, the pilot taxied for six minutes to the end of and off the runway.

All 301 passengers and crew perished in the fire. A spokesperson for Lockheed said there were eight doors and escape hatches, all of which could be manually operated from inside or outside in an emergency.

The pilot was faulted for not using a minimum stop emergency landing procedure, taxiing too long, and not instructing the passengers properly for emergency exit. Presumably, onboard panic was a major issue. Keep in mind that all airplane doors, unlike the doors of public buildings, open inward, making escape difficult during panic. Also, dangerous fumes can quickly incapacitate everyone inside.

This disaster contrasts significantly with Air France Flight 358, which safely evacuated 309 passengers and crew from an Airbus 340 on August 2, 2005. Although the official report is not yet available, evacuation is said to have occurred in less than two minutes despite passengers stopping to retrieve carry-on luggage and to take pictures with their cell phones. Also, passengers were evacuated safely despite the fact that four exit doors were blocked or had their emergency chutes fail to open. (There are eight doors available for emergency evacuation.) The plane was totally gutted by fire moments later (Figures 7.7A and 7.7B).

Federal certifications of new aircraft designs require emergency evacuation of the plane within 90 seconds with half of the exits blocked. The certification process requires demonstration of emergency exit with unrehearsed passengers. The test must be performed with limited lighting and distribution of minor obstructions (blankets, carry-on luggage, etc.) and ran-

Table 7.2. Timeline of Flight SV163

18:15	First indication of fire in cargo hold 7 minutes after takeoff
18:19	Decision to return to Riyadh, 60 miles away
18:25	Fire spreads into cabin with ensuing panic
18:36	Plane lands
18:42	Taxiing finally ends; engines shut off
19:05	Emergency workers enter cabin

Figs. 7.7A and 7.7B. Air France Flight 358. All 309 passengers and crew evacuated
safely. *Photos:* Transportation Safety Board of Canada; used by permission.
Source: A05H0002-Air France Flight 358 photos and
www.tsb.gc.ca/en/media/photo_gallery.asp, Transportation Safety Board of Canada,
2005. Reproduced with the permission of the Minister of Public Works and
Government Services Canada, 2006.

dom blockage of half the doors. At least 35% of the "test passengers" must be over 50 years of age. Three dolls are carried by passengers to simulate evacuation of infants. In addition to testing cabin layout and door access, crew training is also being tested. Presumably, the crew is extensively rehearsed and "at the top of their game" for the certification test.

A minor fire does not have to end in disaster or near disaster. A more typical example occurred in November 2000. The NTSB reported that a lightning strike caused arcing in a ceiling wire and ignition of adjacent material. In this case, aggressive discharge of a handheld halon extinguisher toward the blistered ceiling panel seemed to stop the fire. The plane then performed an emergency landing and evacuation without incident.

Engine Fires

Engine fires have historically been quite common and are usually a non-event. Modern engine fire detection equipment and fire extinguishers mounted inside the engine are expected to easily solve any fire problems. Additionally, the planes are designed to fly safely without one engine. Still, unusual circumstances can occur. (See, for example, the Kegworth crash described in Chapter 2, in which a Boeing 737 with two engines had an engine fire, and the flight crew shut off the wrong engine.)

There are many possible ways for an engine to catch on fire. Fundamentally, fuel leaks somewhere that it should not and ignites.

In one of the more unusual stories, in 1965, an engine fire on a 707 ignited an explosion in the wing fuel tank. An engine and 25 feet of the wing tip fell off. Amazingly, an emergency landing was successfully completed with no injuries to the passengers or crew.

Titanium Engine Fires

Titanium is used inside the cooler sections of a jet engine. Titanium is used in aircraft engines because of its low density, high strength, and corrosion resistance. Unlike most other structural metals, titanium ignites at a lower temperature than it melts at, and it has a lower ability to conduct or transfer heat. Because heat is not readily conducted away from its source, the titanium accumulates heat and more rapidly reaches its ignition temperature. Hard rubs are the most common source of heat. Rubs may result from foreign object damage, secondary damage, a stall, bearing failure, or unbalance.

There have been over 140 known titanium fires in aircraft turbine engines in flight and during ground tests. A few of these fires have been serious from a flight safety point of view. In almost all these instances, the titanium fire was a secondary event: something else failed first and resulted in a titanium rub. Often, the initial failure was a titanium compressor blade that failed from foreign object ingestion, vibration, a heavy rub, or some other occurrence. Once ignited, titanium combustion continues until the titanium is depleted, the air pressure falls below some critical value required to sustain combustion, the combustion progresses to a thick section, or the ignition energy source is removed. Titanium fires are fast burning—20 seconds or less—and are extremely intense. The molten particles in titanium fires generate highly erosive hot sprays which have burned through compressor casings with radial expulsion of molten metal.

Engine certification requires many unusual tests including ingestion of a 4-pound bird (recently, artificial birds have been used) and release of a fan, compressor, or turbine blade. All of these events will destroy the engine. To pass the test, the engine must not create a titanium fire, or any other type of fire, and the damage must be contained within the engine.

Post-Crash Fires

Post-crash ground fires are a significant source of passenger injury. A 1999 study by the FAA identified 17 instances since 1967 in which the fuselage was penetrated by fire, resulting in injuries to passengers on board. Typically, the plane is sitting in a pool of leaked fuel. The fuselage may be ruptured or intact after the crash. A ruptured fuselage may allow the external fire to enter the cabin more easily. Sometimes, the fire enters the cabin through open emergency exits. In other cases, the wind blowing the flames against the fuselage is a major factor.

Assuming that the fuselage is intact after the crash, the major barriers to a fire penetrating the cabin interior are the aluminum skin and the underlying thermal acoustic insulation. The burn-through resistance of the aluminum is about 30 to 60 seconds, depending on its thickness, and the insulation adds an additional 1 to 2 minutes.

A DC-10 sat in a burning pool of fuel for 2 to 3 minutes before being extinguished by airport fire crews. This fire did not penetrate the fuselage, providing evidence that wide-bodied aircraft (DC-10, 747, and L-1011) with thicker fuselage skin can resist burn-through longer. For example, a 747 has fuselage thickness of 0.071 inches versus 0.036 inches for a 737.

Since a relatively modest impact force can easily produce a fuel leak, a fire can start almost immediately upon impact. The source of fuel is a ruptured fuel tank or engine fuel line. Ignition results from hot engine parts, arcing of electrical wiring, electrostatic discharge, friction sparks,[4] and overheated brakes.

Compared to propeller planes, jet aircraft provide additional sources of ignition and are more susceptible to post-crash fires. On impact, propellers strike the ground and stop rotating. Jet engines continue to rotate after impact and suck in large quantities of air. If a fuel mist (liquid droplets suspended in air) is sucked into the engine, ignition can occur in the combustion chamber or on any other hot metal surfaces above the ignition temperature. (Jet fuel will self-ignite at 435–485°F; ignition from a spark can occur at much lower temperatures.) This can produce flames at the engine inlet or exit and ignite other spilled fuel.

Fire Forensics

Because crashes often result in fuel leaks and hot engine parts, post-crash fires are quite common. In-flight fires are relatively rare but always have the potential for catastrophic damage. Therefore, it is important to distinguish between an in-flight and a post-crash fire.

Most in-flight fires eventually burn through the structure and are exposed to the air moving past the plane, known as the slipstream. The added oxygen results in two important forensic clues:

1. The slipstream increases the intensity of the fire and raises the temperature significantly.
2. The fire pattern follows the flow of the slipstream.

A ground fire of pooled hydrocarbon fuel generates temperatures around 2,000°F. A local "chimney effect," where rising hot air adds oxygen that fans the flames, can raise the temperature of a ground fire to some extent. An in-flight fire burning the same fuel with added oxygen will burn in excess of 3,000°F. Therefore, if melted components with a melting point above 2,000°F are found, an in-flight fire should be suspected.

Melting temperatures above 2,000°F for common aircraft materials are given in Table 7.3.

Table 7.3. Melting Temperatures for Common Aircraft Materials

Material	Temperature (°F)
Steel*	2,620–2,795
Stainless steel	2,550–2,740
Titanium	2,820–3,000
Copper	1,981

*Variations occur depending on the precise alloy. Copper wire is almost pure copper and has a single melting temperature.

During a ground fire, molten aluminum will fall down under the influence of gravity and collect in puddles. The pattern could be altered by ambient wind conditions.

With an in-flight fire, the molten aluminum will spray and splatter downstream of the slipstream and adhere to the colder surfaces. The slipstream will blast molten aluminum into very small droplets. (Aluminum droplets formed by gravity during a ground fire are much larger.) The metal spray may resemble dirt on the surface. An experienced investigator will quickly recognize (and feel) the grain-sized bits of molten metal splatter by rubbing a hand against the now roughened surface.

Burning of hydrocarbon fuels creates soot (particles of carbon) that will deposit on surfaces less than 700°F. Soot patterns can give important clues. A soot pattern on the ground flows upward (hot air rises) or in the direction of the prevailing wind. An in-flight soot pattern is dominated by the slipstream. Often a fire will appear as an expanding or V-shaped pattern aft with the slipstream and emanating from the fire origin. Sometimes, clean areas occur from a shadowing effect behind obstructions (lap joints, rivet heads, antennae) and leave a clean, soot-free area downstream of the obstruction.

A ground fire frequently starts at a point and spreads upwards in a V-shaped pattern. The fire can also spread out along a pool of fuel and destroy the V-shaped pattern but puddled molten aluminum will often survive as evidence.

Which Came First: Fire Damage or Impact Damage?

If wreckage (not in a post-crash fire area) is found burned, it would suggest an in-flight fire. However, it is possible that a part picks up fire damage as it passes through the impact fireball and lands elsewhere.

Heavy objects can penetrate the ground or roll, bounce, and skid in the direction of flight. This motion may be faster than the propagation of a ground fire. Buried wreckage with soot often indicates an in-flight fire. Still, every rule of thumb has an exception. For example, a heavy object could also have been soaked in fuel and burst into flames upon impact.

Scratches, fractured edges, and crumpled metal may provide suggestive (but not definitive) clues. Soot on a fracture surface indicates impact first, followed by post-impact fire. The impact tears the metal and makes a clean fracture surface. Exposure to a post-impact fire results in soot on the fracture surface. Soot deposits deep in the folds of metal that crumpled during impact also imply an in-flight fire.

Metal structures often tear at a line of rivets. If the part is exposed to fire before fracture, the area under the rivet heads and at the edges of the rivet holes should be clean.

If a fire occurs in-flight, any mud should be on top of soot. For a post-crash fire, the part picks up mud first and then soot. This can be misleading, however, as the part can tumble through an initial fireball and pick up soot, keep tumbling, and pick up mud on top of the soot before coming to rest.

Soot from an in-flight fire can be re-coated with soot during a post-crash fire thus making it difficult to sort out. Also, a post-crash fire can be so extensive that all in-flight fire evidence is destroyed.

Combustion: The Chemical Reaction

Fire (combustion) is a chemical reaction between oxygen and a fuel. Liquids and solids do not actually burn, but instead they give off flammable gases that do. The liquid or solid fuel must be vaporized before it can burn. Vaporization occurs at the surface.

The molecules at the surface of a liquid fuel are attracted to each other and to the molecules underneath. Heat energy added to the surface molecules overcomes this attraction, allowing the molecules to escape into the air. Most solids turn into vapor by a process called pyrolysis. Heat energy breaks down the atomic bonds of bigger molecules thus releasing smaller molecules or molecular fragments that escape and become flammable vapors.

Upon heating, a fuel releases gases and/or smoke. The temperature at which volatile gases are released varies greatly, occurring at about 300°F for wood and –40°F for gasoline. Smoke consists of compounds of hydrogen, carbon, and oxygen. (Carbon particles deposit on surfaces as soot.) At this point, burning has not yet begun. When the volatile gases are heated above their ignition temperature (about 500°F for wood), the atomic bonds in the compounds break down, giving off considerable heat, and form new bonds with oxygen. This is the chemical reaction of combustion. The remaining unburned material of solid fuel forms char, a nearly pure form of carbon. The heat given off by the combustion creates more volatile gases and heats them above their ignition temperature. The process becomes self-perpetuating. When surrounding fuel is heated to its ignition temperature, the fire spreads.

The process can be observed with a magnifying glass, paper, and sunlight. Bright sunlight is focused to a small point on the paper with the magnifying glass. The heated paper will initially smoke and char before suddenly bursting into flame when the gases reach their ignition temperature.

Only some compounds break apart and re-form with oxygen and burn. With sufficient heat, all solids will do at least one of the following:

- Burn, the chemical reaction of combustion
- Char or thermally degrade with heat
- Melt

A steel forging will melt before it burns. Titanium burns before it melts (potentially resulting in nasty fires inside jet engines). Many materials, including plastics, char at lower temperatures before they ignite. Compounds containing carbon and hydrogen, including all liquid fuels and most plastics, give off carbon dioxide when burning. During burning and charring, many of the plastics used in cabin interiors also produce thick smoke and noxious vapors, creating additional hazards.

Explosions

An explosion is a rapid increase in volume with release of energy in a violent manner. Explosions can be chemical, mechanical, or nuclear. Examples of mechanical explosions are a boiler exploding with a sudden release of pressurized steam and a fuselage exploding with a sudden release of compressed air in a process known as explosive decompression (see Chapter 4).

A chemical explosion is a type of combustion and is classified as either a deflagration or a detonation depending on the rate of combustion.

Deflagration

The rate of combustion depends on the size and shape of the fuel. A pile of toothpicks catches fire and burns faster than a solid block of wood because of higher surface area available to absorb heat and initiate the chemical reaction of combustion. A dust cloud or a mist (suspended droplets of any combustible liquid, including jet fuel) has a high surface area and will combust very rapidly in a subsonic explosion known as a deflagration. The fireball that occurs after some airplane crashes is a deflagration of jet fuel mist.

Large pressures result only if the deflagration occurs inside a container. Typically, the container ruptures with an explosive release of compressed gas energy. The expanding gases of a gunpowder explosion confined in a gun barrel by an accelerating musket ball is another example of a deflagration.

TWA Flight 800: Post-Crash Analysis and Testing

Combustion inside a fuel tank is extremely dangerous and is likely to destroy the plane. Such was the fate of TWA Flight 800 in 1996. The breakup sequence, recovery, and reconstruction were discussed in Chapter 3. The focus here is the combustion process inside the fuel tank. Specifically, investigators addressed the following questions:

- Was the fuel tank explosive at the time of the accident? Yes.
- Was the explosive pressure pulse large enough to destroy the plane? Yes.
- Was an ignition source or location of ignition identified? Not definitively.

Combustion inside a fuel tank is a deflagration—and a slow one at that. Analysis and testing of the circumstances of Flight 800 estimate the flame propagation speed at only 1 to 1.65 feet per second. Expansion of the combustion gases inside the fuel tank increased the pressure until the container burst. With the energy of compressed gas, the overpressurized fuel tank mechanically exploded.

Boeing engineers estimated that a pressure of only 20 to 30 psi would burst the fuel tank. The pressure applied over a relatively large area of the fuel tank wall, about 20 feet wide × 6 feet high, resulted in a pressure thrust force (force = pressure × area) almost equal to the weight of the plane. (See Chapter 3 for a description of the fuel tank including diagrams.)

The tank wall also serves as a main structural member reinforcing the wing-to-fuselage connection. In Flight 800, destruction of the tank wall split the plane in two.

A specific ratio of fuel to air is needed for an explosion. The explosive limits of a gas are the upper and lower fuel-to-air ratios required for ignition and rapid combustion. At the lower limit of fuel concentration, there is not enough fuel to sustain combustion, because the fuel molecules are too far apart. At the higher limit, there is not enough oxygen. Regardless of the fuel-to-air ratio, an ignition source is still required.

The investigators first addressed the question, Was there an explosive fuel-to-air ratio in the fuel tank? After arriving from Athens, Flight 800 left New York City's JFK Airport with an empty center wing tank. This actually means that there was about a 0.19-inch film of residual fuel, about 50 gallons, in the 13,000-gallon tank.

Previous data set the lower flammability ratio limit at 0.032 to 0.035. Therefore, for every 1,000 gas molecules in the vapor space, combustion can occur if 32 to 35 of them are fuel molecules.

When a liquid first partly fills a container, the liquid immediately begins to evaporate into the vapor space of the container. Evaporation continues and increases the amount of evaporated liquid in the vapor space until an equilibrium point is reached. This process is described on a molecular level below. A molecular model is useful to explain how changes in temperature and pressure alter the fuel-to-air ratio (and explosive characteristics) of the vapor space in the fuel tank.

When a liquid fuel is placed in a fuel tank, some percentage of the liquid molecules on the surface will have sufficient energy to overcome the attraction toward each other and evaporate into the vapor space above the surface (Figure 7.8). The liquid fuel molecules that have evaporated into

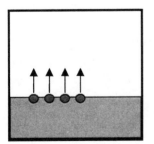

Fig. 7.8. When liquid is placed in a container, some molecules have enough energy to escape into the vapor space above the liquid in a process known as evaporation.

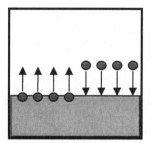

Fig. 7.9. Eventually, equilibrium is reached when the number of liquid fuel molecules evaporating equals the number of gas fuel molecules returning (condensing) to the liquid.

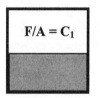

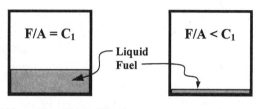

F/A = Fuel/Air Ratio

Fig. 7.10. Except for very small liquid levels, the fuel-to-air ratio in the vapor space is constant (at a given temperature and pressure) for any amount of liquid fuel.

the vapor space are so widely spaced that they are no longer in the liquid state. They are now gas molecules.

As the liquid molecules accumulate in the vapor space above the liquid, some percentage will jump back into the liquid surface. According to Boltzmann,[5] the process is statistical. The fuel molecules, liquid or gas, have a distribution of energies. Some percentage will have the required energy to jump one way or the other. Eventually, equilibrium is reached when the gas fuel molecules return to the liquid at the same rate as the liquid fuel molecules escape into the vapor space (Figure 7.9).

TWA Flight 800: A Variety of Conditions Were Studied

In a container at a given pressure and temperature, the fuel-to-air ratio in the vapor space is the same for all liquid levels, except for the very smallest. The liquid fuel molecules evaporate until saturation is reached, at which point evaporation ceases. For very small amounts of fuel, there is not enough liquid to evaporate into the vapor space and reach saturation (Figure 7.10).

The practical limits of draining a fuel tank leave behind a residual level of jet fuel that is sufficient to obtain saturation. However, very small amounts of fuel will affect evaporation of the fuel molecules.

Relative Humidity: A Related Concept

When evaporation in a container reaches equilibrium, the vapor is said to be saturated. A similar situation occurs when water evaporates into air. The water-to-air ratio for air saturated with water vapor at 70°F is about 2.7%, but it rises to 4.7% at 90°F.

The relative humidity of air is measured in relation to air saturated with water vapor. Fifty percent relative humidity means the air holds 50% of the water vapor it is able to hold at a given temperature. The air is saturated, or holds as much evaporated water as possible, at 100% relative humidity. Usually, 100% relative humidity means it is foggy. High humidity affects human comfort by making evaporation of moisture on the skin difficult. Conversely, low humidity results in dry skin, as the water in the skin cells more easily evaporates out of the body. Low humidity is also associated with passenger discomfort on long flights. The air at 40,000 feet is very cold and dry and tends to dehydrate people, partially explaining the discomfort of long plane rides. The new Boeing 787 adds humidifiers for passenger comfort.

The fuel is actually a distribution of hydrocarbon molecules with various molecular weights. The lighter molecules evaporate more easily. For very small levels of liquid fuel, there are not enough light molecules to evaporate into the vapor space. Since the heavier fuel molecules are harder to evaporate, the net effect is a reduced fuel-to-air ratio for fuel tanks with low levels of fuel. The effect is known as mass loading (the ratio of the mass of liquid fuel to the volume of the tank). For very small mass loadings, the fuel-to-air ratio decreases. The effect is shown in Figure 7.11.

The density of the most common jet fuel, known as Jet A, is 783 kg/m³. A mass loading of 783 kg/m³ corresponds to the tank being 100% filled with fuel. A mass loading of 78.3 kg/m³ corresponds to the tank being 10% filled with fuel. Flight 800 had an estimated 50 gallons, or about 3 kg/m³. Extensive testing was done to establish the effect of mass loading on the conditions present during Flight 800.

Effect of Altitude

When a plane flies to 13,700 feet (the altitude of the Flight 800 accident), the air pressure decreases from 14.7 psi to 8.7 psi. To equalize pressure in the fuel tank with the outside atmosphere, the vapor space is vented (Figure 7.12). To simplify the discussion, assume that venting occurs all at once

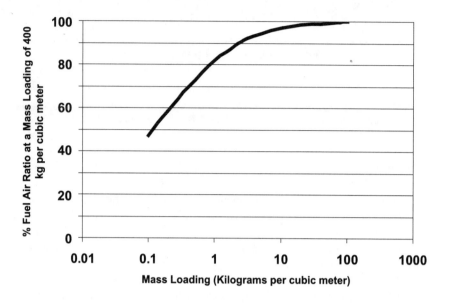

Fig. 7.11. Effect of mass loading on fuel-to-air ratio. *Source*: FAA, AC 25.981-2

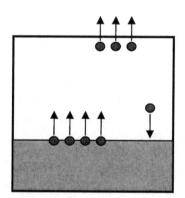

Fig. 7.12. At 13,700 feet, the atmospheric pressure is reduced. Fuel molecules in the vapor space are vented out of the tank to equalize pressure with the outside atmosphere.

in a single venting "blast" event.[6] To pick a realistic number, assume the fuel-to-air ratio before venting is 3.5% and remains so immediately after venting (i.e., before the liquid fuel molecules have a chance to evaporate). However, after venting there are fewer fuel molecules in the vapor space.

Even though the fuel-to-air ratio remains at 3.5% immediately after venting at 13,700 feet, this is a transient condition. There are now fewer fuel molecules in the vapor space to maintain equilibrium with the constant transfer into and out of the liquid surface. Fuel molecules begin to accumulate in the vapor space until a new equilibrium fuel-to-air ratio is reached at perhaps 4.5%.

Just like water boils at a lower temperature on a mountain, evaporation is also easier at higher altitude. Consider air pressure to consist of impacts from individual air molecules. The pounding of air molecules on the liquid surface hinders the evaporation process and keeps the liquid in its place. At higher altitude, the air is less dense, so there are fewer air molecules pounding on the liquid surface.

At 13,700 feet, the temperature varies with ambient weather conditions; +10°F is a realistic value. However, common to most Boeing designs, immediately beneath the center wing tank is the air conditioning equipment (Figure 7.13).

Air from the jet engine compressors is used to ventilate the plane. If the plane is sitting on a runway in the summer, the air may need to be cooled. The air conditioning equipment (described in more detail in Chapter 4) cools, heats, mixes, compresses, and expands various air streams as needed to provide the necessary cabin air temperatures. This

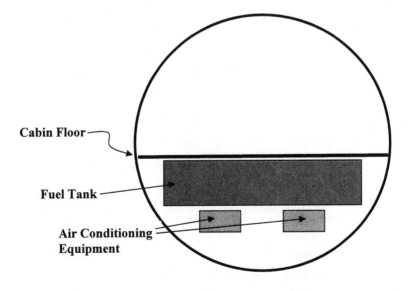

Fig. 7.13. Approximate layout of center wing fuel tank

equipment generates thermal energy and heats up the center wing tank vapor space.

If the liquid fuel is heated, the molecules have additional energy and are more likely to evaporate into the vapor space. The liquid might be heated to perhaps 122°F and the fuel-to-air ratio might increase to 5.5%.

This modest temperature change affects the ignition energy required for combustion and gives insight into subsequent testing done during the investigation. Before reviewing those tests, there is one more significant variable to discuss: the fuel. The most common jet fuel used is known as Jet A, aviation kerosene very similar to the kerosene used in lamps.

Although there are 23 specifications that the fuel must meet, chemical composition for jet fuel is not defined. Fundamentally, crude oil comes out of the ground as a mixture of various molecular weights of hydrogen and carbon molecules. Jet fuel mostly contains molecules with between 4 and 20 carbon atoms. The purpose of an oil refinery is to produce products (gasoline, jet fuel, heating oil, kerosene, etc.) with uniform properties. This is done by separating out molecules with a similar range of molecular weights to obtain uniform performance of the end product.[7] Separation is done by boiling; the lower molecular weights boil at a lower temperature.

Chemical analysis of Jet A fuel considered similar to that used in Flight 800 identified over 200 different chemical compounds. The composition of Jet A varies greatly from company to company, depending on the refining process and even on the particular oil well.

The "flash point" is the lowest temperature at which a liquid gives off vapor in sufficient concentration to combust when exposed to a "standard" ignition source. For purposes of safe handling, minimum flash points of 100°F or higher are required for jet fuel. For lower flash points, the fuel is considered too volatile and likely to combust from stray static electricity during normal handling. The flash point is relevant to explosiveness, but it is not an accurate predictor of explosiveness because of the many variables involved (temperature, ignition energy, etc.). Flash point is considered an important indicator of relative flammability of different fuels.

Refiners have economic incentives to deliver jet fuel with much safer, higher flash points. The different refined streams that contribute to lower flash points are usually more valuable in making gasoline. A survey of flash points for military fuels shows a variation from 106 to 127°F.

As part of the Flight 800 crash investigation, jet fuel with a flash point between 113 and 118°F was tested under a variety of conditions.

Flight 800 originated in Athens, Greece, which was the source of the residual fuel in the center wing tank. A sample of fuel was obtained from Flight 881, which left Athens shortly after Flight 800. The sample is believed to be similar to the fuel that exploded. The flash point of Flight 881's fuel was 114°F.

There was significant batch-to-batch variation in ignition energy and blast pressures generated when testing different brands of fuel during explosion tests.

The temperature in the center wing tank vapor space and its effect on fuel-to-air ratio (and combustion) was the next question addressed by the investigators.

Flight 800 took off on a hot day in July, flew to 13,700 feet into colder air, and exploded. Six test flights were flown with an instrumented 747 to measure 48 temperatures in the fuel tank vapor space, 42 temperatures on the fuel tank surface, and 33 temperatures on the air conditioning equipment surface and surrounding air space. The test flights occurred in July 1997, approximately one year after the accident. Heated by the air conditioning equipment, the highest vapor space temperature measured was 145°F just before takeoff. At 13,700 feet, the vapor space cooled to 102 to 127°F and shortly after cooled out of the combustion range.

Temperatures were measured in the different bays, or sections, of the fuel tank. The temperature range (at 13,700 feet) of the three hottest bays is shown in Figure 7.14.

The fuel-to-air ratio was sampled in Bay 2 during one of the test flights. The fuel-to-air ratio data are summarized in Table 7.4. The flash point of the fuel tested from the test flight was 116°F.

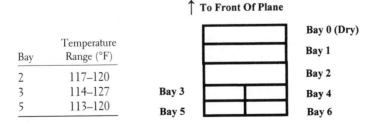

Bay	Temperature Range (°F)
2	117–120
3	114–127
5	113–120

Fig. 7.14. Location of bays in the fuel tank and temperatures measured during the test flight

Consider something more familiar: humid air. Air at 72°F is saturated with water vapor when the water-to-air ratio reaches 2.7%. Warmer air will evaporate more water, and lower air pressure will saturate with a higher water-to-air ratio. Air pressure decreases at higher altitudes and evaporates water more easily. It "feels" drier and more comfortable in Denver at 5,000 feet, because the mile-high city has lower air pressure. Comfort is increased because moisture evaporates off the skin more easily.

The investigators were also interested in identifying the required ignition energy. Will higher fuel-to-air ratios ignite with smaller sparks? Will lower ratios ignite with larger sparks?

The lower flammability fuel-to-air ratio established by other researchers in the past is relative to the ignition source used. It was difficult to reconcile previous data because everyone did the test differently and used a different ignition source. Ignition energies for different altitudes and fuel temperatures are estimated in Figure 7.15.

A joule is the amount of work done by turning on a one-watt light bulb for one second. Physicists would say a joule is the amount of work done by one watt in one second. This compares to the static spark from walking on a rug (1–10 mJ), automotive spark plugs (50–100 mJ), and a turbine engine igniter (about 12 J).

Combustion versus Explosion

The distinction between combustion and explosion is an important one. Combustion is propagation of a flame. If confined, the combustion gases will accumulate and quickly create an explosive pressure. Actual explosion (a violent release of energy) does not occur until the container bursts with a rapid release of compressed gas energy. It is possible that the pressures never reach the fuel tank burst pressure, a topic also explored by the investigators of Flight 800.

Table 7.4. Summary of Fuel-to-Air Ratio Data

Test Condition	Temperature (°F)	Fuel/Air Ratio
Accepted lower flammability limit		0.032–0.038
Test flight: Taxi	123	0.034
10,000 ft	115	0.046
14,000 ft	117	0.054
Fuel sample from Athens	104	0.036
	122	0.066

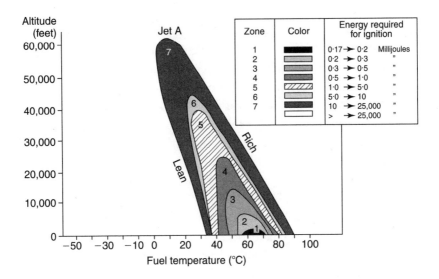

Fig. 7.15. Estimated minimum ignition energies for Jet A fuel. *Source*: FAA, Report DOT/FAA/AR-98/26

Hundreds of tests were conducted by the crash investigators to identify the conditions under which jet fuel combustion in the tank would occur. The variables studied included the chemical properties of the fuel, minimum ignition energy, temperature, pressure, and mass loading. Other goals were to map flame propagation speeds and explosive pressures and to obtain data suitable for computer simulations.

The lowest fuel-to-air ratio found to support combustion was 0.038 at a pressure simulating 13,800 feet. This mixture ignited with spark energy of 80 joules (J).

Other results from the laboratory test cell experiments are as follows:

- The lowest ignition temperature at simulated 13,800 feet pressure was 96.4°F.
- Peak explosive pressures ranged from 39.2 psi at 104°F to 52.2 psi at 122°F. This range should be compared to Boeing's estimate of the fuel tank burst pressure at 20–30 psi.
- As the temperature goes down, the explosive pressure decreases and eventually the vapor does not even ignite. This point is important. If the empty tank had been only one-quarter full, it would have absorbed the heat from the air conditioning equipment and the

vapor space temperature would have decreased, probably out of the explosive range. Recall that a hotter vapor space evaporates more fuel and has a higher fuel-to-air ratio.

- The ignition energy decreased from 0.5 J at 104°F to 0.5 mJ at 122°F, a decrease of 1,000 times. The power supplied to the fuel gage wiring is limited such that any spark energy is less than 0.02 mJ, about 10% of the recognized minimum ignition energy of hydrocarbon fuels. (The "rule of thumb" ignition energy of hydrocarbon fuels, about 0.25 mJ, is based on higher fuel-to-air ratios than studied here.)

- At temperatures less than 86°F, ignition was not possible with 100 J. Flight 800 was actually delayed at the gate for one hour while the ground crew tracked down a suspected passenger-baggage mismatch. A passenger's bag had been pulled because the air line suspected it was unattended; eventually, it was confirmed that the passenger had been on the plane all along. Because the explosiveness of the fuel-air mixture is sensitive to temperature, the on-ground delay could have been a factor in the disaster.

The leading, but still unconfirmed, theory for ignition on Flight 800 remains that, somehow, a higher voltage wire arced into a low-power fuel gage wire.

One-Quarter Scale Explosive Test of Fuel Tank

The NTSB-sponsored test program continued with a series of one-quarter scale model combustion experiments on the center wing tank. Seventy-two tests were conducted over a two-year period. Initially, a simulated fuel mixture consisting of 1.4% propane and 7% hydrogen (remainder air) was used. The simulated fuel was adjusted until the explosive pressures and flame speed at atmospheric pressures matched the Jet A fuel with the conditions experienced by Flight 800. These tests initially saved investigators from the complex, time-consuming, and expensive task of building equipment to match the pressure and temperature cycles present during the flight.

With the simulated fuel, the computer models were quite accurate. However, when actual jet fuel was used and a complex test rig was constructed to heat the tank and lower the pressure below atmospheric pressure, the comparison between the experiments and computer models showed greater variability.

Flame propagation varied as it passed through communicating holes[8] in the fuel tank partition plates and usually involved some degree of quenching (flames go out), reignition, and partial combustion. Quenching was less predictable with the jet fuel and caused the experimental results to be inconsistent with computer simulations. Quenching was also very sensitive to location of the ignition source.

The phenomenon of turbulence also adds to the difficulty of accurate computer modeling. Turbulence can be illustrated with a faucet. At very low flow rates, the water flows in a laminar fashion. Laminar flow is glasslike, quiet, and predictable. For example, if ink is introduced into a laminar flow stream with a hypodermic needle, it will flow in a steady, straight line. As the flow velocity and mass increase, the flow becomes chaotic and turbulent with many small internal vortices and eddies making the flow noisy and opaque. If ink is introduced into a turbulent flow, it will also quickly break up into a chaotic and random series of vortices and eddies. Although the overall average direction and velocity of the water flowing out of the faucet is very predictable, the path of an individual water particle is not.

Weather, being turbulent, is difficult to predict with certainty. Turbulence in the atmosphere occasionally surprises the pilot by slamming the plane and injuring anyone caught without their seat belt fastened. It may be a frightening experience but not a dire threat to a large commercial aircraft. Turbulence is not impossible to model with a computer, but it adds greatly to the model's complexity.

Because of the variation in turbulent flows between the scale model tests and the full-scale fuel tank, as well as the problems with flame quenching, the scale model tests were not meant to be a definitive predictor with respect to the full-scale fuel tank. The scale model tests were designed to calibrate and test the full-scale computer models and give insight into the multichamber combustion process.

Peak pressure pulses varied significantly depending on the fuel and location of ignition. Sample pressure pulse measurements from the one-quarter scale tests are given in Figure 7.16.

Researchers reached the following conclusions from the one-quarter scale tests:

- Jet fuel ignited and combusted during all tests, approximating conditions that existed in Flight 800.

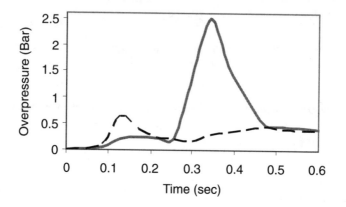

Fig. 7.16. Example results from one-quarter scale fuel tank combustion tests. Overpressure of 1 bar equals atmosphere (14.7 psi). *Source*: Baer et al. "Extended Modeling Studies"

- In certain tests, the pressure levels exceeded those required to rupture the center wing tank.

Additionally, 32 full-scale computer simulations were performed. Investigators considered eight ignition locations, two temperatures, and various time delays between failure of the two baffles in the tank. This analysis found several ignition locations that predicted pressures consistent with observed damage. Because of the uncertainties in the computer modeling, however, a probable ignition location could not be identified. The most likely scenario was a short circuit outside the fuel tank that allowed excess voltage to enter through the low-voltage fuel tank wiring.

Other Ignition Sources Considered

Many other potential ignition sources were considered. These sources included hot surface ignition, fire in the landing gear wheel well, fire in the air conditioning pack, uncontained engine failure, turbine burst in the air conditioning pack, malfunctioning pumps, static electricity, lightning, meteor strike, and radiated electromagnetic energy from transmitters outside the airplane coupled to the fuel gage wiring.

Except for meteor strike and radiated electromagnetic energy, there is prior experience with fuel tank explosions for the other ignition sources listed. Design changes over the years have reduced the risks of ignition from these sources, also there was no fire damage indicating these failure modes.

Radiated electromagnetic energy. Radiated electromagnetic energy could couple to the fuel gage wiring. Forty ground-based emitters were identified (Doppler, Coast Guard, Navy, Air National Guard, space tracking, airports, etc.). All transmitters were studied by NASA and the Department of Defense. The maximum energy inside the plane, but not on the wires, was less than 0.1 mJ. The study concluded the energy on the fuel gage wires was at least 100 times less than the energy required for ignition.

Lightning. No weather anomalies were reported.

Missile, missile fragment, and meteorite strike. A meteorite would be expected to leave the telltale features of a high-velocity penetration similar to a missile strike. Because of controversial eyewitness reports of a missile strike, this possibility was addressed extensively, but it was eventually dismissed; the rationale is addressed in Chapter 3.

Auto ignition and hot surface ignition. Auto ignition of the fuel (spontaneous ignition without a spark or other ignition source) could occur at 460°F, and hot surface ignition could result if any part of the metal fuel tank walls reach 900 to 1,300°F. However, other types of thermal damage would be apparent if these temperatures were somehow reached.

Investigators considered failure of various pumps and engine failure modes that might result in hot surfaces, but no evidence was found.

Static discharge. The U.S. Air Force Wright Patterson Research Laboratory was contracted to assess static buildup of several components inside the fuel tank. Testing revealed that some clamps could build up a voltage potential of up to 650 volts, but they were estimated to provide discharge spark energy of only 0.0095 mJ.

Arcing or Short Circuiting

Based on evidence from reconstruction (see Chapter 3), the NTSB concluded that the center wing tank exploded. The source of the ignition could not be determined with certainty, but of the sources evaluated, the most likely was a short circuit outside the center wing tank that allowed excessive voltage to enter the tank through the fuel gage wiring. This could happen if exposed wire from nearby higher voltage wires arced into low-voltage fuel gage wiring. There was no direct evidence to support this theory, but there was a great deal of indirect evidence. In addition to ruling out all other possible sources of ignition in the fuel tank, the following points support the theory.

- *Evidence of an electrical anomaly on Flight 800*
  The captain commented about a "crazy" engine fuel flow indicator two-and-a-half minutes before complete loss of power. Additionally, the cockpit voice recorder registered two acoustic
  indications one second before power loss. Subsequent testing
  showed that these indications could result from power interruptions in adjacent circuits. Also, the recovered fuel gage indicator
  from Flight 800 read a value that differed from the value
  recorded by the ground refueler. Testing demonstrated that this
  could result from a voltage spike.
- *Evidence of damaged wire*
  Evidence of arcing was found on wires that could have created
  excess voltage in the fuel tank, but it is difficult to say if the arcing triggered the accident or resulted from the accident. The investigators examined wiring on 26 aircraft of varying ages. All of
  the aircraft contained numerous examples of mechanically damaged, chafed, cracked, and contaminated wires. Also found were
  sharp-edged metal drill shavings and fluid stains.
- *Damaged wire arced during testing*
  A short circuit can occur if the wire's internal conductors are exposed and there is direct contact between bare copper or if a bridge
  is created by contaminants such as metal shavings or fluid.

Three types of short circuits were studied.

1. Wet short circuits, in which 1% saline solution (lavatory fluid) was
   dripped (6–10 drops per minute) on wire with damaged insulation
   and exposed bare copper
2. Dry, nonabrasive short circuits, in which metal shavings were
   placed between wires with damaged insulation
3. Dry, abrasive short circuits, in which metal shavings were placed
   between vibrating wires with intact insulation

The tests showed that peak currents of more than 100 amps on a par-
ent wire released up to 400 mJ of energy to a lower voltage victim wire.
This occurred even if the victim wire was protected by circuit breakers.
The 400 mJ of transferred energy greatly exceeds the 0.002 mJ energy nor-

mally supplied to the fuel gage wiring and the smallest experimentally measured ignition spark of 0.5 mJ.

Some explanation is in order concerning the probability of exploding
a fuel tank from external arcing. The following four events have to occur
simultaneously.

1. The fuel tank must contain a flammable fuel-to-air ratio; which is
   estimated to occur 7% of the time. The vapor space of a plane
   taking off, even on a hot day, is quickly cooled at altitude below
   the combustion range.
2. Damaged wire on the low-voltage fuel gage wiring must occur in
   close proximity to damaged high-voltage wiring.
3. Electrical contact must be made between the high- and low-voltage
   wires either by contact or by a short distance spanned by fluid or
   metal shavings.
4. The circuit breakers must fail to protect the circuit.

Researchers re-created damaged wire and facilitated arcing with metal
shavings and fluid drips. Even then, arcing was a random event that did
not occur most of the time. Often, the circuit breakers did trip with the
most severe arcs, but not always.

A stray arc of sufficient energy inside a fuel tank with a combustible
fuel-to-air ratio will ignite with near certainty. This differs from arcing
near combustible material. The testing conducted for Swissair Flight 111
demonstrated that ignition of the insulation was sporadic.

Fuel Tank Design

Assuming that the fuel-to-air mixture will be explosive at least occasionally,
engineers go to great lengths to eliminate all ignition sources. From previous crash investigations, the following potential sources of ignition inside
a fuel tank have been identified.

- Electrical arcs
- Overheated hot wire
- Friction sparks
- Hot surface ignition

Electrical arcs and filament heating are hopefully eliminated by reducing all voltage into the fuel tank so that any potential arcing is limited to one-tenth the ignition energy of jet fuel (normally considered to be 0.25 mJ). Limiting all transient electrical voltages to less than 40 volts is considered adequate.

Friction sparks sometimes occur from rubbing steel parts inside pumps. Design features that eliminate this possibility include, for example, layout so that the pump is always submerged in fuel.

Wing tanks are considered too cold to support combustion. However, center wing tanks require thermal design evaluation to prevent all possibility of hot surface ignition. This evaluation must consider the possibility of nearby equipment failure and overheating.

One of the most serious risks is static electricity building up in the fuel from friction as the fuel flows through pipes and pumps.

All materials have a different affinity for electrons. Static electrical charge can be generated when dissimilar surfaces move across each other and electrons transfer from one material to the other. Rubbing causes an occasional electron to jump from one material to the other. Continuous rubbing builds up a charge or voltage potential between the two materials. For example, when fuel moves through a pipe, hose, or valve, electrons transfer from the fuel to the other material. The rate at which the static charge dissipates is proportional to the fuel's ability to conduct electricity. Pure hydrocarbons are essentially nonconductors, but trace contaminants do provide a small amount of conduction.

Rapidly pumping a liquid that is a relatively poor electrical conductor can result in a static charge being created much faster than it dissipates. When the accumulated voltage exceeds a critical value, a spark can jump from the liquid into the air above the liquid.

All pumps and valves that enter the fuel tank from the outside have grounding wires to prevent buildup of static electricity. Also, fuel flow rates are limited to prevent electrostatic buildup. The U.S. Air Force research lab studying Flight 800 considered static buildup from the flow from a hypothesized fuel tank leak. Their analysis concluded that a spark of only 0.0095 mJ—not enough to ignite the fuel air vapor—could form from static buildup.

The use of a fuel additive reduces the hazard of charge accumulation in less than optimum handling situations. The additive does not prevent charge generation; rather, it increases the rate of charge dissipation by increasing fuel conductivity.

Lightning

The fuel tank is also designed to withstand lightning strikes. Lightning is a frequent yet benign occurrence. On average, every plane is hit by lightning once a year without causing significant damage. Lightning disasters are extremely rare. The last known aircraft destroyed by lightning in the United States[9] was in 1963. A Boeing 707 exploded when lightning ignited combustible fumes in the fuel tank.

Generally, lightning strikes an extremity (nose, wing tip), passes through the metal structure, and exits another extremity. If the structure is properly bonded and the bond is maintained, there is neither arcing nor sparks as the electrons pass through. If arcing occurs in the presence of flammable vapors, a fire or explosion can occur. The location at which this occurs is seldom the point at which the lightning strike enters or exits the aircraft. This is helpful to the investigator, because the evidence of the lightning strike will not necessarily be destroyed by any subsequent fire.

The aircraft skin around the fuel tanks must be thick enough to withstand a burn-through. All of the structural joints and fasteners must be tightly connected to prevent arcing. Access doors, fuel filler caps, and any vents are particularly vulnerable and must be designed and tested to withstand lightning. All the pipes and fuel lines that carry fuel to the engines (and the engines themselves) must be protected against lightning. In addition, new fuels used since the 1970s are safer. Jet A, essentially kerosene, is less volatile (evaporates less) compared to the fuels similar to gasoline that it replaced. Jet A needs a higher temperature to be explosive and is estimated to be nonexplosive 93% of the time on a typical flight.

Although passengers and crew may see a flash and hear a loud noise if lightning strikes their plane, nothing serious should happen because of the careful lightning protection engineered into the aircraft and its sensitive electronic components. Pilots occasionally report temporary flickering of lights or short-lived interference with instruments.

Electrostatic theory states that negatively charged electrons will repel each other and move as far apart as possible. This movement places all excess electrons from the lightning strike on the outer surface of the metal fuselage, which explains why people inside the aircraft are not shocked. The ground crew, touching the plane on the outside, may not be as well protected. Sometimes, a plane struck by lightning will park at the end of the runway until the excess charge bleeds off.

Rule Changes as a Result of Flight 800

As a result of the TWA Flight 800 disaster, numerous rule changes, mainly increased inspection of various components, were mandated to minimize the possibility of fuel tank ignition. A one-time reassessment of fuel tank ignition sources was also required. The ignition sources that were evaluated are shown in Figure 7.17.

Detonation

A final cause of combustion that we should consider is detonation of an explosive. This possibility was ruled out early on in the Flight 800 investigation, but explosions have destroyed airplanes in other air disasters.

A detonation is combustion that occurs at extremely high speeds. The gases from C-4, a military explosive (only available illegally on the black market), expand at over 26,000 feet per second, far in excess of the speed of sound (about 1,100 feet per second). This rapid propagation compresses the air faster than it can move out of the way, resulting in a shock wave and much potential damage even without confinement in a container.

In gunpowder, a "low explosive," the oxidizer (the chemical that supplies oxygen), and the fuel are mixed like salt and pepper shaken together. Saltpeter (potassium nitrate, KNO_3) gives off oxygen as it decomposes and the rate of combustion accelerates. The propagation of the flame is limited by thermal conduction (how fast the material heats up). With a "high-explosive" detonation, the oxidizer and fuel are premixed on a molecular level. A high-explosive detonation derives its energy from a chemical reaction, just like gunpowder does, but the energy transfer occurs not by heat conduction but instead by a much faster high-speed compressive shock wave.

A few drops of nitroglycerol will ignite and flame without detonation or explosion. A drop or two of nitroglycerol dripped on an anvil and struck with a hammer will detonate. Most modern explosives require percussive shock or other triggering devices (e.g., a detonator or a blasting cap) for detonation. When the shock wave moves through the explosive material at supersonic speeds, the entire block of material is activated to combust nearly simultaneously. Ninety percent of the chemical reaction of a detonation is completed in 10^{-6} to 10^{-9} seconds. The blasting cap breaks up the molecular structure of the high explosive, and enormous amounts of chemi-

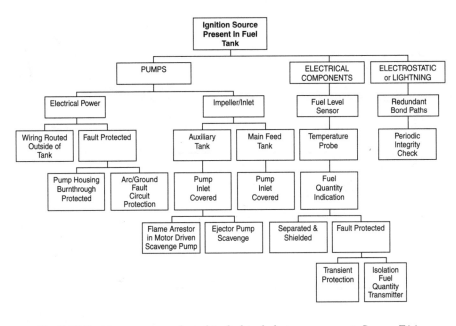

Fig. 7.17. Ignition sources evaluated in fuel tank design assessment. *Source*: FAA, Report AC 25.981-1B

cal bond energy are released when the explosive molecules are smashed into ions and loose atoms.

Blast damage from high explosives leaves distinctive markings easily identified by an experienced investigator.

Normally, if aluminum is overloaded, it will deform extensively before fracturing. Because of the high speeds associated with a detonation, the metal does not have time to deform in its usual manner. This results in fragmentation of the fuselage opposite the blast, usually in the form of a blast hole. In addition to a blast hole, the metal around the blast hole is deformed outward and away from the blast. (A detonation inside the plane blasts out, and a detonation outside the plane blasts in.)

Also present is radiating surface pitting from minute particles blasting out in all directions and embedding in nearby surfaces. This results in a visibly roughened surface. Explosions also result in high-speed penetration of metal fragments into adjacent objects and structures. If explosions are suspected during a crash investigation, all human remains, seat cushions, baggage, and other items are x-rayed and examined for penetration of fragments. If a fragment hits a metal structure, it may leave a rimmed crater caused by instant melting and splashback of material.

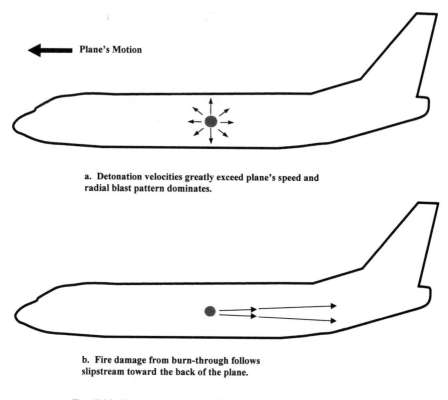

Plane's Motion

a. Detonation velocities greatly exceed plane's speed and radial blast pattern dominates.

b. Fire damage from burn-through follows slipstream toward the back of the plane.

Fig. 7.18. Detonations and in-flight fires have distinctive patterns.

There is nothing on an airplane that will detonate under any circumstances. Any evidence of a detonation immediately indicates a bomb.

Because the velocities of a detonation are much greater than the speed of the plane, damage radiates in all directions from the detonation. An in-flight fire, on the other hand, follows the slipstream, creating a different pattern (Figure 7.18).

Detection of Explosive Residue

In addition to the visual evidence of a detonation, modern equipment is extremely accurate at detecting minute quantities of explosive residue.

Explosive residue is detected in a two-step process. The chemicals to be analyzed are separated inside a gas chromatograph. The compounds are vaporized and passed into a column. Different compounds pass through the column and into the detector at different rates, depending on their boiling point.

The separated and pure compounds are then ionized with electromagnetic waves. One method involves blasting electrons off unidentified molecules to create positive ions. The charged ions are deflected by passing them between charged plates. Ions with different charges and masses are deflected at different angles. The number of deflected ions is counted by the current that occurs when the ions strike at a detector plate. Each residue will have a "signature" of charged ions, similar to a fingerprint, that can be compared to a library of known spectra.

An updated version of the same technology claims to detect one trillionth of a gram of explosive compounds in an air sample. Similar technology is used to inspect passenger luggage for bombs.

An earlier laboratory version of the device was used in the 1988 Pan Am Flight 103 Lockerbie crash investigation. The explosive center was easily located from the fracture patterns (see Chapter 3), but it was also verified by measuring the explosive residue.

8...

Crash Testing

The most observed, instrumented, photographed, and measured crash in the history of aviation occurred in the Mohave Desert on December 1, 1984. Although the test crash ended in a spectacular fireball, the results were disappointing. After years of careful planning and research, the intention was to avoid a post-crash fire.

An unofficial rumor has the Secretary of Transportation, the project's sponsor through the FAA, abruptly leaving the observation room muttering "Don't ask me to fund this again!" A total lack of any research on the subject after that date supports the rumor.

The remotely piloted Boeing 720 plane was also crashed to test the structural response of energy-absorbing seats. Because of piloting "issues," the impact forces were lower than expected, and the seat tests were inconclusive (Figure 8.1).

Controlled Impact Demonstration

In 1980, the FAA initiated planning to conduct a controlled impact of a typical jet transport aircraft. This project became known as the Controlled Impact Demonstration (CID).

A jet fuel additive had been developed to prevent misting and thereby inhibit ignition and flame propagation and ultimately prevent a post-crash fire. Numerous smaller-scale and laboratory tests were very promising. The stated purpose of the CID was to test the additive under realistic full-scale crash conditions.

Antimisting kerosene adds a long-chain polymer molecule to the jet fuel. The polymer prevents formation of tiny droplets or mists. However, because of potential clogging problems inside the engine, the molecule must be bro-

Fig. 8.1. The Controlled Impact Demonstration (CID) Boeing 720 had a cabin full of crash dummies. *Photo:* NASA Dryden

ken down into shorter molecules in a process known as "degradation." Each of the four engines had a "degrader" attached for this purpose.

A Boeing 720 owned by the FAA and ready for retirement (after 20,000 flight hours and over 54,000 takeoff-landing cycles) was chosen for the crash test. Although this plane was considered obsolete, it was consistent with design and manufacturing practices used for other more current aircraft of the day. Most modern fuselages are structurally very similar.

The FAA invited NASA to participate in crashworthiness experiments including

- experimental determination of crash forces and failure mechanisms;
- crash testing of new energy-absorbing seat designs;
- improved cabin fire safety.

Planned CID Crash Scenario

In the late 1970s, the FAA and NASA contracted the three major U.S. transport airframe manufacturers of the day: Boeing, Lockheed, and McDonnell Douglas. The purpose was to review all company and government accident data to define a range of survivable crash scenarios. One hundred and seventy-six well-documented survivable accidents were selected for detailed study. The CID crash scenario evolved from these studies.

The planned crash sequence required the aircraft to descend symmetrically along a glide slope at a controlled sink rate with the orientation shown in Figure 8.2.

An extremely small impact footprint, 30 feet wide and 200 feet long, was the target. A slight nose-up attitude was chosen to obtain initial ground contact in the back of the plane.

After primary impact with nose up and wheels retracted, the aircraft would slide between a corridor of eight obstructions spaced 50–100 feet apart and designed to cut the wing tanks and spill fuel. The plane was expected to continue sliding along a gravel surface striking six pairs of light poles spaced every 100 feet. Each light pole had five 300-watt lights. The 60 shattered light bulbs and their exposed filaments served as 60 potential ignition sources. It was hoped that the antimisting kerosene would prevent any post-crash fire.

The remote piloting techniques were developed with a series of 14 test flights, including 69 approaches, to within 150 feet of the crash site.

Actual CID Crash Events

On final approach and at an altitude of 200 feet, the aircraft was off center to the right. Instead of going around for another attempt, the remote pilot decided to try a last-minute correction. The decision was difficult. Once below 150 feet, high-speed (and limited-duration) onboard cameras and other data-recording equipment would automatically be activated. If the crash was aborted, the recording equipment would miss the crash event.

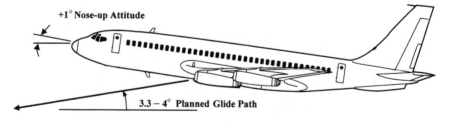

+1° Nose-up Attitude

3.3 – 4° Planned Glide Path

Fig. 8.2. Planned orientation of CID impact

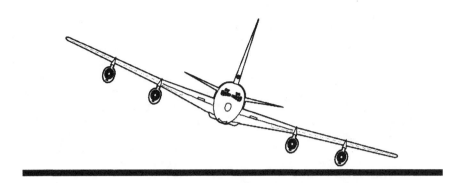

Fig. 8.3. Unplanned left wing tilt (down 13 degrees) resulted in first contact under left wing. *Image*: NASA Dryden

Fig. 8.4. Initial contact was with no. 1 engine. *Photo*: NASA Dryden

At 150 feet, the pilot made a fairly sharp left aileron control input to center the aircraft on the target zone. This resulted in a rolling oscillation. Despite attempts to damp the oscillation, the engine under the left wing hit with a 13-degree roll at about 172 mph and a 17-ft/sec sink rate (Figure 8.3). The plane was 285 feet short of the planned impact area (Figure 8.4).

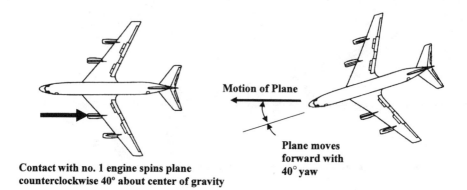

Contact with no. 1 engine spins plane counterclockwise 40° about center of gravity

Motion of Plane

Plane moves forward with 40° yaw

Fig. 8.5. Contact with no. 1 engine under left wing caused the plane to spin or yaw.

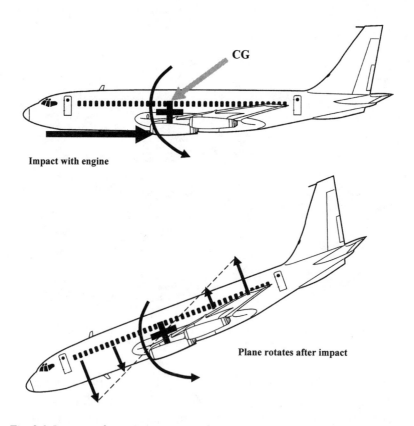

CG

Impact with engine

Plane rotates after impact

Fig. 8.6. Impact with no. 1 engine caused aircraft to pitch or rotate down about its center of gravity (CG).

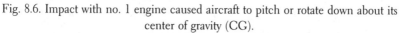

The unplanned asymmetrical first contact resulted in three effects:

1. The contact to the left of its center of gravity caused the plane to yaw to the left (Figure 8.5).
2. The initial contact reduced the sink rate from 17 ft/sec at the center of gravity to about 12 ft/sec.
3. The impact with no. 1 engine below the center of gravity pitched the aircraft down (Figure 8.6).

The rotational velocity added about 6 ft/sec to the front of the plane and subtracted a similar speed from the rear of the plane. (The front of the plane impacted the ground at about 12 ft/sec + 6 ft/sec = 18 ft/sec. The back of the plane impacted at about 12 ft/sec − 6 ft/sec = 6 ft/sec.) The actual rotational velocity depends on the distance from the center of gravity and equals $V = r\,\omega$, where r is the distance from the center of gravity and ω is the angular velocity of the plane. For the test plane, ω was observed to be about 5.7 degrees of rotation per second. Because of this added rotational speed, the impact forces were larger in the front and smaller in the back. The crash instrument readings verified this trend.

The second impact occurred on the fuselage. The fuselage, angled nose down about 2.5 degrees, pitched (rotated) into the ground and impacted about 0.46 seconds later near the nose wheel well. The plane slid forward for 11 seconds (Figure 8.7).

During the slide, the no. 3 engine on the right side was destroyed by impact with an obstruction (designed to open up the wing fuel tanks) 1.8 seconds after first contact (Figure 8.8A). The destroyed engine became an overwhelming ignition source that ignited the fireball (Figure 8.8B). The entire plane erupted in a fireball immediately after the destruction of the no. 3 engine (Figure 8.9).

Damage to the fuselage allowed fire to enter the cabin. After another 5–10 seconds, visibility was near zero throughout the aircraft, as recorded by the onboard cameras. Although the fireball only lasted about nine seconds, it took firefighters almost two hours to douse the flames in the fuel-soaked gravel under the plane.

From the standpoint of the antimisting fuel, the test was a major setback, but the data collected on crashworthiness was still significant. The recorded film contained unique information on the development of fire and smoke in the interior of the aircraft. Except for the fireball, this crash

Fig. 8.7. Right wing no. 3 engine about to impact obstruction and ignite a fireball.
Photo: NASA Dryden

was survivable. The FAA estimates that 19 of the 53 simulated passengers might have survived the fire.

If the obstructions had ripped open the wings as originally planned, instead of destroying an engine, the test would have had a much better chance of success. The destroyed engine released flammable lubricants and hydraulic fluid that were immediately ignited by hot engine parts. Inside the engine, the fuel additive is "degraded" (i.e., the flammability is restored). Even if the polymer is still intact, the fuel will instantly vaporize and ignite when exposed to an open flame.

CID Instrumentation

A total of 350 transducers were scattered throughout the plane including 305 accelerometers and 45 strain gage transducers (see accelerometer and strain gage boxes). To increase reliability, the transducers were connected to two independent 176-channel data acquisition systems. The data were recorded redundantly onboard while simultaneously being transmitted via four telemetry transmitting systems.

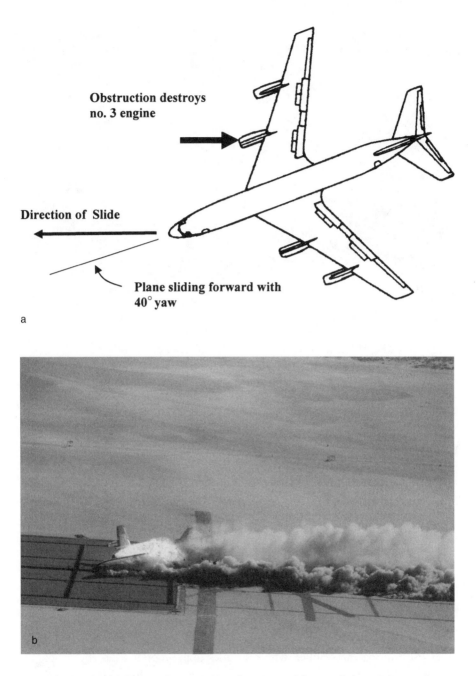

Figs. 8.8A and 8.8B. No. 3 engine on right wing was destroyed about 1.8 seconds after initial contact. *Photo:* NASA Dryden

Fig. 8.9. The plane erupted in a fireball. *Photo:* NASA Dryden

---

All the seats were filled with dummies of appropriate size and weight, but only 11 were fully instrumented anthropomorphic crash dummies (crash dummies are discussed later in this chapter). Each of the 11 instrumented dummies had a triaxial accelerometer in the pelvis. Three dummies had additional accelerometers in the head and thorax. Each lap belt restraint had a load cell to measure the tensile loading in the seat belt.

WHAT IS AN ACCELEROMETER?

Accelerations are measured with a piezoelectric crystal that converts mechanical motion into measurable voltages. A similar crystal was discussed in Chapter 5 in which the inverse operation occurred (a piezoelectric crystal was used to convert electric energy into mechanical vibrations or ultrasonic waves used to inspect for fatigue cracks).

There were 17 different passenger seat configurations tested. The seats were modified to absorb more crash energy. An example of a seat modified to absorb energy is shown in Figure 8.11, the diagonal seat brace is designed to collapse at a constant force, thereby absorbing energy. The modified seats were in the back of the plane to coincide with the planned impact in the rear of the fuselage. Instead, fuselage impact occurred in the front. The lower than expected sink rate (and, therefore, G load) in the rear did not activate the energy-absorbing features of the innovative seat designs.

CID Crash Data

Because the left-wing impact reduced the sink rate from the planned 17 ft/sec to 12 ft/sec, measured accelerations were substantially lower than an-

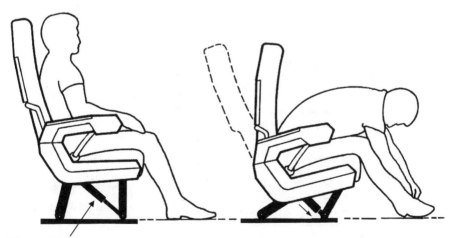

Diagonal Seat Brace

Fig. 8.11. Energy absorbing diagonal brace and illustration of forward flail of upper torso during impact. Adapted from FAA, Report DOT/FAA/CT-85/25

ticipated. The highest vertical accelerations along the floor were in the cockpit near the nose wheel well (location of first fuselage contact) at about 14 G. The remaining cabin floor was typically at about 7 G or less.

The peak longitudinal floor accelerations showed a similar distribution with the highest, 7 G, occurring in the cockpit. The longitudinal accelerations were generally around 4 G over the rest of the floor. The peak transverse floor accelerations ranged from about 5 G in the cockpit to 1 G in the fuselage.

Normal and longitudinal G loading along the length of the fuselage is shown in Figure 8.12 at various body station locations. (Body station 540 is 140 inches farther from the nose than body station 400.) G loading on the floor is particularly important, because it can be used to calibrate seat crash testing, which is discussed later in this chapter.

The accelerometers in the test fuselage were located on the fuselage bottom, subfloor, floor, and inside the pelvis of crash dummies. These locations were chosen to determine how the crash loads "flow" through the structure from the ground, through the floor, into the seat, and ultimately into an accelerometer mounted in the pelvis of a crash dummy. The peak normal G readings at various locations are shown in Figure 8.13 for body station 540.

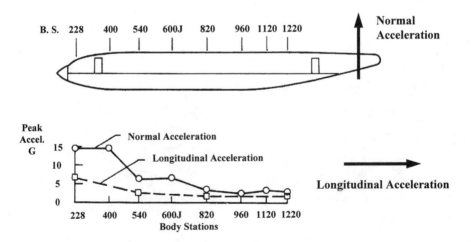

Fig. 8.12. Peak floor accelerations at various body stations (B.S.) showing effect of higher impact velocity in front of plane where first contact of fuselage occurred

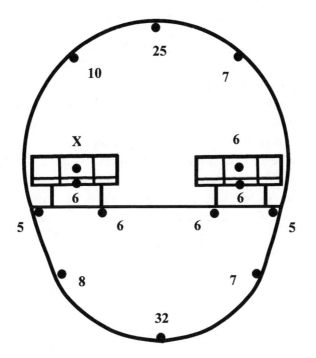

Fig. 8.13. Cross section of fuselage showing peak normal accelerations in G's at
body station 540. (X indicates no reading taken.) *Source*: Hayduk, NASA
Conference Publication 2395

The highest accelerometer readings always occur with first contact on
the bottom of the fuselage. Hopefully, the fuselage will crush and absorb
much of the crash forces, thereby protecting the crash dummy. In the CID
test, the fuselage was crushed vertically by 6–7 inches in the forward cabin
area near the nose wheel, by 9 inches forward of the wing, and by about 6
inches behind the wings.

Crash Testing: Historical Perspective

Compared to automotive crash testing, there have been relatively few airplane crash tests. Airplane crash tests are considerably more difficult and
expensive for a variety of reasons.

A car is limited to moving in one plane, so front, offset front, and side
crash tests significantly cover the crash envelope. Additional crash variables
exist for an airplane, which can rotate in space and has both forward movement and sink rate. Very small changes in the orientation of the plane completely change the location of impact. Impacting on the nose is very

different than initial contact on the tail or wing or elsewhere on the fuselage. Data are needed to identify an envelope of survivable impact speeds (vertical and horizontal), flight path angles, plane orientation (attitude, roll angles, yaw angle), and impact location. Simply put, there are too many experimental variables to sort out with such expensive testing.

Obviously, destroying a plane costs more than destroying a car, but that is just part of the total expense. By now, a car crash test is routine and needs limited engineering intervention. On the other hand, teams of engineers will spend months designing the instrumentation for a plane crash test and many more months interpreting the data. There is more structure to be instrumented for a plane test. Also, years might pass between plane crash tests, so the software and electronics must be updated and verified. Usually, there is a specific focus of the test, requiring a unique layout. The engineers more or less have to reinvent the wheel for each plane crash test conducted.

Early Crash Testing

Except for a few test crashes by the National Advisory Committee for Aeronautics (NACA), NASA's predecessor, in the 1950s and a few military crashes of smaller fighter planes, there were no test crashes of consequence until 1964. In 1964, the FAA conducted two test crashes—a DC-7 and an L-1649 Constellation—on planes of significant size with respect to modern-day aircraft.

The data system failed during the DC-7 crash test and no data were collected.

The Constellation crash tests were actually a sequence of crashes on the same plane as follows.

1. The airplane was accelerated to the climb-out speed at the maximum takeoff weight.
2. All three landing gear were knocked off with barriers to simulate a hard landing.
3. The right wing impacted a pole.
4. The fuselage impacted a 6-degree earth mound at 172 ft/sec (117 mph). This is equivalent to a plane with a 6-degree negative nose-down attitude contacting level ground. The vertical component of velocity creates an effective sink rate of about 18 ft/sec. The fuselage was intact after this impact.

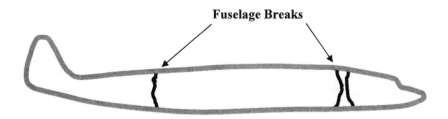

Fuselage Breaks

Fig. 8.14. Fuselage breaks in L-1649 Constellation crash test. Adapted from Wittlin and Lackey, NASA Contractor Report 166089

Table 8.1. Twenty-degree Impact of Constellation

Location	Peak Longitudinal G's
Forward cabin	±10
Mid-cabin	−5 to −10

5. Seventy-five feet after contact with the 6-degree mound, the fuselage contacted a second earth mound with a slope of 20 degrees at a speed of 103 ft/sec (70 mph). The airplane bounced off the first mound and became airborne in between the two mounds. The effective sink rate was 35 ft/sec. The fuselage broke into three sections during impact with the second mound (Figure 8.14). The impact and fuselage breaks were very similar to the Kegworth Flight 092 crash discussed in Chapter 2.

Sample crash data for the 20-degree impact are shown in Table 8.1. This crash was considered highly survivable. In fact, the wreckage was used for subsequent evacuation studies with volunteer passengers.

Drop Tests

A plane about to contact the ground has forward kinetic energy and downward kinetic energy. During a crash, the forward kinetic energy is often dissipated by a long ground slide. However, because the rigid ground is unforgiving and there is nothing equivalent to a ground slide for vertical impact, many crashes result in significant vertical impact damage. Unless the plane flies or slides into a barrier, most crash damage results from vertical impact. (Vertical sink rate data during crashes are presented in Chapter 2.)

Table 8.2. Fuselage Sections Dropped in Preparation for CID

Location	Body Stations	Length (ft)	Empty Weight (lb)	Loaded Weight (lb)
Forward of wings	600–600J	12	1,870	5,100
Center, including wheel wells	820–960	13	5,400	7,964
Aft	1,020–1,140	12	?	6,395

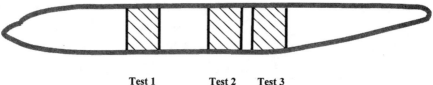

Test 1   **Test 2**   **Test 3**
Forward   **Center**   **Aft**

Fig. 8.15. Location of fuselage sections used in drop tests performed in preparation for the Controlled Impact Demonstration

The damage associated with vertical impact is also distinct from longitudinal impact. In a vertical impact, the fuselage crushes, the seats collapse, and the occupants' spines are compressed, potentially resulting in spinal injuries (see Chapter 9).

For all these reasons, numerous drop tests—a plane is literally lifted and then dropped—have been performed over the years, usually on sections of fuselage and occasionally on an entire plane.

Drop Tests for CID

In preparation for the Controlled Impact Demonstration, several drop tests were conducted. These tests were done to determine load and deflection data for various fuselage sections (for use in computer models) and to identify impact requirements for the CID's data acquisition hardware.

The dropped fuselage sections of a Boeing 707 (a Boeing 720 is a shortened 707 with identical fuselage construction details) are described in Table 8.2. The location of the sections is shown in Figure 8.15.

The empty weight identified in the table is structure only (fuselage and reinforcing ribs), while the loaded weight includes storage bins, paneling, insulation, ducting, and instrumentation, as well as numerous 165-pound crash dummies.

Each section was dropped 6.2 feet to obtain an impact velocity of 20 ft/sec. The drop height of 6.2 feet is a rather small number compared to a cruising altitude of 40,000 feet. However, one has to consider how gently a jet aircraft is supposed to land, perhaps 2 to 3 ft/sec. The maximum design landing sink rate of 10 ft/sec is equivalent to dropping 18.7 inches (see Chapter 2 for other landing gear design requirements).

Truly falling from 40,000 feet is an impossible situation in terms of survivability, whereas relatively modest design improvements can be expected to save lives during a hard landing. Also, a hard landing is considerably more likely to happen than falling out of the sky from high altitude.

The difference between 20 ft/sec and the design landing sink rate of 10 ft/sec may not sound like much, but recall that kinetic energy = ½ mass × velocity^2. A plane landing at 20 ft/sec has four times the kinetic energy as a plane landing at 10 ft/sec and will significantly damage the fuselage.

Test section 1. As expected, the three sections performed differently in each drop test because of differences in structural stiffness. Test 1, the "soft" forward section, had crushing of 22 to 23 inches in the front and 18 to 19 inches in the back. High-speed motion picture analysis showed a maximum deflection of 26 inches at 0.21 seconds after impact. The maximum deflection was followed by an elastic rebound or springback of approximately 3 inches, resulting in a final permanent crush of 22 to 23 inches in the front.

The bottom of the fuselage resisted crushing through deformation of the frames (circumferential reinforcing ribs spaced every 20 inches inside the fuselage). The longitudinal reinforcing ribs and skin offered little resistance to crushing. All seven frames ruptured near the bottom contact point during development of a "snap-through" collapse (Figure 8.16). The energy absorbed by crushing of the bottom fuselage greatly reduced the G loading recorded by the accelerometers on the floor and inside the crash dummies.

Test section 2. Test 2, the center section (including the reinforcement of the wheel wells), was the stiffest section. Test 2 had no crushing of the bottom of the fuselage and absorbed very little energy during bottom contact. As a consequence, high loads were transmitted from the bottom of the fuselage to the floor and into the seats and crash test dummies. Maximum normal accelerations were 70 G at the fuselage bottom, 70 to 95 G at the cabin floor, and 40 to 60 G at the dummy pelvis, eas-

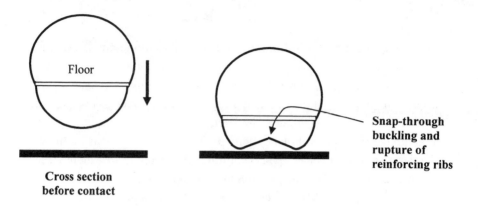

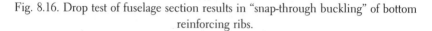

Snap-through buckling and rupture of reinforcing ribs

Cross section before contact

Fig. 8.16. Drop test of fuselage section results in "snap-through buckling" of bottom reinforcing ribs.

ily exceeding human tolerances. Figure 8.17 shows before and after photos of test section 2.

Stronger is not always better, a concept that violates our commonsense notion of how things should work. When it comes to absorbing crash energy, flexibility and deformation are better. This is best demonstrated with a bubble gum "thought experiment."

Consider a solid rod placed against your head and struck with a hammer (Figure 8.18). The effect of the hammer blow on the head greatly depends on the material of the rod. If the rod is made of steel, the hammer blow on the end of the steel rod is passed directly to the head without any alteration. No protection is provided to the head, because the steel rod has little or no energy-absorbing deformation. The weaker bubble gum deforms and absorbs energy, thereby protecting the head. The soft fuselage sections crush and act like bubble gum, while the stronger reinforced section does not crush and acts like the solid steel rod.

Figure 8.19 shows acceleration data from the top of the fuselage of test section 2. The metal between the roof and the impact zone on the bottom acted like a spring resulting in many up and down oscillations. The fuselage deflected with linear elastic motion similar to a spring mass system. The first negative peak was at about 0.06 seconds, representing the time it took for the impact shock to travel from the contact point to the top of the fuselage. Frictional damping reduced the amplitude of the spikes.

Because this motion on the fuselage ceiling is linear elastic, the oscillations are unusually well defined and very similar to a damped sine wave.

Figs. 8.17A and 8.17B. Center section before (lifted 6.2 feet) and after drop test. Note partial collapse of seats. Bottom crush zone is very small because of reinforcement for wheel well. *Photo:* NASA Langley

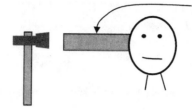

Solid rod made of bubble gum or steel

Fig. 8.18. A solid rod made of bubble gum will protect the head from a hammer blow better than a stronger steel rod.

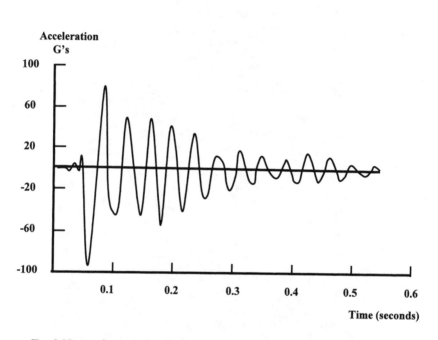

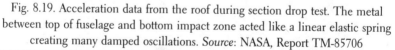

Fig. 8.19. Acceleration data from the roof during section drop test. The metal between top of fuselage and bottom impact zone acted like a linear elastic spring creating many damped oscillations. *Source:* NASA, Report TM-85706

Most crash data involves highly nonlinear crushing and tearing of metal. For example, the sudden fracture of a component results in the rapid release of its stored spring energy causing secondary spikes.

Similar data from an accelerometer at the bottom of the wheel well showed one violent spike followed by lesser oscillations (Figure 8.20). In this case, there was little metal between the impact zone and the accelerometer to create the spring-like oscillations.

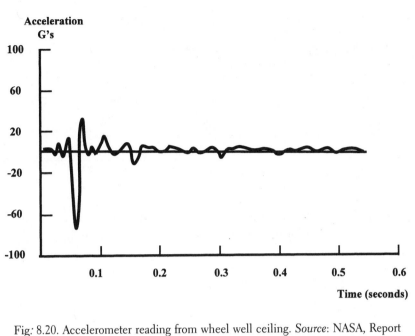

Fig: 8.20. Accelerometer reading from wheel well ceiling. *Source*: NASA, Report TM-85706

Test section 3. Test 3 was a section behind (aft) of the wing with a tapering longitudinal cross section. This "soft" section experienced approximately 14 inches of crushing in the front and 18 inches in the tapered back end. The crushing absorbed a lot of energy and provided protection for the crash dummies. The peak G loads on the bottom of the fuselage varied from 109 G in the front section to 64 G in the back. The peak floor accelerations varied from 14 G near the walls to 25 G in the center. The peak accelerations in the dummies' spines were 19 G.

The tapered cross section resulted in the impact force not passing through the center of gravity. This resulted in a pitching motion that caused the back end to hit harder and crush more.

Somewhat in defiance of common sense, the fuselage section with the least visible damage (no crushing), Test 2, resulted in the highest impact loads on the crash dummies. The peak accelerations for the three tests are shown in Table 8.3.

In addition to the three section drop tests, an entire 707 was also dropped in preparation for the CID. The plane was dropped with a 1-degree nose-up attitude and 17 ft/sec sink rate, the same as the planned (but not obtained) CID impact. The crush zones varied from 3 inches in the hard

Table 8.3. Comparison of Peak Accelerations (G's) for Three 707 Fuselage
Sections Dropped at 20 ft/sec

Test Number and Section	Fuselage Bottom	Floor	Crash Dummy Pelvis
Test 1, forward	56–overload*	8–22	6–15
Test 2, center (wheel wells)	60–overload*	60–90	40–60
Test 3, aft	64–109	15–25	19

*Overload indicates that the instruments maxed out.

zones to 12 inches in the softer sections. These crushes compare to up to
24 inches of crush for individual sections. Obviously, the individual parts
are not as strong as the connected whole.

Other Drop Tests

Except for small planes and helicopters, there have only been a few
other drop tests conducted by the FAA since the Controlled Impact
Demonstration.

In 1993, 1999, and 2000 drop tests were done on a tapered section
of a Boeing 707 and two sections of a Boeing 737. In all three cases,
a 10-foot-long section of fuselage was dropped 14 feet to impact at
30 ft/sec. These tests were conducted to assess the crashworthiness of
different configurations of overhead bins and auxiliary fuel tanks
mounted under the floor boards. The 707 cross section is shown in
Figure 8.21.

Giving added insight into potential injury to passengers, limited data are
available from these tests for crash test dummies with load cells and accelerometers mounted in the lumbar and pelvic regions. These dummy
crash data are given in Table 8.4.

There was also an extensive effort by NASA to model these drop tests
with computer simulation. This effort is discussed later in the chapter.

Seat Crash Testing

One of the requirements for survivability is maintaining the tie-down chain
between the floor, the seat rails, and the seats. Seats have broken free during crashes, causing serious injury and death (see Kegworth Flight 092
crash in Chapter 2). Well-designed seats will also reduce spinal injuries by
absorbing energy during crashes with high sink rates.

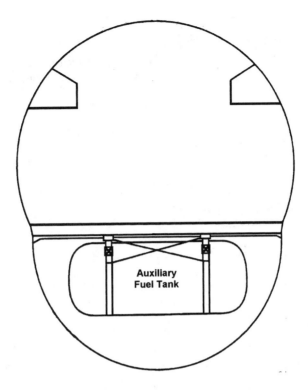

Fig. 8.21. Cross section of Boeing 707 fuselage section dropped 14 feet to assess overhead bins and auxiliary fuel tank

Typical Seat Configuration

The seats slide back and forth along the cabin floor on tracks. Seat spacing is varied by the airlines for a variety of reasons. For example, on longer flights the seats might be set farther apart for passenger comfort. Seats in first class, at exit rows, and behind bulkheads all have different spacing. The airlines may also gradually move the seats closer together for all the obvious economic incentives.

Table 8.4. Crash Dummy Spinal Loading from 14-ft Drop Tests

Test Section	Number of Crash Dummies	Spinal Loading (lb)	Maximum G Loads	Duration (sec)
707	2	1,404–1,461	26	0.052
737	5	638–922	11.0–15.1	0.043–0.122

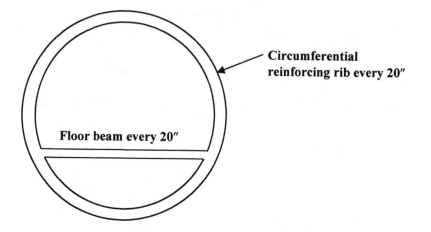

Fig. 8.22. Floor beams support seat tracks and connect to circumferential reinforcing ribs.

The seats connect to tracks with special attachment fittings. The tracks, which run the length of the fuselage, sit on floor beams that connect to the circumferential ribs spaced every 20 inches to reinforce the fuselage (Figure 8.22). (The 20" spacing is standard on almost all large commercial jets.)

The underlying seat-support structure might appear flimsy when considering that crash forces sometimes cause seat detachment. However, the seats are designed to minimize weight and to collapse and absorb crash energy. A stronger seat structure will transmit larger forces to the floor, but the reinforcement to accomplish this requires stronger tracks, floor beams, ribs, and fuselage, thus affecting the plane's weight. An example seat structure is shown in Figure 8.23.

The FAA's required seat strength has been increased recently, but the topic is controversial as some safety experts believe the seats should be stronger still. Unfortunately, for the most violent (and fatal) crashes, seat strength makes little difference. The Kegworth Flight 092 crash, in which the plane hit an embankment as it moved forward and the seats detached and crashed into the front of the plane, is an exception. Most survivable crashes occur with excess sink rate (i.e., crashing, vertically into the ground) where seat detachment is a lesser factor.

Strength requirements for seats were first mandated in 1952 for the following inertial loads: 9 G forward, 3 G sideways, 2 G upward, and 6 G downward (Figure 8.24).

Fig. 8.23. Example of a passenger seat attached to tracks, left end faces forward

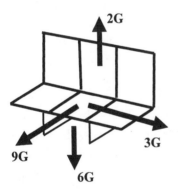

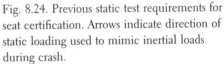

Fig. 8.24. Previous static test requirements for seat certification. Arrows indicate direction of static loading used to mimic inertial loads during crash.

A static test of 9 G forward involves pulling on a seat with a steady and continuous force equal to nine times the weight of the seat plus a 170-pound passenger. If a triple seat weighed 100 pounds, a static load would be applied with a cable and winch system to the seat belts equal to

$$9 \times (3 \text{ passengers} \times 170 \text{ lb/passenger} + 100 \text{ lb}) = 5{,}490 \text{ lb.}$$

Failure during these tests usually occurred just above the required load with subsequent release of the seat. Similarly, the floor structures were designed to barely sustain the seat loads.

In 1988, new rules were adopted for seat strength in all planes designed after 1988. At the time, the new rules only applied to the Boeing 777 and Airbus 330, both introduced since 1988. Airlines, in anticipation of a phase-in of the new requirements, were installing stronger seats during major refurbishing. Seat manufacturers stopped manufacturing the old seats in the 1990s. The rule was modified in October 2005 to include all planes manufactured after October 2009 regardless of when they were first designed. (For example, the Boeing 737, still in production, was first designed and certified in the 1960s.)

The static loads were increased to 3 G in the vertical direction and 4 G in the sideways direction. A backward load of 1.5 G was added for the first time for rebound impact after an initial forward impact. To minimize seat detachment, the attachments are designed for inertial loadings equal to 1.33 times the loadings specified above.

Also required for the first time is dynamic testing or simulated crash testing. The difference between static testing and dynamic testing is analogous to standing on a scale (static load) versus jumping on the scale (dynamic load). If you jump on, the reading overshoots your actual weight. During dynamic testing, the crash dummy will flail forward creating new and different (but more realistic) loading patterns. Head-path patterns are also evaluated behind bulkheads making certification unique to each plane design.

One goal of all crash testing and computer simulation over the years was to identify how the impact forces for a survivable crash "flows" through the fuselage to the cabin floor and eventually interacts with the passenger. After careful study of crash test data, identification of survivable accidents, and years of computer simulation, survivable crash pulses were identified for use with seat design. The crash loads developed represent a compromise between passenger survivability, excessive loading on the floor, and excessive weight in the supporting structure. They became the basis for the Federal Aviation Regulation titled "Dynamic Evaluation of Seat Restraint Systems and Occupant Protection on Transport Airplanes."

The seat to be tested is mounted on a special sled. The FAA facility (there are others) often used by seat manufacturers uses a sled mounted on two precision rails. The sled is propelled to 44 ft/sec and then rapidly decelerated.

Two tests are required: one that simulates vertical ground impact and one that simulates horizontal impact with a ground-level obstruction.

Test 1: 14 G vertical-longitudinal crash. The vertical-longitudinal test is meant to simulate a ground impact with a high rate of descent. The ori-

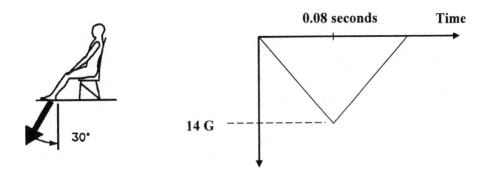

Fig. 8.25. Crash pulse for current 14 G vertical crash test required for seat certification. Arrow indicates direction of sled motion. Aircraft is moving down and deceleration is upward, so inertial loading tries to crush the spine downward.

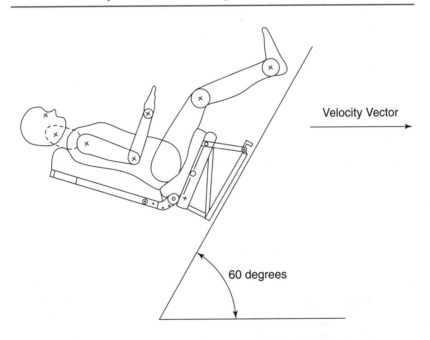

Fig. 8.26. One common configuration for current 14 G vertical seat crash test on a horizontal sled

entation of the crash dummy mounted on the crash sled is shown in Figures 8.25 and 8.26. The 14 G crash pulse ramps up in 0.08 seconds, as shown in Figure 8.25. For this mostly vertical impact, the seat must absorb energy and protect the passenger from spinal injury.

Test 2: 16 G horizontal crash. The horizontal test is meant to simulate impact with a ground-level obstruction. The orientation of the crash dummy mounted on the crash sled and the crash pulse are shown in Figure 8.27. The 16 G crash pulse ramps up in 0.09 seconds. If the seat attachments break, the seat will be flung forward, as happened in the Kegworth Flight 092 crash. Another consideration is the impact load on the crash dummy's head as it hits a seat in front, a bulkhead, or a galley wall.

Because of crash testing at 16 G's, the new seat designs are commonly referred to as 16-G seats, whereas the old seats are referred to as 9-G seats.

Floor deformations often occur during survivable crashes. Therefore, Test 2 must be done with simulated floor distortions. The floor rails attaching the seat must be misaligned, as shown in Figure 8.28.

Requirements for Passing Seat Crash Tests

A seat must meet a variety of requirements to pass a crash test.

- The maximum compressive load measured between the pelvis and the lumbar column of the anthropomorphic dummy must not exceed 1,500 pounds.
- The seat must remain attached at all required points, although the structure may distort.
- Permanent deformations, measured after the test, must not exceed three inches. This requirement minimizes the possibility of trapping passengers within deformed seats during emergency exit.

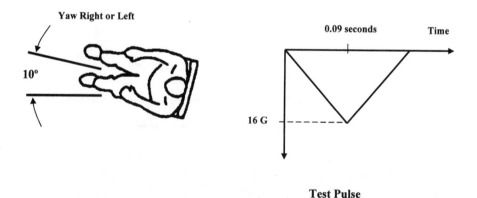

Fig. 8.27. Crash pulse for current 16 G longitudinal seat crash test required for certification

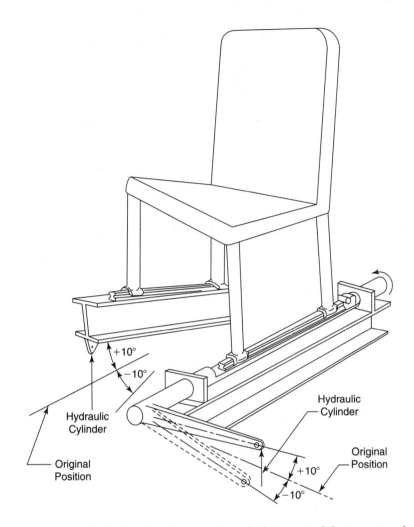

Fig. 8.28. A method of applying floor warping, which is required during seat crash testing. *Source*: FAA, Report DOT/FAA/CT-82/69

There are also limitations on permanent seat rotation for similar reasons (Figure 8.29).

- The head injury criterion (HIC), described in detail in Chapter 9, must be below 1,000. The HIC is a calculation based on the area under force versus time plots of the accelerometers mounted in the crash dummy's head.

Airlines have problems meeting the new head impact requirements behind bulkheads, galleys, and lavatories. One option is to remove these seats, an un-

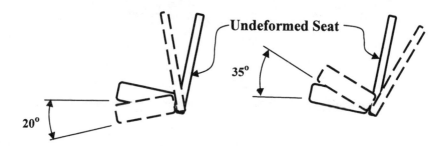

Fig. 8.29. Post-test seat rotation must not exceed 20 degrees down or 35 degrees up.
Source: Desjardins et al., SAE Paper 881378

desirable choice that affects revenue. Another option, being phased in by some airlines, is to equip the seat belts on the troublesome seats with air bags.

The FAA has studied the bulkhead and HIC problem. Table 8.5 shows sample data for some of the crash variables associated with a HIC test for a horizontal sled test of 16 G.

The HIC value will vary greatly depending on the bulkhead material, thickness, and crush strength. The values in Table 8.5 show sample rela-

---

HEAD INJURY AND THE BOWLING BALL TEST

Flat surfaces within striking radius of the head (tables, bulkheads, partitions) require padding with energy-absorbing material during taxi, takeoff, and landing. A test suggested in FAA's "Transport Airplane Cabin Interiors Crashworthiness Handbook" involves simulating a 9 G impact to the head of a 50th percentile male using a bowling ball. The bowling ball must strike the surface with at least 2,780 inch pounds of kinetic energy and rebound with only 375 inch pounds of energy to pass the test. If the bowling ball rebounds with kinetic energy greater than 375 inch pounds, more energy-absorbing material must be added to the surface.

Assuming that the bowling ball weighs 10 pounds and using the equation for the kinetic energy = ½ mass × velocity², required velocity of the bowling ball can be calculated.

$$2{,}780 \text{ in lb} = \tfrac{1}{2} \; \frac{10 \text{ lb sec}^2}{32.2 \text{ ft}} \; \frac{\text{ft}}{12 \text{ in}} \times \text{velocity}^2.$$

Solving the equation, the required velocity of the bowling ball for initial impact is 463.5 in/sec or 38.6 ft/sec (26 mph).

Table 8.5. Sample Data for HIC Test Results
of a 16 G Horizontal Crash into a Bulkhead

Seat Setback Distance (in)	Head Peak Acceleration (G)	HIC	Head Impact Velocity (ft/sec)
35	132	783	54
33	165	1,259	43

Table 8.6. Design Floor Loads

Aircraft	Up Load (lb)	Forward Load (lb)
737	5,360	Unavailable
767	3,650–6,100	2,200–3,660
DC-9	4,500–5,170	4,500
DC-10	4,000–6,000	3,600–5,500
A300	7,700	5,950
A320	5,500	4,400

Source: Desjardins et al., SAE Paper 881378

tionships among peak acceleration, head impact velocity, and HIC value. With all other variables held constant, Table 8.5 also shows the effect of an increased setback distance on HIC.

There are also requirements during seat testing for femur injuries; these are not discussed here.

Seat Design

The seat design has to optimize the following requirements:

- Provide seat structure adequate to pass the crash tests without failure
- Not exceed floor loading limits specified by the aircraft manufacturer (Table 8.6 and Figure 8.30)
- Limit permanent distortion of the seat so that emergency evacuation is not impeded

A rigid seat can be designed to withstand the new 16-G dynamic testing (without structural failure of the seat itself), but the resulting loads will exceed the floor strength. The seat must be designed to absorb some of the crash energy and limit the loading applied to the floor. The seat manufacturers have devised a number of energy-absorbing designs to limit the inertial loading and the loading on the floor.

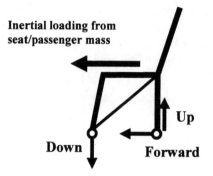

Inertial loading from seat/passenger mass

Up

Down

Forward

Fig. 8.30. The maximum floor loads occur at the attachment points of the seat. Shown are the forces the chair exerts on the floor.

Dynamic Seat Response: Load Attenuation and Magnification

The seat will interact with the input crash pulse from the cabin floor and respond in one of several different ways that illustrate many basic principles of physics.

1. The crash load can be passed directly to the passenger without any alteration.
2. The seat can elastically deform, absorb energy, and rebound like a spring.
   a. The elastic rebound may occur such that inertial crash loads are attenuated.
   b. The elastic rebound may occur such that inertial crash loads are magnified.
3. The seat can respond elastically, absorbing energy that is recovered by rebound, followed by a plastic deformation absorbing energy that is not recovered. This is the most common response for a strained metallic structure, and the response for which seats are designed.
4. The seat can plastically deform to absorb energy without any rebound.

The first and last points represent extreme cases, are the easiest to explain, and are relatively intuitive.

If the seat is rigidly attached to the floor and moves with it, the crash loads are passed through directly without any changes. This is similar to the hammer blow on the steel rod held against the head in Figure 8.18. The steel rod transmits the hammer blow directly to the head without alteration. However, if the seat structure is made of a flexible material, like bubble gum, it plastically deforms and absorbs much of the crash energy. Of course, there are limits to the amount of plastic deformation any material can receive before it breaks or fractures. Bubble gum, although flexible and capable of absorbing a lot of energy, is not sufficiently strong to withstand crash forces. The seat structure design also determines the crash response. Very thin support legs will flex more than solid legs with a massive cross section. The two extremes, points 1 and 4 above, are a very rigid material and structure versus a very flexible material and structure.

The concept of load attenuation or magnification is more abstract. If the plane crashes to a stop in 0.3 seconds, but the seat continues to deform such that the occupant stops in 0.4 seconds, the rate of change of velocity (acceleration) is reduced on the occupant. Load attenuation has occurred, and the G load on the passenger is less than what occurs on the cabin floor. However, if rebound of the seat is such that the occupant is flung up before the plane finishes crashing down, the acceleration and G load on the passenger will actually increase. G loading on the passenger is greater than on the floor, because load magnification has occurred.

Load Deflection Requirements for Seat Design

The energy absorbed by the seat (the area under the seat-force deflection curve) must be large enough to pass the impact test without exceeding the maximum loads specified on (1) the crash dummy's spine, and (2) the aircraft's floor structure. If one of these limits is exceeded, the seat must be redesigned to absorb more energy by deflecting more and thereby transmitting lower forces. The loads on the crash dummy's spine depend on the energy-absorbing capability of the entire system. This includes fuselage crush and seat deformation.

Horizontal and vertical flexibility of the seat is a good thing and will provide increased passenger protection, but there are other practical limits. Basically, there is only so much space available. Permanent deformations of the seat after a crash are limited by FAA regulation so as not to impede emergency evacuation.

The tests must be conducted with an occupant simulated by a 170-pound anthropomorphic test dummy. A 170-pound dummy was chosen as representative of a 50th percentile adult male. Obviously, the structural response and failure of an 80-pound triple seat frame will be greatly affected by the inertia of three 170-pound crash dummies.

Development of the Crash Dummy

Crash dummies were first used by the military in the late 1940s, mostly for ejection seat development and testing. Crash testing came into its own for automotive applications when Congress passed the National Traffic and Motor Vehicle Safety Act of 1966. Initially, the National Highway Traffic Safety Administration insisted on testing constraint systems with an Alderson's VIP-50, one of the earlier dummy designs. The automotive companies complained about lack of repeatability in the crash results, eventually winning this point in federal court. General Motors took a leadership role in developing the Hybrid I, quickly followed by the Hybrid II, which satisfied the courts, the government, and automotive manufacturers and became the standard for frontal automotive testing.

The Hybrid II anthropomorphic test dummy (ATD) is also known as the Title 49, Part 572, Subpart B, 50th percentile male, in honor of the U.S. Code of Federal Regulations[1] that covers automotive crash testing. There is now a Hybrid III, but a modified Hybrid II remains the FAA standard for seat crash testing and the dummy used in all crash tests described here.

Six other crash dummies were eventually developed and used for mandated automotive crash testing:

Three infants: newborn, six months, and nine months
Two children: three years and six years
5th percentile female

The ATD has two fundamental requirements:

1. It must give consistent results with repeated tests.
2. It must mimic the human body in terms of dimensions, mass distribution, joint location, stiffness and motion, load deflection properties, etc. This is an area of ongoing research and continued progress for automotive researchers.

In spite of the many efforts to make the ATD humanlike—the whole point being to identify the forces that damage humans—testing a crash dummy is not quite the same. Although there is an extensive history of cadaver and animal[2] crash tests and even limited volunteer testing, there is something to be said for the consistency of using what is essentially an instrumented machine.

Even if it was practical to somehow destructively test human subjects to evaluate crash forces, the inherent variability of the human body would make it very difficult to use the results for design standards. Ten times more tests would be required to correctly account for human strength variability and to define a lower limit of human injury. The forces that damage the human body are simply not as well defined as the forces that break, for example, a piece of steel.

In one sense, all crash tests are comparisons to all other crash tests. Did the design change make the component more crashworthy or less? There are many advantages to addressing this question with machinelike precision and the repeatability of a crash dummy. Ideally, there should be a calibration curve which relates crash test data obtained from the dummy to the percentage of human occupants who will survive. These data, to some extent, have slowly evolved and exist for head and spinal injuries and other body parts (see Chapter 9).

Crash Dummy Dimensions and Weights

In designing ATDs, approximately 200 measurements of segment length, circumference, masses, moments of inertia (explained later), joint location, and joint ranges of motion were taken, largely from segmented cadavers. Standardization of all these values contributed greatly toward the repeatability of dummy crash tests. The Hybrid II was the first dummy to use segmented parts that easily disassembled for measurement and quality control. This simple improvement allowed more accurate control of individual part mass, center of gravity, and moments of inertia.

Particularly important for computer simulation is the link length, or distance between joints, which requires a special kind of motion study. A human shoulder joint, for example, can lift and rotate. It is not a single fixed joint as constructed in a crash dummy. The best location for the dummy joint so that the motion during a crash matches between the dummy and a human is not a trivial determination.

Sample dimensions and weights of the Part 572 crash dummy are given in Tables 8.7 and 8.8.

Crash Dummy Mechanical Properties

To obtain repeatability, mechanical properties of the individual crash dummy segments are specified in the Federal Regulations for crash dummy performance. For example, when the dummy's head is dropped 10 inches

Table 8.7. Example Crash
Dummy Dimensions

Position on Body*	Length (in)
Top of head	35.7
Shoulder pivot	22.1
Hip pivot	3.9
Bottom of foot	17.3

X axis

*All measurements from top of seat (X axis in figure)

Table 8.8. Example Dummy
Segment Weights

Segment	Weight (lb)
Head	11.2
Upper torso (including lumbar spine)	41.5
Lower torso (including visceral sac and upper thighs)	37.5
Foot	2.8

onto a steel block, accelerometers mounted in the head must read between 210 and 260 G's. These values were determined by drop tests using detached cadaver heads with embedded sensors.

Strength and flexibility are also very important. As just one example, the head and neck substructure must be swung from a rigid pendulum until it strikes a surface at 23.5 ft/sec with an impact force between 20 and 24 G's. The head is to rotate (relative to the neck attachment) 68 ± 5 degrees with additional requirements on rotation versus time during the impact sequence.

All the mechanical requirements trace back to extensive cadaver testing and limited volunteer testing. The military, for example, tested the flexibility of pilots' necks with four different motions (Table 8.9):

A. Forward (the downstroke of the "yes" motion)
B. Backward (the upstroke of "yes")
C. Front to side (looking forward to one half of the "no" motion)
D. Front to shoulder (touching head to shoulder)

Table 8.9. Flexibility of Neck

Motion	Voluntary Degrees of Motion	Forced Degrees of Motion
A	61	77
B	60	76
C	78	83
D	41	63

Source: Coltman et al., Report USAAVSCOM
TR 89–D-22B

Table 8.10. Spinal Flexibility Required in Test Dummy

Rotation (degrees)	Force (lb)
0	0
20	28
30	40
40	52

When subjected to continuously applied force, the crash dummy spine assembly must flex by an amount shown in Table 8.10 for each specified force level. When the force is removed, the dummy's spine should straighten to within 12 degrees of its initial position. (The spine is not a linear spring.)

The Hybrid II design paid special attention to joints, with the result being increased repeatability of the torques resisting rotation in the shoulders, knees, neck, etc. The dummy joints operate on a clutch principle. The clutch force is given by a spring pressing against two plastic pressure plates.

Under certain circumstances, humans attached by only a seat belt (no shoulder harness) tend to "submarine" under the belt. In the 1970s, the motion of the crash dummy was compared to fresh cadavers during sled and car crash tests. Also, much was learned by comparing the dummy's response to the response of human volunteers up to the limit of injury-free crash impacts. Initially, the dummy tended to submarine or slide under the seat belt more than the cadavers. After modification to the dummy's abdominal material and pelvic shape and careful attention to joint stiffness, the overall motion agreed fairly well.

During a crash, the head and torso flail forward. Every part of the crash dummy must reach the same point at the same velocity as a person exposed to the same crash forces. This is done by having an ATD that matches the human's mass distribution and joint stiffness.

Distribution of Mass

Not only is the mass of the individual segments important, but also the distribution of mass. The distribution of mass will affect how the segments rotate. If an object is mounted on a shaft and loaded with a twisting force or torque, the object will spin on the shaft with increasing angular acceleration. There is a rotational version of Newton's Second Law ($F = ma$) that describes rotational acceleration (Figure 8.31).

The mass moment of inertia is a property of the object, similar to mass, that describes the object's rotational acceleration. Just like mass describes an object's resistance to acceleration, a mass moment of inertia describes an object's resistance to rotation.

Unlike mass (which remains constant for an object), the mass moment of inertia changes when the axis of rotation changes. This difference can

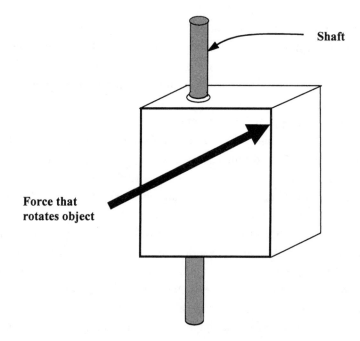

Fig. 8.31. Applying a torque to an object mounted on a shaft results in angular acceleration of the object (it spins faster and faster). The angular acceleration changes if the shaft is moved to a different location.

Table 8.11. Example Moments of Inertia

Item	Moment of Inertia (in lb sec^2)
Head	0.145
Torso	9.625
Basketball	0.055
Bowling ball	0.455

be felt by swinging a bat over one's head. The motion of the bat will feel different depending on where the bat is held. Holding the small end will be different than holding the large end or holding it in the middle. In each case, the distribution of mass, relative to the pivot point, is different.

A few sample mass moments of inertia are given in Table 8.11 for body parts and, for comparison, a bowling ball and basketball. All values are for rotation about the center of gravity.

The crash dummy used by the FAA is actually a modified automotive dummy. For a dummy to be used in aircraft testing, consideration must be given to the vertical component of the crash force—the force that tries to crush the spine. This vertical load is unique to airplane crash events. To measure this vertical impact force, a load cell is placed between the pelvis and bottom of the spine. If the peak load value is less than 1,500 pounds, then injury is unlikely. The military has a similar need for their crash dummies used to evaluate ejection seat inertial loading (see compressive strength of spine in Chapter 9).

Computer Modeling: Simple Theory

Except for the earliest tests, a major goal of any aircraft crash testing has always been to collect data usable for computer simulation research.

Ideally, crash data are used to verify and gain confidence in a computer model. The computer model can then be used for parametric studies to evaluate many changes of the crash variables (velocity, orientation of the plane, and impact location) and a multitude of changes to the aircraft structure. Computer crash simulation is very difficult. A precise prediction of crash conditions remains elusive for a variety of practical reasons. However, computer simulations are helpful to aircraft designers to compare design changes and their effect on crashworthiness. In spite of the difficulties, very good and useful predictions can be obtained.

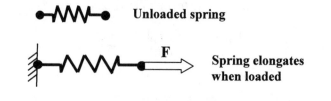

Unloaded spring

F

Spring elongates
when loaded

Fig. 8.32. A simple spring

A computer simulation of a structure loaded with simple dead-weight loads is considered accurate if it agrees within perhaps 10%. There are many sources of error and variation in the computer model and the experiment that prevent perfect agreement. For example, dimensional agreement is always off a bit.

Crash simulation has additional difficulties. As individual components begin to collapse, tear, or fracture, they will occur close to the ±10% accuracy of prediction. If the sequence and timing of five small components collapsing and fracturing are off slightly, the load will "flow" through the remaining structure with a different load pattern and accumulate errors.

Another problem is crashing into dirt, which presents a whole new set of hard-to-predict variables. Fundamentally, the task is at the limits of computer technology. Given the difficulties, comparisons between computer simulations and experimental data are quite good.

Many of the basic concepts of computer simulation can be explained by understanding a force versus displacement plot for a simple spring (Figure 8.32). If a simple, linear helical spring is fixed on one end and extended with a force on the other, a plot can be made of force versus displacement, shown in Figure 8.33. The slope of the line is commonly referred to as the spring constant, K. The equation of this line is

$$\text{force} = (\text{spring constant}) \times \text{displacement}.$$

Given the spring constant and either force or displacement, the remaining variable can be solved for.

The analysis of two springs is a little more complicated. Figure 8.34 shows a model of two spring elements and three "nodes." Elements and nodes are fundamental concepts for computer simulation of structures.

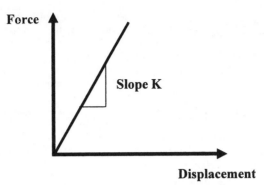

Fig. 8.33. The relationship between force and displacement for a linear spring

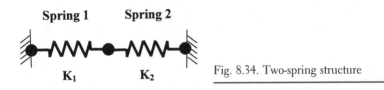

Fig. 8.34. Two-spring structure

The two springs may have different spring constants. This would occur if the coil diameters or number of coils differ or if the springs are made of wire with different diameters.

If the center node is loaded with 200 pounds, for example, the spring on the left elongates and the spring on the right shortens (Figure 8.35). The forces can be evaluated individually with a "free body diagram" of each spring (Figure 8.36). The sum of the forces on each spring must equal zero. An equation describing the deflection of each spring can be written. Also, the two unknown spring forces, F_1 and F_2, must equate to 200 pounds. The system of equations can be solved to determine the force in each spring and the displacement of the center node.

Airplanes are modeled with many springs and many equations. Two different approaches have been developed in a method known commercially as "finite element analysis." The first computer method uses perhaps a few dozen stick "elements" to model the plane. The springiness of each element is determined experimentally by testing (one of the purposes of drop testing fuselage sections). The second method uses a three-dimensional network of tens of thousands of elements to model the springiness of the underlying material and how it changes as the

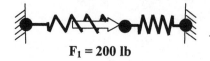

Fig. 8.35. Two-spring structure is loaded with 200 pounds on the center node.

$F_1 = 200$ lb

Spring 1 elongates

Spring 2 shortens

Fig. 8.36. The forces in each spring can be evaluated.

aluminum sheet metal is shaped into fuselage skin and attached to re-inforcing ribs.

The models become even more complicated when displacements in three-dimensional space are considered, when the model becomes highly "nonlinear" as crushing and tearing occur in the fuselage, and when iner-tial effects are added. (Instead of the forces summing to zero, an inertial term must be added—the mass × acceleration term in Newton's Second Law.)

The history of airplane crash computer simulation closely follows the development of computers. As computers grew in size, speed, and mem-ory, larger computer models to simulate crashes followed.

The earliest computer simulation of the Controlled Impact Demon-stration used 17 nodes and 16 stick elements. In 1999, a 10-foot section of a 737 was dropped. The model simulation contained 9,759 nodes.

An entire plane, an ATR42-300 (a European commuter plane that seats up to 50 passengers) approximately 74 feet long, was dropped 14 feet in 2003. The computer simulations, completed by NASA, included 57,643 nodes. About eight months were spent carefully measuring the airframe geometry. The coordinates of 25,917 points were identified with direct measurements. The 0.3-second crash event ran for 130 hours on a com-puter workstation. The simulation accurately predicted the major structural failures, including collapse and failure of the fuselage reinforcing ribs sup-porting the wing structure (Table 8.12).

Computer Modeling of Seat Crash Tests

Modeling seat crash tests is inherently easier and more accurate than mod-eling the impact of an entire plane. A seat has fewer components and less

Table 8.12. Comparison of Experimental
Data and Computer Simulation for
European Commuter Plane Drop Test

Experimental Results* (G's)	Computer Simulation (G's)
28	24
24	30
34	32

*Floor accelerations at three locations

metal to model, so considerably less time is required for both engineers and computers.

In a plane, there are many components that fracture or collapse. As discussed in the section on computer simulation of crash testing, there is statistical scatter associated with a single fracture event. Multiple fractures will accumulate error. When simulating a seat crash, fewer components fracture, collapse, crush, or buckle. There is simply less error to accumulate.

The FAA has stated its intent to put greater emphasis on analytical computer modeling of crash dynamics for transport category aircraft seats. In 2003, the FAA issued guidelines for acceptable methods to demonstrate compliance by computer simulation for certification of 16-G seats in lieu of crash testing. The computer simulation can only be used for incremental design changes, and computer models must be verified by crash testing.

The FAA expects the following crash data to agree with 10% of the computer simulation.

- The peak floor reaction loads
- The seat belt loads
- The spine load time-history and maximum spine load value

In addition, the Head Injury Criterion (HIC) simulation should agree within 50 HIC units of the test results. However, unless the computer model demonstrates decreased head velocity with any proposed design changes, the FAA does not support computer simulation of HIC values above 700.

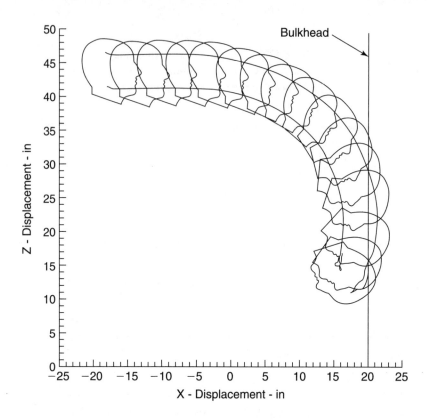

Fig. 8.37. Computer simulation of head impact into a bulkhead. Interference between head and bulkhead represents deflection of the bulkhead. *Source:* FAA, Report DOT/FAA/AR-02/103

Table 8.13. Comparison of HIC Values for Computer Simulation
and Sled Crash Test (16 G Horizontal Test)

	Seat Setback Distance (in)	Head Impact Velocity (ft/sec)	Head Peak Acceleration (G)	HIC
Sled test	32	45.7	168	853
Simulation	32	44.9	166	918
Sled test	37	48.1	81	552
Simulation	37	47.6	80	560

Crash dummies greatly simplify the computer simulation. The computer merely has to model a mechanical device that can be poked, prodded, and destructively tested as needed. The crash dummy is simply easier to model and gives more repeatable results than human tissue. Also, 14 joints are far easier to mathematically model on a computer than the complexities of an actual person. The crash dummy measures head and chest G loads and forces on the neck, spine, and femur. The computer model calculates similar accelerations and forces.

FAA Bulkhead Design Studies, Computer Simulation, and HIC Requirements

The FAA has done extensive testing and computer simulation to address the HIC requirement behind bulkheads. The variables are extensive, including bulkhead material and thickness, seat belt and seat flexibility, seat setback distance, and bulkhead panel crush and collapse strength. Head impact results will vary greatly depending on how these variables are chosen. Figure 8.37 shows the path of a head as it impacts a bulkhead, and Table 8.13 compares computer simulation data with crash test data for impact with a bulkhead. Table 8.13 also demonstrates decreased HIC with increased seat setback. All tests were done with a 16 G horizontal impact.

9...

Human Tolerances to G Loads
and Crash Forces

On September 14, 2003, Air Force captain Chris Stricklin opened a show for the U.S. Air Force Thunderbirds. He decided to dazzle the crowd with a daring act in his F-16. He would climb steeply, roll the plane, dive, and at the last second, pull out.

Seconds after takeoff, Stricklin was in the third part of his maneuver, the dive. He quickly sensed that something was wrong. The ground was approaching too fast; he wasn't going to make it. There was no time to make sense of what the instruments were telling him. Instinct took over. Stricklin could do one of two things: eject or stay in the plane. Either way, he would be risking his life.

Ejections have always been violent, resulting in many injuries and fatalities. Although design improvements have reduced the frequency of injuries, ejection remains literally a bone-jarring experience.[1]

Training had taught Stricklin that hesitation was the greatest cause of failed ejections. But he couldn't eject yet. He had to get his plane away from the crowd, and he had to reduce his sink rate. At his current sink rate, even ejection wouldn't save him.

The sink rate—how fast the plane is speeding towards the ground—must be reduced below some critical value; otherwise, the ejected pilot continues on this downward path, descending into the ground. The critical sink rate depends on the altitude, speed, and heading of the plane. Stricklin pulled hard on the controls to try and get his plane into a better position.

Pulling the ejection handle initiates a complex and rapid sequence of events. For Captain Stricklin, the sequence would occur 140 feet off the ground and just 0.8 seconds before the plane crashed. In milliseconds, the

Fig. 9.1. Photo of Captain Stricklin's ejection. Although the plane appears
to be flying horizontally, there is a considerable vertical component
of velocity. The plane crashed an instant later.
Photo: Staff Sgt. Bennie J. Davis III, U.S. Air Force

pilot's harness tightened around him, forcing him into a position that prevents flail injuries. The plane's canopy blew off and an initial blast shot the ejection seat 40 inches up a set of rails. A second rocket blast removed the pilot from the plane (Figure 9.1).

An ejection seat has two parachutes: the main chute and a smaller drogue chute sometimes used to help deploy the main chute. Depending on flight conditions, these chutes are deployed differently. As the seat travels up the rails, altitude and pressure sensors are used to select the mode of ejection. For low-altitude, low-speed ejections (with little time before the plane strikes the ground), the main chute deploys immediately. For very high-altitude ejections, the drogue chute deploys first, because the pilot needs to quickly descend to a safe altitude for breathing. After a considerable delay, the main chute is triggered. In a third mode, the main chute deploys at an intermediate rate, as compared to the two extremes described above.

Actual G loads (discussed later in this chapter) during ejection depend on the motion of the plane, the weight of the pilot, and the force of the rocket blast. In his ejection, Stricklin was estimated to be "pulling" 10 to 15 times normal gravity.

Captain Stricklin landed disoriented and surrounded by flames. It took him a second or two to realize he was alive. Although he landed near the flaming wreckage, Stricklin wasn't burned. His survival kit, dangling by a cord, hit the ground before him and sent dust into the air, creating a 10-foot-diameter flame-free zone. Still dazed, Stricklin decided to "just lay down in that happy place and wait."[2]

Two days later, Stricklin got out of his hospital bed to hug his wife, but again, something was wrong. "We've been married for 10 years," he said, "and she wasn't the same height she usually was." In the ejection, Stricklin's spine had been compressed by two inches. He gradually regained his height, all but half an inch within a few months.

The Air Force investigation concluded that the maneuver was erroneously initiated at 1,670 feet instead of the usual 2,500 feet. New regulations now require Thunderbird pilots to climb to 3,500 feet before performing this daring dive.

Acceleration

Stepping off a platform results in the acceleration of free fall from Earth's gravity. For every second of free fall, the velocity increases by 32 ft/sec, corresponding to an acceleration (or rate of change of velocity) of 32 ft/sec per second, sometimes written as 32 ft/sec/sec or more commonly as 32 ft/sec^2. The increasing velocity and distances associated with free fall are shown in Table 9.1.

It is common to express acceleration as factors of Earth's gravity. For example, if an ejection seat accelerates the pilot up at 480 ft/sec^2, the ejected pilot will experience $480/32 = 15$ G's. The pilot's spine is being compressed with 15 times the load of normal gravity.

Table 9.1. Velocity and Distances
Associated with Free Fall

Time (sec)	Velocity (ft/sec)	Total Distance (ft)
0	0	0
1	32	16
2	64	64
3	96	144
4	128	256

Inertia

A pilot ejecting up experiences the same compressive loading on the spine as a passenger in an airplane crashing down. To better understand this phenomenon, the property of matter known as inertia must be explored.

To paraphrase Newton, an object in motion (or rest) tends to remain in that state of motion (or rest) unless an external force is applied to it.

The tendency of motion to maintain itself is also know as the principle of inertia. Inertia is a property of matter that allows it to resist changes in motion. When pushing a block at rest on a frictionless surface, the equal and opposite force felt by your finger is supplied by the inertia of the block.

Inertial loads can best be explained by standing in an elevator with a scale underfoot. If the elevator is not moving, the scale will read the normal weight, W. If the elevator accelerates up at one half the acceleration due to gravity (16.2 ft/sec^2), the person's mass will resist this acceleration, and the scale will read $W + W/2$, or 150% of the person's normal weight. If the elevator accelerates down at half the acceleration due to gravity, the scale will read $W - W/2$, or 50% of the person's normal weight. If the elevator is accelerating down at a rate equal to the acceleration due to gravity (i.e., free fall), the scale will read zero. The effect of acceleration on a given mass is commonly referred to as G force, G load, or simply G's. Eight G's, for example, is eight times the normal force of gravity. G loads for a few common activities are given in Table 9.2.

If an object is isolated in outer space, away from any gravitational force, there is no up or down for this weightless mass. If a platform is placed under the mass and accelerated at 32.2 ft/sec^2, the normal acceleration of Earth's gravity, the object will experience an inertial loading equivalent to

Table 9.2. Example Accelerations

Activity	G
Accelerating a Porsche*	0.7
Carrier catapult launch	2.5
Roller coaster[†]	3.5–5
Shuttle launch	3

*0–60 mph in 3.9 seconds
[†]Evaluation of 110 amusement park rides

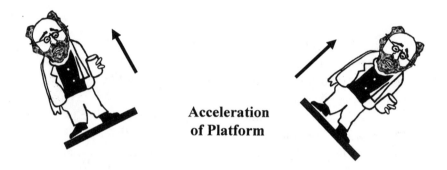

Acceleration of Platform

Fig. 9.2. An isolated body in space (isolated from any gravitational attraction) is weightless. If the body is accelerated at 32.2 ft/sec^2, it experiences inertial loading opposite the direction of the acceleration.

Earth's normal gravity. The sensation of up will be the direction of acceleration, regardless of which direction it is.

Consider a person standing on a platform in deep space, holding a cup of coffee, and accelerating at 32.2 ft/sec^2, as shown in Figure 9.2. The force to accelerate the platform 32.2 ft/sec^2 equals the normal weight of the body on Earth.[3] The inertial loading feels like normal gravity, and the coffee stays in the cup. As explained by Einstein, a body (human or otherwise) cannot tell the difference between the presence of Earth's gravitational field and being in space with no gravitational field but accelerating at 32.2 ft/sec^2. Think of inertial loading as being a virtual gravity always acting in the opposite direction of acceleration.

Pilots, especially fighter or acrobatic pilots, can experience G loads for many seconds or even minutes. A crash results in increased G loads for briefer periods of time.

G loading affects the body in two ways. A pilot can temporarily lose vision or even consciousness, because the circulatory system cannot properly pump the "heavier" blood. The G loads during a crash can cause physical damage to the body.

Standard Convention for Inertial Loading on the Body

It is important to understand the convention used with arrows on a G loading diagram. The convention used here is that the arrow points in the direction of inertial loading (i.e., the direction of the virtual gravity). The inertial loading arrow points down, from head to foot (+Gz), for the loading associated with normal gravity.

Inertial loading is always in the opposite direction of the acceleration. For example, an elevator accelerating up results in +Gz inertia loads (Figure 9.3). In this case, the +Gz inertial load arrow points down (adding to normal gravity), increases the reading on a scale under a person's feet, and tries to compress the spine.

The eyeball, connected by soft tissue, will move around relative to the head. Fighter pilots pulling enough G's are able to feel their eyeballs move. It is common pilot jargon to call out G loading direction by the movement of the eyeballs relative to the rest of the head. For example, +Gz is also known as G loads in the "eyeballs down" direction. In addition to showing the direction of inertial loading, the arrows in Figure 9.3 show the direction of eyeball movement.

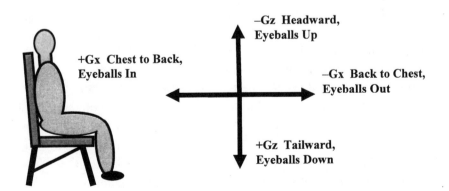

Fig. 9.3. Convention for inertial loads. Arrows point in direction of inertial loads (opposite the direction of acceleration) and in the direction of eyeball movement.

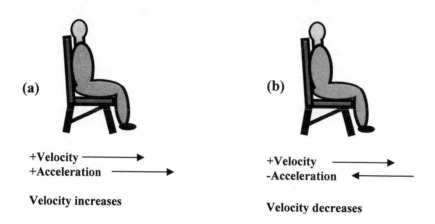

(a)

+Velocity ⟶
+Acceleration ⟶

Velocity increases

(b)

+Velocity ⟶
−Acceleration ⟵

Velocity decreases

Fig. 9.4. Sign convention for forward velocity and positive and negative acceleration. (a) Car is accelerating forward from a stop. (b) Car is moving forward but braking or crashing into a barrier.

Before continuing, one needs to understand the sign convention for velocity and acceleration. If moving forward (+velocity) and accelerating forward (+acceleration), the velocity increases—the case of an accelerating car. If moving forward (+velocity) and accelerating backwards (−acceleration), the velocity decreases—the case of a car slowing down or a plane crashing into an embankment (Figure 9.4).

If a plane is flying forward with a vertical downward sink rate, the most likely crash forces are −Gx (eyeballs out) and +Gz (eyeballs down). Eyeballs down, or +Gz, results from a hard landing. The plane, moving down, crashes into the ground and causes an upward acceleration. The resulting eyeballs-down G loads try to compress the spine. If the same plane flying forward crashes into an embankment, the acceleration points backward towards the plane's tail. The inertial loading is −Gx or eyeballs out. Of course, if the plane is tumbling when it contacts the ground, or if tumbling results from asymmetric contact with an engine or wing tip, inertial loads can be in any direction relative to the plane.

A carrier landing, car crash, and airplane crash all result in chest-to-back acceleration and eyeballs-out inertial loading. The motion is in the +Gx direction, but the acceleration is in the −Gx direction and the inertial loading is +Gx. The seat belt keeps the passenger from flying forward out of the seat. The eyeballs, being loosely connected to the head, continue to move forward, lagging behind the deceleration of the rest of the body. This results in the eyeballs-out sensation. If the seat belt had load cells, a tensile force would be recorded.

An ejection seat launch and the vertical component of a plane crash with high downward sink rates both result in headward acceleration and +Gz inertial loading that tries to compress the spine. If the crash is survivable, spine injuries often occur.

Figure 9.5 shows common examples of accelerations and inertial loading.

In addition to compressing the spine, longer-term exposure to +Gz acceleration will affect the heart's ability to adequately pump G-loaded (heavier) blood to the eyes and brain. This effect was first noticed by pilots in World War I when pulling out of a dive.

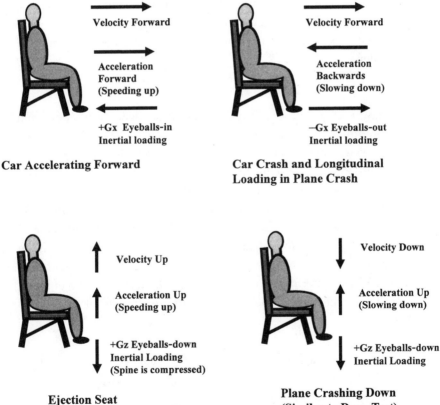

Fig. 9.5. Common examples of accelerations and inertial loading

A loss of visual acuity caused by diminished blood flow to the eyes, some-times known as grayout, usually happens first. With increased G loads, a person totally loses sight or may black out.

Eventually, with sufficient reduction of blood flow to the brain, a person loses consciousness. At this point, many people have jerky, convulsive movements. In centrifugal studies, 50% of pilots lost consciousness without any warning symptoms. Data collected from 1,000 Navy pilots with onset rates of 1 G per second are shown in Table 9.3. At 1 G per second, the data in the table could also be expressed in seconds. The average pilot experiences grayout in 4.1 seconds.

For G loading pressing into the chest (+Gx loading), tolerance is 10 to 12 G's for 2 to 3 minutes. All movement becomes burdensome, including movement of the chest cavity required for breathing. With impact or crash loading, 50 to 100 G's will result in damage to the aorta, the major artery taking blood from the heart to the body. At first, the aorta is torn, and eventually it can be completely severed, a common injury in car accidents.

Grayout, Blackout, and Blood Pressure

The pressure associated with a column of blood (between the heart and brain) is a simple hydraulic model that explains many of the basic cardiovascular features of G loads and blood flow to the brain. The brain is approximately 350 mm above the heart. (Consistent with conventional blood pressure terminology, metric units are used.)

First measured with a column of mercury, atmospheric pressure and blood pressure are traditionally described as the pressure at the bottom of a column of mercury. Normal atmospheric pressure, about 14.7 psia, is the pressure at the bottom of a 760-mm column of mercury (760 mm of mercury = 14.7 psia). Recall that pressure at the bottom of a column of liquid equals the weight of the liquid divided by the area the weight acts on (see Chapter 4).

Table 9.3. Navy Pilots' Average Response

Symptom	Average Threshold (G)	Range (G)
Grayout	4.1	2.2–7.1
Blackout	4.7	2.7–7.8
Unconsciousness	5.4	3.0–8.4

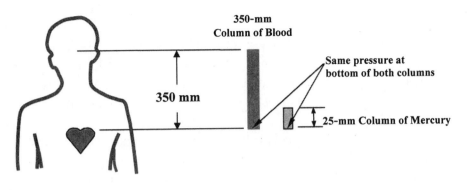

Fig. 9.6. To pump blood to the brain, the heart must overcome the pressure from 350 mm of blood. Because of differences in density, a 350-mm column of blood exerts the same pressure as 25 mm of mercury.

Since blood only weighs about 7.3% as much as mercury, a column of blood 350 mm high has the same pressure as a column of mercury 0.073 × 350 mm = 25.5 mm of mercury (Figure 9.6). In other words, for the heart to pump blood to the brain, it must supply a pressure of at least 25.5 mm of mercury to overcome the weight of the blood.

At 5 G, the column of blood between the heart and the brain weighs five times as much and exerts five times as much pressure at the bottom of the column. At 5 G, the pressure of the column of blood equals 5 × 25.5 mm = 127.5 mm.

When the pressure required at 5 G is compared to the normal blood pressure supplied by the heart, there is not enough pressure to supply blood to the brain. The brain is starved of blood and the pilot passes out. Essentially, there is zero blood pressure in the brain in 5 G.

The pressure at the heart after contraction, on average, equals 120 mm of mercury. The pressure during contraction of the heart is called systolic pressure. The common usage is to say the systolic pressure is 120 mm of mercury, or 120 mm, or just 120. The heart pressure of 120 mm easily overcomes the 25 mm pressure resulting from the weight of blood at 1 G. However, it does not overcome the required pressure of 127.5 mm at 5 G.

At about 4 G grayout (and eventually blackout or loss of sight) occurs, because blood is not being adequately supplied to the retina in the eyes. At about 4 G, the column of blood has a pressure of 4 × 25.5 mm = 102 mm. A pressure of at least 102 mm is required to lift the blood to the eyes at 4 G. It takes an additional 20 mm of pressure to adequately push the blood into the retinas.

If the heart supplies 120 mm at 4 G, the blood pressure at the eyes equals 120 mm − 102 mm = 18 mm. The required pressure to adequately supply the eyes (20 mm) is not met, so grayout has probably occurred. Blackout or total loss of sight, which sometimes precedes unconsciousness, is explained by complete lack of blood flow to the eyes.

This model does not include the cardiovascular response of a gradual onset of G loading. The heart rate can compensate somewhat for increased G loads, and every person is different, as shown in the results in Table 9.3 for 1,000 pilots.

Fighter pilots are taught the anti-G straining maneuver. They strain their legs, abdomen, and buttocks to constrict the blood vessels in the lower body. This helps prevent blood from pooling in the lower body, thereby making it easy to pump blood to the upper body. Fighter plane maneuverability is limited in part by the pilot's resistance to G forces.

Anti-G suits use pressurized air bladders inside special pants to provide the same constricting effects. The pilot must still strain until the air bladders become pressurized, or for maximum effect, the pilot will strain while the bladders are pressurized. After learning the straining techniques, 99% of fighter pilots can withstand 9 G's for 15 seconds in a centrifuge.

G-Suits

G-suits were developed to help pilots resist G loads. The first G-suit, developed during World War II, is an interesting story of historical note.

Wilbur Franks, a Canadian cancer researcher, kept breaking his test tubes when spinning them in a centrifuge. The spinning liquids would become heavier with increased G loads and break the glass. His solution was to float the tubes in water to provide an equivalent pressure on the outside of the tube.

Franks heard about pilots blacking out during dives and wondered if the same idea would work for humans. First, though, he picked a smaller test subject. Franks filled a condom with water and inserted a mouse into the first tiny G-suit. The mouse tolerated 240 G's, an extremely encouraging result.

In May 1940, a tailor was hired to secretly sew the first human G-suit. The suit consisted of a rubber bladder filled with water and covered with non-stretchable material. As the blood got heavier with G loads, the water in the suit would automatically get equally as heavy and press against the body to minimize pooling of blood in the lower extremities.

The concept worked and was quickly tried by British pilots. Although the suit worked for its intended purpose, it was very uncomfortable to wear.

It was soon decided that testing the suit in planes was dangerous and a human centrifuge, the first one built by the Allies, was developed in 1941. Centrifuge testing determined that only the bottom part of the body needed coverage by the G-suit.

The G-suit was successfully used in combat over North Africa in 1942. The pilots reported favorably on their increased maneuverability, and the Royal Air Force recommended adoption of the suit for operational use by fighter pilots. The RAF believed it would provide a significant advantage over their German counterparts. Despite having stockpiled over 8,000 suits, it was decided not to use them and to preserve secrecy until the suits could be used for maximum advantage—during the invasion of Europe.

By the time of the invasion, the nature of the air war had changed. Instead of short, furious dogfights, the fighter pilots were engaged in bomber escorts for extended periods. It was difficult to keep the water in the suits warm, and the pilots resisted being soaked in water for 6 to 8 hours at a time. Comfort and fit problems remained.

As fighter planes became jets in the 1950s, the jet engine's compressor had plenty of compressed air available to pressurize the G-suits. Air-filled suits, the current standard, are more complicated, needing control systems to provide appropriate pressure, but they are considerably more comfortable. None the less, water, unlike air, responds instantly to increased G loads without any control piping and instrumentation. More recently, the Swedes have revisited water-filled suits.

Disorientation Due to Inertial Effects

Pilot disorientation is another serious problem caused by inertial loading. One of many scenarios involves a false sense of up and down. Historically, this happened with early attempts at instrument flying. Pilots had to learn to trust their instrument readings more than their attachment to mother earth and stop flying "by the seat of their pants," as they had learned by sitting under normal 1 G. Disorientation and learning to trust their instruments remain problems for inexperienced pilots.[4] When a plane enters a turn, it generates centrifugal inertial loads (Figure 9.7) similar to swinging

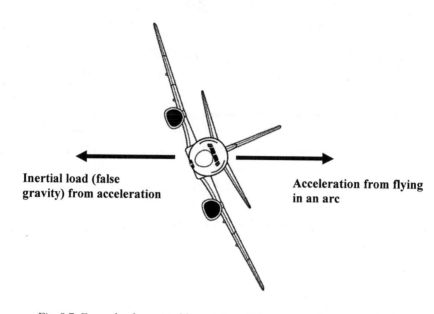

Inertial load (false gravity) from acceleration

Acceleration from flying in an arc

Fig. 9.7. Excess bank creates false gravity, adding to normal gravity and pilot confusion.

---

a pail of water overhead or driving a car around a turn. The equation that describes this acceleration[5] is

$$\text{acceleration} = \text{velocity}^2/\text{radius}.$$

An instrument failure and pilot disorientation contributed to the crash of a Boeing 737 on June 6, 1992. Twenty minutes after Flight 201 left Panama City, the plane entered a routine turn at 25,000 feet. When the artificial horizon instrument failed, the crew became disoriented and the aircraft rolled through 90 degrees, flipped over, and entered a steep dive, breaking up at approximately 13,000 feet. The aircraft was flying over featureless Panamanian jungle at night and the crew were unable to determine the aircraft's orientation. Inertial loading from the turn was suspected of contributing to their disorientation.

NASA's *Vomit Comet*

If an airplane is flying an arc in the vertical plane, the inertial load can add to or subtract from normal gravity (Figure 9.8). If downward acceleration from flying in the arc equals normal G, the plane is in a temporary state of free fall and the passengers will experience weightlessness. The same effect occurs when cresting "over the top" in a roller coaster. In total free fall, the downward acceleration of 1 G creates an upward inertial force that balances gravity, resulting in the weightless sensation of free fall.

NASA's famous Vomit Comet—an airplane for training astronauts and conducting research in zero-G conditions—flies in an arc on purpose to obtain about 25 seconds of weightlessness before the plane runs out of horizon and has to climb. As an aside, Tom Hanks spent a lot of time in the Vomit Comet during the filming of Apollo 13.

Inertial Loading Tricks the Inner Ear

Another pilot disorientation scenario involves inertial loading tricking the inner ear. In addition to sensing up and down motion, the inner ear contains organs that act as an amazing feedback control system and help us to

1. maintain and regain balance from a variety of situations;
2. maintain an awareness of where our body is even when moving with our eyes shut;
3. keep our eyes focused on an object even when our head is moving.

The system is designed to work with 1 G and with that 1 G always being down.

The disorientation effect can be approximated by spinning in a circle and then trying to walk an unmarked straight line. The tendency is to walk a curved line in the direction of the spin because of the continued motion of the fluid in the inner ear. If asked to walk a marked straight line, the vi-

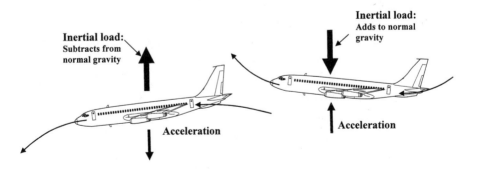

Fig. 9.8. Inertial loads from flying in an arc subtract or add to normal gravity, depending whether the arc curves down or up.

sual input will usually compensate for the fluid in the inner ear, demonstrating the complex interaction between vision and the body's motion detectors.

When an inexperienced pilot enters a turn under instrument conditions, the initial sensation of turning will disappear after two or three turns. If the pilot takes the appropriate corrective action by responding to the instrument readings to stop the turn, the pilot will experience a sensation of turning in the opposite direction. If the pilot incorrectly reacts to this sensation, the turn will be reentered, possibly with disastrous results. This illusion is known as the "graveyard spiral."

There are also other illusions related to various flying motions that are well known in the flying community.

Human Tolerance to Impact

The ability of the human body to withstand deceleration has been studied extensively by a variety of means including parachuting, ejections, survivors of long free falls, and human accidents. Better data have come from rocket-powered sleds and animal and cadaver studies. As technology has improved, anthropomorphic dummies and mathematical computer models have also been used. Still, it must be remembered that estimates of human tolerance to impacts are just that—estimates.

Another important, but limited, data source of human tolerance to impact loads comes from volunteers. Whole-body human tolerance limits result from tests with voluntary human subjects who are exposed to increasingly severe impacts. The severity of impact is increased until the subject feels that further tests would be unacceptable. Minor accidental injury may occur.

It is known that human tolerance to deceleration depends on the following:

- The acceleration duration
- The acceleration magnitude (peak G)
- The rate of acceleration
- The orientation of the acceleration with respect to the body
- The type of seat and restraint
- The physical characteristics of the person
- The distribution of force over body parts

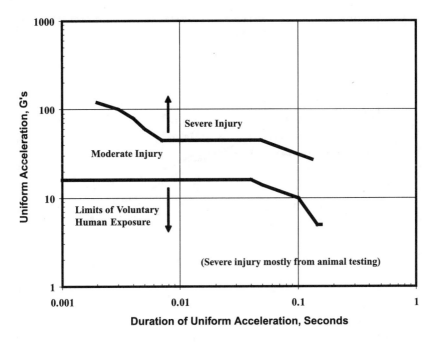

Fig. 9.9. Eyeballs down, +Gz. *Source*: Pesman and Eiband, Technical Note 3775

Note that the following discussion assumes a survivable volume exists after the crash. In other words, the occupant was not crushed mechanically. Also, trauma from other blunt objects is not considered. Human tissue can, of course, be damaged by much lower forces than presented below if the impact force is concentrated on a small area. Uniform deceleration of the entire body is assumed, but hammer blows and sharp cuts are not considered. Depending on how contact is made or how the person is constrained, the injury from inertial load versus impact force can be difficult to separate. For example, jumping off a five-story building onto concrete pavement will result in rapid deceleration and possible internal injuries from inertial loads. Separate from inertial loading, the concentrated impact crash forces will obviously damage human tissue. As another extreme example, moving into a knife edge at very low speed will also damage human tissue without any inertial loading taking place.

As explained in Chapter 2, aircraft move forward and downward (with horizontal and vertical motion). If a plane crashes into a barrier, eyeballs-out or –Gx inertial loading is created. If a plane crashes down into the ground, eyeballs-down or +Gz inertial loading occurs. Because planes tend to crash in these two directions, human tolerance to –Gx and +Gz have

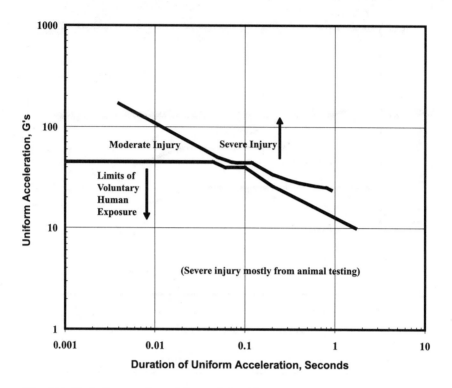

Fig. 9.10. Eyeballs out, –Gx with lap and shoulder restraint system. *Source*: Pesman and Eiband, Technical Note 3775

been studied the most. Because crash test dummies have become increasingly sophisticated, the most significant cadaver and live human test data available remain from the 1950s and are summarized in Figures 9.9 and 9.10. Most of the data for eyeballs out (–Gx) comes from Colonel John Stapp of the U.S. Air Force and his famous rocket-powered sled experiments from the early 1950s.

Colonel John Stapp and His Rocket Sled

Imagine sitting in an amusement park ride and accelerating faster until your bones begin to break, your brain gets rattled with concussion, and blindness sets in. Now imagine doing this on new test equipment being developed for the first time with a great deal of uncertainty about exactly what will happen. Further imagine doing this not once, but repeatedly, after painful injury. This was Colonel Stapp's reality. Colonel Stapp, a test pilot who did all his testing four feet off the ground, was the original rocket man, and a man who deserved all his medals.

Shortly after World War II, Chuck Yeager broke the sound barrier. Fighter planes were being designed for Mach 2 (twice the speed of sound or approximately 1,500 mph) and the U.S. Air Force desperately needed data on human deceleration and windblast tolerance to design ejection systems.

At the time, conventional wisdom confidently stated that 18 G's would be lethal to humans. However, there were hints from accidents and crashes that, given the proper circumstances, the human body could tolerate much greater impacts. For example, a 16-year-old boy plunged 244 feet from the Golden Gate Bridge, hit the water at 73 mph feet first, and survived with a broken shoulder and back. Stapp and his staff studied hundreds of similar stories.

Located at Edwards (then called Muroc) Air Force Base in California was a 2,000-foot-long straight rail track with 45 feet of hydraulic brakes at the end. Originally set up to test captured Nazi V-1 "buzz bombs," the track now tested a 1,500-pound "rocket sled." Each rocket bottle would produce 5,000 pounds of thrust. By varying the number of bottles (up to four) and the brake forces, a variety of stopping G forces could be created.

To work out the bugs in the equipment, initial testing was done with a crash dummy. During one of the earlier runs, Oscar, the test dummy, was wearing only a light safety belt. The breaks locked up producing "only" 30 G's. Oscar ripped the belt, flew through a one-inch-thick wooden windscreen, and finally landed 710 feet away. As the writer of a recent article put it, "Clearly some damnable forces of physics were at work."[6]

After testing with the crash dummy, testing continued with animals. With enough G's, the animals simply came apart. Eventually, it was learned that an animal would survive injury free at roughly one-quarter the G loads that caused fatal injury. In other words, if a bear or hog died at 200 G's, it would not sustain an injury at up to 50 G's with the same position and restraint system.

Everyone assumed the testing was too dangerous for humans. Stapp (a captain at the time) was also a medical doctor assigned to supervise the research project. Captain Stapp startled everyone on his first day by announcing that he was going to be the test subject. A human is designed to hold his spine straight and walk upright, reasoned Stapp. Also, an animal cannot tell you if the straps are too loose or too tight or rubbing in the wrong places.

Fig. 9.11. Colonel Stapp in his rocket sled. *Photo:* Edwards Air Force Base History Office

After 8 months and 35 test runs, Stapp felt they were ready for the first manned run. When the mechanic and flight surgeon had finished inserting Stapp in the rocket sled, he was totally immobilized by straps around his chest, elbows, wrists, knees, thighs, and ankles (Figure 9.11). (Animal tests had shown that limbs pulled into the wind stream would be broken.) Using just one rocket bottle, Stapp experienced a modest 10 G's on his first run in December 1947.

During the next few years, Stapp established his reputation as a fearless test subject by enduring 28 test runs and numerous injuries. He endured runs of

35, 38, and 46 G's. In addition to several concussions, he cracked his ribs, broke his wrist twice, and lost a few dental fillings (Figure 9.12). In one famous story, Stapp set his own broken wrist on his way back to the office.

The eyes were quickly determined to be the most vulnerable part of the body with respect to G loading. With forward $+Gx$ acceleration (eyeballs in), grayout would occur around 18 G's as the blood left the eyeballs and pooled in the back of the head. For reversed acceleration ($-Gx$, eyeballs out), Stapp would temporarily experience redouts caused by breaking capillaries and hemorrhaging as blood rushed into the eyeballs.

Stapp initially refused to allow others to ride the rocket sled. There was one obvious benefit: Stapp could write extremely accurate medical reports on his favorite test subject, Colonel Stapp. He also feared that the "can-do hotdog" attitude of the average test pilot might produce a disaster. Even-

Fig. 9.12. Colonel Stapp decelerates in the rocket sled. *Photo:* Edwards Air Force Base History Office

Fig. 9.13. Rocket bottles on the back of Colonel Stapp's rocket sled
Photo: Edwards Air Force Base History Office

tually, seven other volunteers did ride the sled, but whenever a new test profile was developed, Stapp was the first test subject.

Being concerned about the windblast in the latest Mach 2 fighter planes, the U.S. Air Force built a new, faster, more powerful sled at Holloman Air Force Base in New Mexico. The new track, 3,550 feet long, ended in a water tank. Water scoops on the sled could provide any desired braking force by simply varying the depth of water in the tank. The new sled would use up to 12 rockets (Figure 9.13).

As long as the skin was protected and the head and limbs were secured from flail, windblast turned out to be a non-event compared to G forces. With protective clothing, Stapp easily withstood windblast pressures of up to 1,100 pounds per square foot. Windblasts of up to 3,300 pounds per square foot resulted in second- and third-degree burns on chimpanzees when the helmet and clothing were torn off exposing the skin.[7]

Stapp was famous in his time with numerous interviews (in 1955 he was on the cover of *Time* magazine). Also famous for his wit, Stapp is given credit for popularizing, of all things, Murphy's Law (if anything can go wrong, it will). In an interesting aside, Murphy was a real person—a West Point graduate and army test engineer. Because he designed a new and better sensor to measure acceleration, he was briefly associated with Stapp's test program. The strain gages in Murphy's sensor were wired backwards by mistake, resulting in zero readings. At this point, the details become murky, with everyone telling a slightly different version of the story. However, they all agree that someone said something that got repeated as Murphy's Law, and Stapp is given credit for presenting Murphy's Law to the world in a news conference.

Colonel Stapp rode his 29th and final ride in December 1954 at Holloman Air Force Base. He was shot to 623 mph in five seconds and came back to rest in just over one second. This ride earned him the title "fastest man on Earth." Subjected to 46 G's, Stapp's blood rushed into his eyes (eyeballs out) during the stop and turned his world pink. He was rushed to the hospital, and his eyesight gradually returned. An hour later, he was eating lunch.

The Air Force grounded Colonel Stapp because he had become too valuable to risk losing. After Stapp performed the first-ever automotive crash tests,[8] the Air Force loaned him to the Department of Transportation where he did pioneering work on automotive crash studies.

The ultimate G-load record on a rocket sled occurred on May 16, 1958. Captain Beeding survived an accidental exposure to 83 G's for 0.04 seconds. Beeding lost consciousness 10 seconds after impact and briefly went into shock. He was hospitalized for three days and recovered from severe back and head pain. This crash was not fatal because of the extremely brief exposure.

Many design changes and improvements were made to airplanes because of the rocket sled tests. Military seats on aircraft were upgraded from 18 G's to 35 G's. Military transport seats were often reversed to face backwards (this distributes crash loads over the back instead of on the straps). Improvements were also made in harnesses and helmets. (Some safety advocates argue that all passenger seats should face backwards. Others argue that this would entail major design costs and that passengers would resist because of increased motion sickness.)

Inertial Loading Studies During WWII

During World War II, the Germans worked on solving a different inertial loading problem: the safe ejection of a pilot.

As Germany developed a new generation of higher-performance military aircraft, the need for aircrew safety became apparent. By 1939, the Luftwaffe's aviation medicine branch was actively experimenting with ejection systems and using instruments that measured G forces on the human body. The fundamental problem with ejection is to "shoot" the pilot up out of the cockpit fast enough to avoid being struck by the plane's tail. Blasting the pilot up creates the same spinal loads as crashing down. In both cases, the spine is being compressed (see Figure 9.5). The Luftwaffe's tests roughly defined the human tolerance at about $+20$ G for a duration of about 0.1 second.

In early 1942, the first known emergency use of an ejection seat occurred during a test flight of the Heinkel He-280 jet fighter. By late 1942, all German experimental aircraft being flight tested were equipped with some form of ejection seat.

By the end of World War II, more than 60 emergency ejections had been made by Luftwaffe pilots. These facts were unknown outside Germany until after the war ended. Captured German research stimulated new interest in Allied escape technology.

Initial German designs used compressed springs, compressed gas, or explosive charges. One early experimental compressed gas system used 1,700 psig pressure, resulting in $27 +Gz$. The older systems used a very high force for a brief period of time, resulting in larger accelerations, 17 to 22 G's not being unusual. Newer rocket-propelled systems could provide propulsion over a longer period of time, thus lowering the required acceleration. The newer seats are typically in the range of 12 to 14 G's.

During an ejection, the seat must climb at a high enough speed for the plane to pass underneath it. For a large plane, such as the B-1, the pilot is 15.6 feet below the tail but also 109.8 feet away from it. A slower seat ejection is acceptable, because the tail is far back. In a fighter such as the F-4, however, the 7-foot-high tail is much closer to the pilot and requires a much faster seat. Examples of ejection G loads for a B-1 and F-4 at various speeds are given in Table 9.4. The time to tail assumes that the aircraft moves at the airspeed listed in Table 9.4, while the seat, accelerating up at the required G's, will just clear the tail.

Table 9.4. Required Ejection Conditions for Bomber versus Fighter

Airspeed (mph)	B-1 Time to Tail (sec)	Required G Force	F-4 Time to Tail (sec)	Required G Force
345	0.22	9.0	0.08	11.0
460	0.16	12.0	0.06	14.7
575	0.13	15.0	0.05	18.4

Source: Herker, "Physics Primer"

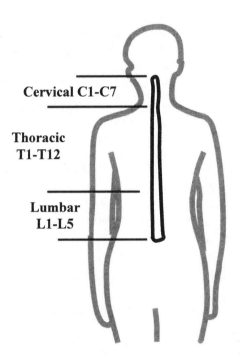

Cervical C1-C7

Thoracic T1-T12

Lumbar L1-L5

Fig. 9.14. Vertebrae of the spine

The Luftwaffe's aviation medicine branch also did active research on the strength of the human spine and the limit of ejection accelerations. Results from their tests were eventually written into FAA rules for passenger seat design.

Strength of the Spine

The spine, or vertebral column, is made of many different bones (called vertebrae) stacked on top of each other. Excluding the tail bone, the vertebrae

Table 9.5. Strength of Vertebrae in a 160-lb Man

Vertebra	Body Weight Supported per Vertebra (lb)	% of Body Weight Supported	Range of Fracture Strengths (lb)*	Average Fracture Strength (lb)	G Loading Strength
T1†		9			
T2		12			
T3		15			
T4		18			
T5		21			
T6		25			
T7		29			
T8	52.8	33	1,190–1,410	1,315	24.9
T9	59.2	37	1,345–1,587	1,493	25.2
T10	64.0	40	1,455–1,764	1,632	25.5
T11	70.4	44	1,587–1,896	1,700	24.2
T12	75.2	47	1,521–1,984	1,757	23.4
L1	80.0	50	1,587–1,984	1,790	22.4
L2	84.8	53	1,764–2,183	1,925	22.7
L3	89.6	56	1,984–2,425	2,161	24.1
L4	92.8	58	1,984–2,425	2,168	23.4
L5	96.0	60	2,205–2,646	2,366	24.6

Source: Stech and Payne, AMRL Technical Report 66-157

*Sample size small; only 3–5 measurements made.

†T1 to T7 not tested for all variables; these vertebrae support a relatively low percentage of the body's weight.

are grouped into three types: 7 cervical, 12 thoracic, and 5 lumbar (Figure 9.14). The vertebrae support the body and also protect the spinal cord.

Vertebral strength data were obtained by Siegfried Ruff in Germany during WWII.[9] Ruff also estimated the percentage of body weight supported by each vertebra (Table 9.5). A few trends are apparent from the table. First, the lower vertebrae support more of the body's weight and are correspondingly stronger with respect to mechanical fracture. Second, the G loading strength (equal to the average fracture strength divided by the weight supported in a 160-pound man) is surprisingly constant at around 25 G. This data can be compared to ejection seat data gathered by the U.S. Air Force. Their statistics from 175 ejections showed that accelerations from 17.5 to 18.4 G had a 7% incidence of vertebral fractures.

The average fracture strength in Table 9.5 ranges from 1,315 to 2,366 pounds. Thoracic vertebra T9 has a strength of almost 1,500 pounds. This strength, which undercuts almost all the fracture strength data for the ver-

Table 9.6. Deceleration Damage to Individual Body Parts

Injury	G Load
Pulmonary contusion (bruise)	25
Nose fracture	30
Vertebral body compression (body position dependent)	20–30
Fracture dislocation of cervical vertebra C1 on C2	20–40
Mandible (lower jaw bone) fracture	40
Maxilla (upper jaw bone) fracture	50
Aorta intimal tear (internal tear)	50
Aorta transection (completely severed)	80–100
Pelvic fracture	100–200
Vertebral body transection (completely severed)	200–300

Table 9.7. Survivable Whole-Body Impacts

Position	Limit (G)	Duration (Sec)
Eyeballs out (−Gx)*	45	0.1
Eyeballs out (−Gx)	25	0.2
Eyeballs in (+Gx)	83	0.04
Eyeballs down (+Gz)	20	0.1
Eyeballs up (−Gz)	15	0.1
Eyeballs left/right (+/–Gy)	9	0.1

*Fully restrained subjects exposed to whole-body impacts at up to 250 G/sec onset rates. Injuries are known to occur if limits are exceeded. With lap belt restraint only, –Gx tolerance may be reduced by two-thirds.

tebrae that support a significant percentage of the body's weight, was incorporated into the requirements for a seat to pass the seat crash test (see Chapter 8). In other words, to pass the crash test a dummy in the seat must have a spinal crush load of 1,500 pounds or less, as measured by a load cell between the pelvis and lumbar column.

Naval Flight Surgeon's Pocket Reference

The Naval Flight Surgeon's Pocket Reference to Aircraft Mishap Investigation contains a government summation of all known data on human impact injuries, including studies of military crashes. All the military data are from fit young males wearing 3-inch-wide shoulder straps and seat belts, including thigh straps. It is roughly estimated that for eyeballs-out (–Gx) testing, the human tolerance to G loads when not wearing shoulder straps is only one-third the tolerance when wearing shoulder straps.

Table 9.6 shows the G loads that cause a variety of injuries to the body, and Table 9.7 shows survivable whole-body impacts. Again, it is cautioned that these numbers are rough estimates and not absolute values. Even if consistent data could be collected, they would show tremendous statistical scatter, because all humans do not respond the same.

Chest Compression

Human volunteers could sustain 20% chest compression in static loading without injury. Tests with cadavers showed that ribs were often fractured above 20% compression at impact velocities of 16 to 23 ft/sec. Compression of 40% caused multiple rib fractures and is considered to have a 50% probability of inducing a life-threatening injury. At low-impact velocities of less than 1 m/sec, injury is limited to crushed tissue. At very high velocities, greater than 30 m/sec, the impact results in blast injuries in the lungs and then in other hollow organs. At velocities between 1 and 98 ft/sec, the response is viscous in that damage is related to the product of the velocity and compression.

Concussion

Concussion can result from a nonimpact motion of the head. Consciousness can be maintained up to 150 G's if the duration of exposure is very short. However, nonfatal concussion may reduce the chance of survival during post-crash hazards such as fire or submersion in water. Fatal brain injuries have been shown to result from lacerations of brain tissue or from shear of the brain stem.

Head Injury Criterion

Head impact has also been studied in detail. The Head Injury Criterion (HIC) was developed from cadaver skull testing conducted at Wayne State University in the 1960s. Post-impact injury was determined based on the presence or absence of a skull fracture. Based on this testing, the values in Table 9.8 have been proposed as safe.

Crash testing is now done with an instrumented crash dummy (see Chapter 8), which has accelerometers embedded inside its head. The G

Table 9.8. Suggested Values For Safe Head Impact

Impact G Load	Duration (msec)
180	2
135	3
110	4
100	5
90	6
85	7
80	8
74	10
57	20

Source: Synder, "Human Impact Tolerance," *SAE Transactions*

load readings from these accelerometers during a crash test could be compared to Table 9.8 to determine pass or fail of the test.

The numbers in Table 9.8 are approximate at best for a variety of reasons. There is tremendous person-to-person variation; for example, skull thickness varies. Also, it has been shown that damage depends on the duration of the crash impulse.

In an attempt to reconcile the damage done by a higher G load for a shorter duration with a lower G load for a longer duration, crash investigators researched various correlations of the data. The area under the acceleration versus time plot was found to predict skull fracture best. Eventually, it was decided that the area under the curve raised to the power of 2.5 (along with other mathematical manipulations) was the best predictor. Why was this done? It is somewhat arbitrary and simply a matter of what manipulation best predicts skull fracture. For our purposes, the important point is how well HIC values predict skull fracture.

When cadaver drop test data were reinterpreted with the HIC calculation, it was found that skull fractures occurred over a wide range of HIC values, from 450 to 2,351. The thickest skulls had the highest values, while the thinnest skulls had the lowest. Nine of the 15 cadavers that had HIC values between 1,000 and 1,500 had skull fractures.

Further statistical refinements determined that 15% of the population is expected to sustain a life-threatening brain injury with a HIC of 1,000.

This value increases to 56% with a HIC of 1,500 and 90% with a HIC of 2,000.

As described in Chapter 8, all passenger seats must now be certified with crash testing. To pass the test, the crash dummy HIC value must be 1,000 or lower. This value is also used in federally mandated automotive crash testing. Chapter 8 gives example crash test data relating head impact velocity, G loads, and HIC values for crash dummies.

A Final Note about the Interpretation of Human Impact Data

The best way to interpret the human tolerance data is to compare it to crash test data presented in Chapter 8. A precise comparison is difficult because of the many variables involved but does give some insight into how some crash events compare to defined human G load limits.

Crashworthiness requirements for aircraft structures have slowly evolved over the years. Initially, there were no rules. Later, requirements were put in place and have slowly increased. Today's requirements represent a compromise between excess structural weight and passenger safety. Humans are, of course, the most important components in the whole system and ideally should be the last to break. Unfortunately, under certain rare crash circumstances, this is not always the case.

Closing Note

Again the author wishes to emphasize that flying is safe. As described in this book, the constant attention to detail (every rivet has a paper trail), the billions of dollars spent on testing and design, the never-ending safety improvements, and the "no-stone-left-unturned" crash investigations should reassure the flying public that everything humanly possible is done to maintain safety standards.

Often a crash results from a chain of five or six improbable events beating the odds by occurring together. The response is to fix all six and a few more. Sometimes the investigation initiates national research efforts with participants from NASA, FAA, NTSB, the avaition industry, and many universities.

One of this book's reviewers, a self-described "white-knuckle flyer of long standing," stated it best: "Reading the book did not make a bad situation worse. The book, even as it describes disaster after disaster, manages to leave the reader with the impression that flying, in general, is pretty darn safe."

Notes

1. The Crash Investigation Process

1. The overwhelming amount of evidence reviewed in a modern investigation makes dishonest testimony pointless. The false witness, with access to only a few facts, cannot compete with the abundant facts and documentation available to the investigators, nor can the witness control the testimony of numerous others.

2. Komons, *Bonfires to Beacons*

3. The jet engine is designed to safely shed compressor and turbine blades and pass them out the exhaust (a "contained" failure). An uncontained failure occurs when engine parts exit the engine at high speeds in an uncontrolled manner, posing a danger to passengers and the aircraft. See Chapter 5.

4. Many consider the reduction of safety to dollars and cents and placing a "price tag" on human life to be offensive. Without a protracted debate of the ethical issues involved, consider that 40,000 people die in automotive crashes every year. Perhaps this number could be reduced to 5,000 if safety features were added. The safety features might cost $1 million per car. Every automotive trip involves an implied acceptance of a cost-benefit analysis.

5. Alarms will sound, for example, if cabin pressure or lift on the wings reduces below critical values.

2. How Planes (Often) Crash

1. See Chapters 3 and 7 which discuss the structural breakup and the detonation process relative to the Lockerbie disaster.

2. According to the CAIN research project at the University of Ulster, through 2001 the IRA was responsible for the deaths of 1,706 people, including 497 civilians and 638 British soldiers. Other casulties included rival Protestant groups.

3. *Toronto Star*, "Failure of Both Engines Feared"

4. Ibid.

5. *Toronto Star*, "Jet Crash Kills 40"

6. Air Accidents Investigations Branch, Report No. 4/90

7. Lane and Strichez, "10 Million–to–1 Odds," *Seattle Times*

8. Hanlon, "Villagers Praise Pilot's Heroic Skill," *Toronto Star*

9. *Seattle Times*, "Engine was Shut off Before Crash"

10. There have since been other incidents associated with fuel system problems. In 2001, an Airbus 330 ran out of fuel over the Atlantic and successfully glided just under 100 miles to a safe landing. Gliding from 30,000 feet at 600 mph is easier than the Boeing 737 that crashed near Kegworth trying to glide from 900 feet at a much slower approach speed.

11. Cushman, Jr., "U.S. Orders Checks of 300 Jetliners," *New York Times*

12. Barr, "No Faults Found in Warning Systems," *St. Petersburg Times*

13. Cushman, Jr., "U.S. Orders Checks of 300 Jetliners," *New York Times*

14. Air Accidents Investigation Branch, Report No. 4/90

15. Ibid.

16. The part that extends out of the front of the wing is called a slat, while the part on the trailing edge or back of the wing is called a flap. The flaps and slats move along metal tracks built into the wings. Extending the flaps and the slats increases the wing area and adds lift at the lower speeds during landing and takeoff.

17. Spoilers are flaps that pop out of the wings during landing to "spoil" the lift.

18. Thrust reversers, mounted near the engine exhaust, block the jet blast and turn it to the opposite direction. In this way, the forward thrust of the jet engine becomes a massive brake. Passengers can hear and feel the thrust reversers shortly after touchdown when the pilot throttles up the engines to increase the braking force.

19. NTSB, Report NTSB/AAR-85/06

20. First contact with a wing or engine can also create complex tumbling motions. One such crash is described in detail in Chapter 8.

21. Weight can be converted into mass from $F = ma$ and the definition of weight. A mass accelerating at the speed of gravity, 32.2 ft/sec^2, equals the weight of the mass or Weight = Mass $\times$ 32.2 ft/sec^2.

22. An MD-10 is a Boeing conversion of a McDonell Douglas DC-10 incorporating a new flight deck with six LCD screens. Boeing acquired McDonnell Douglas in 1997.

23. Small general aviation planes are significantly less safe, mainly because the pilots have less experience. Flying into bad weather and getting lost or disoriented are very real dangers for the pilot of a small plane, but they rarely occur with large commercial planes. Adding to reduced safety in small planes are less sophisticated instrumentation and fewer redundant systems.

24. A best-selling book in France was based on the premise that the 9/11 terrorist attack on the Pentagon was a staged hoax, because there were no obvious fuselage remains. There is a discernible fuselage for most planes that crash at 200 mph or less. However, there are numerous examples of complete fragmentation of planes that impact at 500 mph, including the terrorist planes that struck the Pentagon and the World Trade Center in the second attack seen by millions on live TV. Only fragments emerged from the back side of the World Trade Center after impact. The World Trade Center is relatively hollow compared to the Pentagon with its many interior walls.

25. Flight 232 lost all hydraulics. Because of extremely limited flight controls, the plane flipped over right before touchdown, tumbled, and did not follow the "typical" crash patterns described here.

26. There were four breaks in the fuselage but only two of them resulted in complete separation.

27. The seat spacing is adjusted for a variety of reasons. The airline may want to change the number of first class and coach seats, or the spacing is increased for passenger comfort on longer flights and reduced to increase the number of paying customers on shorter flights.

28. Older regulations on seat strength required 9 G's static loading in the longitudinal direction. New regulations require a 16 G dynamic crash certification test. The new rules were being phased in at the time of the Kegworth crash. The seats in the Kegworth crash were designed for 16 G's and were expected to meet the new 16 G requirements, but they were not actually certified with a 16 G crash test. The observed deformation in the Kegworth crash exceeded that which occurs in the 16 G crash tests.

3. In-Flight Breakup

1. Terminal velocity occurs when the air resistance equals the weight of the falling body. At this point, acceleration stops and the body falls at constant speed.

2. The phrase "Pierre Salinger syndrome," the tendency to believe anything read on the Internet, was coined.

3. Being a cynical observer of human nature, I am prepared to accept the willingness of others to distort the truth. However, having never been a fan of conspiracies that require dozens of people to lie in the face of extensive investigation, I will stick to official government conclusions in this book.

4. There are two types of bombers. The lone bomber has a lower probability of destroying the plane. The professional terrorist, though not any smarter, has the backing of an institutional knowledge base. Also, recent bombings tend to be more dangerous because powerful military explosives are more readily available on the black market.

5. The detonation of a military explosive is an extremely high-speed combustion process. The high rate of combustion creates unique blast patterns, some of which are described here. Other characteristics are described in Chapter 7.

6. A transponder on a commercial airplane is a receiver/transmitter that automatically responds to interrogation signals from the air traffic controller's radar, thus supplying continuous information about the plane's location, speed, altitude, etc.

7. After a three-year joint investigation by Scottish authorities and the FBI, indictments for murder were issued in November of 1991. Protracted negotiations with Libyan leader Muammar Gaddafi secured the handover of the accused to Scottish police in neutral Netherlands. In January of 2001, one of two accused men was convicted by three Scottish judges (the second was acquitted).

8. The simplest illustration of air resistance involves sticking a hand out the window of a moving car. The hand will experience the force of air resistance.

9. Radar data of breakups is fragmentary. Large parts usually mask smaller parts.

10. The direction of metal tears or fractures can often be determined by visual examination of the fracture surface. Markings known as "chevron" patterns, a series of

small steps on a fracture surface, indicate the direction of tearing and are very important in identifying the breakup sequence.

11. Acceleration is the rate of change of velocity. If a dropped cannonball accelerates downward at 32.2 ft/sec^2, the velocity is 32.2 ft/sec after one second, 64.4 ft/sec after two seconds, 96.6 ft/sec after three seconds, etc.

12. Momentum = mass × velocity. From conservation of momentum, the momentum of the gun equals the momentum of the bullet. Measuring the much slower motion of the swinging gun permits an estimate of the faster bullet.

13. The wing section speed is very sensitive to wing angle and was probably much lower.

14. A load factor for a fighter plane might be 7 to 9G. As well as structural limits, there are physical limits for the pilot's body, such as the ability to pump heavier blood to the head (see Chapter 9).

15. In a sense, this book is about the tiny details that go awry and result in mechanical failure. Unfortunately, there is no easy way to discover these problems except with years of experience at testing, designing and redesigning components, and studying failures.

16. Wind gusts are another major source of loads.

17. Tails have also broken off from metal fatigue and other mechanical damage.

4. Pressure, Explosive Decompression, and Burst Balloons

1. Flexing of the nose section forward of the break was the source of the springiness. The nose section drooped about one meter from the loss of fuselage structure.

2. The constant random motion of gases is used to explain why a gas expands to fill any container. Without constant motion to keep the gas molecules stirred up and suspended, they would fall to the bottom of the container, as do liquids and solids. The motion results from thermal energy.

3. A molecular spacing refers to the diameter of a water molecule. This assumes a water molecule can be approximated as a hard sphere, as shown in Figure 4.5.

4. If the weight is applied, removed, applied, removed, and repeated tens of thousands of times metal fatigue will result in fracture at much lower stress values than would occur with a single, one-time tensile test.

5. If the elongation per unit length is the same in the straight rod and the rod wrapped into a circle, the stress will be the same in each rod. This allows comparison of the stress in rods of different lengths and shapes.

6. NTSB, Report NTSB/AAR-89/03

7. Fuselages are complex structures. A crack can, in fact, propagate aligned with the axial stress. For example, a manufacturing defect could bias the fatigue crack growth direction, or an attached structure could alter the local stress state. However, fuselage cracks still have a very strong tendency to propagate in the direction of maximum hoop stress (just like a balloon).

8. The pressure nomenclatures psia and psig are introduced here for the first time. All pressure readings are relative to some reference. Psia means pounds per square inch absolute. In this case, the reference pressure is a perfect vacuum or zero pres-

sure. A perfect vacuum is difficult to obtain and technically will only occur if all air molecules are somehow sucked out of a container. Up to this point, all pressure designations of psi have actually meant psia. Psig means pounds per square inch gage. In this case, the reference pressure is normal atmospheric pressure, 14.7 psia. Pumping a pressure vessel up to 4.16 psig actually means pumping it up to 4.16 + 14.7 = 18.86 psia.

9. Wikipedia, "De Havilland Mosquito"

10. Government-owned airline company which became British Airways in 1971.

11. The tensile pull by hand is assumed to be relatively constant. The experiment could be improved by hanging weights to more accurately define the fracture load and stress. Generally, shorter cracks will become "critical" if the hanging weight is increased enough. Results could be matched with fracture mechanics equations.

12. Payne Stewart, the pro golfer, won 11 PGA tournament victories (including the U.S. Open) between 1991 and 1999.

13. NTSB, Report AAR-00-01

14. Colarossi, "Jury Clears Learjet," *Ottawa Citizen*

15. There are other safety mechanisms for lower altitudes depending on the door. The door may have electronic locks similar to a car door. The door may be designed to swing into the wind blast which, at a few hundred mph, will prevent outward movement.

16. As an engineer, I prefer to speak in terms of internal pressure, which of course determines the stresses in the fuselage. The pilot and mountain climber have a more fundamental requirement: breathing. For that reason, cabin pressure is always spoken of in terms of an equivalent altitude above sea level. The pilot would state that the cabin pressure climbed to 14,000 feet. I find it confusing to state the pressure "climbs" to 14,000 feet when the plane is at 32,000 feet and the actual pressure reading is decreasing. I would prefer to state that the cabin pressure reduced to a pressure corresponding to an altitude of 14,000 feet. The pilot's terminology is in more common usage, however.

17. NTSB, Report CHI96IA157

18. WHO, "International Travel and Health"

19. Hellenic Republic, "Aircraft Accident Report, Helios Airways Flight HCY522"

20. Whereas the spacing of the longitudinal ribs varies somewhat from plane to plane, surprisingly the 20" spacing of the circumferential ribs is almost a de facto industry standard and is used on most large commercial aircraft.

21. A torque applied to the fuselage in this manner is normally called a "bending moment" by the structural engineer.

22. The paper fuselage experiment could be improved with measured torques. One end could be held and the other end torqued by hanging weights from a wire wrapped around the PVC elbow. The torque equals the hung weight times the radius of the elbow. The weight could be slowly increased until the "fuselage" collapses on the compressive side. The effects of thicker paper, different diameters, etc., as well as the effect of stapling slit plastic straws to act as reinforcing ribs could be observed and measured.

23. Historically, the first rivets used in aluminum planes were protruded head rivets for simplicity. Countersunk rivets were developed later to minimize air drag.

24. The tape applied after pressurization of the balloon is structurally not the same as reinforcing ribs added to the fuselage before pressurization. The important point is that the balloon was altered from a state of explosive decompression to a state of slow decompression with structural modifications.

5. Jet Propulsion, Burst Engines, and Reliability

1. The Flight 232 crew were on the high end of experience with 30,000, 20,000, and 15,000 estimated hours for the pilot, first officer, and flight engineer. Compare this with smaller general aviation planes: I met a new pilot who purchased a $425,000 plane. He hired an "experienced" pilot to help him fly home, 1,500 miles away. At the airport, we met a university flight instructor. These three pilots had 150, 650, and 250 hours of flight time.

2. Barron, "Engine Fan," *New York Times*

3. Haynes, "The Crash of United Flight 232," Transcript

4. Ibid.

5. NTSB, Report NTSB/AAR-90/06

6. The new Airbus 380, seating 555 passengers, will save about 1 ton of weight by increasing the hydraulic system pressure from the standard 3,000 psi to 5,000 psi. This allows smaller tubing and actuators.

7. The acceleration of gravity is 32.2 ft/sec^2. Acceleration is the rate of change of velocity. After one second, a falling object is traveling at a velocity of 32.2 ft/sec. After two seconds, the velocity has increased to velocity = acceleration × time, or 64.4 ft/sec = 32.2 ft/sec^2 × 2 sec.

8. Subsystems, which include fuel, lubrication, ignition, anti-ice, accessory gear box (runs generators and hydraulics), and bleed air systems (to adjust compressor airflows, cool other engine parts, and pressurize the cabin), add back complexity and account for the many parts actually seen on a jet engine.

9. Mass = weight divided by acceleration of gravity. Angular velocity equals radians per second. The definition of arc length defines a radian where arc length = radius × radians. For a complete circle, circumference = radius × 2π radians, where 2π radians = 360 degrees.

10. The melting temperature is measured in an absolute scale of degrees Kelvin = degrees Centigrade + 273.

11. NTSB, Report NTSB/AAR-90/06

12. An exact count of fatigue striations is difficult, because rubbed and smeared surfaces are being looked at on a microscopic level.

13. NTSB, Report NTSB/AAR-90/06

14. In an example that is perhaps more familiar and easier to understand, the height of 20 adult males is measured and found to range from 4′ 11″ to 6′ 7″. In spite of the data, there is some small probability of finding an adult male shorter than 4′. Statistical methods can be used to process the raw data of the 20 measured heights and

predict that probability. It may be concluded that 99.999% of all males are taller than 4'. Therefore, the reliability of all adult males being 4' or taller is 99.999%.

15. Damage-tolerance design procedures are not used instead of design for strength and design for safe life methods. Instead, in yet another example of how the aerospace industry can never be "too safe," all three design procedures are used.

16. Probability calculations do not work in exactly this way, but the discussion does give insight into how events per billion are arrived at.

17. Flight 232 was the first time Captain Haynes shut down a commercial jet engine in over 30 years of flight experience.

18. For example, gone are controls and instrumentation for the supercharger, propeller-blade angle, fuel-air mixture, carburetor-heat, and cowl flaps (adjusts cooling air to engine).

19. A 330-minute demonstration flight was accomplished in October 2003 over the Pacific.

20. *Wall Street Journal*, "After Engine Blew"

21. Ibid.

22. NTSB, "Jet Engine Problems"

23. The fire-extinguishing system consists of a pressurized vessel filled with an extinguishing agent (CO_2 is but one example). The pressurized vessel is sealed with a precision burst disc that is activated electrically from the cockpit. The extinguishing agent is then distributed throughout the engine with a system of piping and nozzles.

6. Metal Fatigue

1. DC-3 production from 1935 to 1946 totaled 10,654 planes. An estimated 1,000 planes (2004) are still flying. The DC-3 production compares to 1,010 commercial 707s produced from 1958 to 1977. As of early 2006, the Boeing 737 (produced continuously since 1967) had orders for 6,099 planes with 4,966 delivered.

2. Although Aloha Flight 243 was a serious accident, the fact that the plane landed safely and did not break up like the Comets is verification of damage-tolerant designs.

3. Eddy currents are created through a process called electromagnetic induction. When alternating current is applied to a conductor, magnetic fields develop around the conductor. The magnetic fields fluctuate with the alternating current. If another electrical conductor is brought near the changing magnetic field, an electrical current will be induced in this second conductor. Fatigue cracks cause measurable changes in this induced current.

4. Engineers prefer to divide the force by the cross-sectional area of the rod and call this "stress." The displacement is divided by the initial length and is called "strain." The result is called a stress-strain curve.

5. The theory preceded the evidence for a number of years. The invention of the electron microscope permitted direct observation of dislocations.

6. For ductile materials, the reduction of cross section will eventually concentrate at one location, the so-called necking phenomenon. Very soft and ductile materials will neck almost into a point.

7. Combustion

1. The MD-11 is a slightly stretched, updated and re-engineered version of McDonnell Douglas's DC-10. The first MD-11 was delivered the end of 1990. Only 200 were manufactured between 1990 and 2001.

2. Goldberg and Cooper, "Timeless Fishing Village," *New York Times*

3. There are three levels of emergency calls. Pan, Pan, Pan is second only to Mayday, a declaration of grave danger.

4. Aluminum will not spark and ignite spilled fuel. Titanium and steel, found in engine parts and landing gear, will.

5. Ludwig Boltzmann, 1844–1906, Austrian physicist and mathematician. His theories connect the properties of atoms and molecules with the large-scale behavior of gases.

6. The vent outlet is a "breather cap" similar to an automotive fuel cap. It breathes when there is a pressure difference, but otherwise it seals.

7. Kerosene used for lighting in the mid-nineteenth century was very unpredictable because of the difficulty of molecular separation of crude oil. Sometimes it would not light; other times it would light too much and explode. One of John D. Rockefeller's innovations was improved refining techniques and the production of more reliable "Standard Oil," hence the name.

8. There are 89 intercompartment passageways connecting the six fuel bays.

9. Since 1963, two other planes have been destroyed by lightning elsewhere in the world.

8. Crash Testing

1. The Code of Federal Regulations is the codification of rules published in the Federal Register by the executive departments and agencies of the federal government. It is divided into 50 titles that represent broad areas subject to federal regulation. Title 49 is "Transportation."

2. Forced by animal rights activists, General Motors discontinued animal testing in 1993. Other manufacturers followed suit shortly thereafter. Cadaver tests continue.

9. Human Tolerances to G Loads and Crash Forces

1. When asked whether they practice ejections, a Navy pilot told me, "Hell no! It would be like practicing bleeding."

2. Quoted in Freeman, "22 Seconds," *Reader's Digest*

3. Weight equals the force due to gravity on a given mass. The same mass will have different weights on different planets depending on local gravity. From Newton's Second Law, force = mass × acceleration or weight = mass × G where G is the acceleration due to gravity.

4. All commercial pilots have thousands of hours of experience. There are many reasons for this, and high on the list is the many hours it takes for the fundamental sense of up and down to be replaced with an innate trust of the instrument readings.

5. Acceleration is the rate of change of velocity. Velocity is a vector with magnitude and direction. When flying in an arc, the speed is constant but acceleration occurs from constantly changing direction.

6. Spark, "The Story of John Paul Stapp," *Wings/Airpower Magazine*

7. The biggest concern from windblast is flailing. Newer ejection seats lock the limbs in place, but this takes time to fully engage. In 1995, Captain Jon Counsell ejected from an F-15 at Mach 1.4, far in excess of the recommended 690 mph maximum airspeed for ejection. His limbs flailed so violently that he broke his left arm, broke his left left leg in five places, and dislocated both knees. Doctors thought he would never walk again, but seven years later he was back flying F-18s.

8. When his superiors complained about an Air Force officer doing automotive crash tests, Stapp produced statistics showing that more pilots died in automotive accidents than in plane crashes.

9. Similar tests had been done by others, including Wayne State University.

References

1. The Crash Investigation Process

Dominick Pisano. "The Crash That Killed Knute Rockne." *Air & Space Magazine, Smithsonian Institution,* December 1991/January 1992.

Nick A. Komons. *Bonfires to Beacons: Federal Civil Aviation Policy Under the Air Commerce Act, 1926–1938.* Washington, DC: Smithsonian Institution Press, 1989.

New York Times. "Rockne Crash is Laid to Ice-Covered Wings." April 8, 1931.

Charles Barton. *Howard Hughes and His Flying Boat.* Blue Ridge Summit, PA: Tab Books, 1982.

NTSB. "About the NTSB: The Investigation Process," www.ntsb.gov.

NTSB. *Aviation Investigation Manual: Major Teams Investigations.* November 2002.

Jonathan Hars. "The Crash Detectives." *The New Yorker,* August 5, 1996.

2. How Planes (Often) Crash

Toronto Star. "Jet Crash Kills 40 in Britain, New 737's Engine Ablaze as It Plunges near Highway." January 9, 1989.

CAIN (Conflict Archive on the Internet). University of Ulster, Northern Ireland, cain.ulst.ac.uk.

Polly Lane and Vince Strichez. "10 Million-to-1 Odds — 737 Crash Followed Rare Double Engine Failure." *Seattle Times,* January 9, 1989.

Michael Hanlon. "Villagers Praise Pilot's Heroic Skill, Captain of Boeing 737 Lauded For Steering Jet From Houses." *Toronto Star,* January 10, 1989.

Seattle Times. "Engine Was Shut Off Before Crash — Investigators Rule Out Sabotage as Cause of Fatal 737 Accident." January 10, 1989.

Seattle Times. "Britain Seeks Inspection of 737 Circuitry — Crash Probe Turns to Engine-Monitoring System." January 11, 1989.

Robert Barr. "No Faults Found in Warning Systems of Other 737s." *St. Petersburg Times,* January 13, 1989.

Houston Chronicle. "More Wiring Problems Found During Checks of Boeing 757s." January 17, 1989.

New York Times. "New Order on Aircraft Wiring." January 13, 1989.

REFERENCES

John H. Cushman, Jr. "U.S. Orders Checks of 300 Jetliners." *New York Times*, January 12, 1989.

Air Accidents Investigation Branch. "Report on the Accident to Boeing 737-400, G-OBME, near Kegworth, Leicestershire on 8 January 1989." Accident Report No. 4/90, 1990.

Macarthur Job. *Air Disaster, Volume 2*. Osceola, WI: Motorbooks International, 1996.

Airline Accident and Airline Safety and Security Information. www.airsafe.com.

Terry Waddington. *McDonnell Douglas DC-10*. Miami, FL: World Transport Press, 2000.

Aviation Safety Network. "ASN Aviation Safety Database." aviation-safety.net/database/.

NTSB. "American Airlines DC-10-10, N110AA, Chicago-O'Hare International Airport, Chicago, Illinois." Accident Investigation Report NTSB-AAR-79-17, May 25, 1979.

Naval Safety Center, Aeromedical Division. *The Naval Flight Surgeon's Pocket Reference to Aircraft Mishap Investigation*. 5th ed., 2001.

NTSB. "Eastern Air Lines, Inc., L-1011, N310EA, Miami, Florida." NTSB-AAR-73-14, December 29, 1972.

Boeing Commercial Airplanes. "Statistical Summary of Commercial Jet Airplane Accidents, Worldwide Operations, 1959–2003." May 2004.

Air Safety Week. "Landing Overruns—More Common than Commonly Thought." May 7, 2001.

Transportation Safety Board of Canada. "Rejected Take-off/Runway Overrun, Canadian Airlines International, McDonnell Douglas DC-10-30ER C-GCPF, Vancouver International Airport, British Columbia." Report Number A95H0015, October 1996.

NTSB. "Flight 30H, McDonnell Douglas DC-10-30CF, Boston, Massachusetts." NTSB/AAR-85/06, January 23, 1982.

NTSB. "General Aviation Crashworthiness Project: Phase II-Impact Severity and Potential Injury Prevention in General Aviation Accidents." Safety Report NTSB/SR-85/01, March 15, 1985.

R.G. Thomson and C. Caiafa. "Structural Response of Transport Airplanes in Crash Situations." Federal Aviation Administration Technical Center, DOT/FAA/CT-83/42, NASA-TM-85654, November 1983.

Gil Wittlin, Max Gamon, and Dan Shycoff. "Transport Aircraft Crash Dynamics." DOT/FAA/CT-82/69, March 1982.

Aero Magazine. "Updates to 737 Conditional Maintenance Inspection Procedures." Boeing Commercial Airplanes, No. 14, April 2001.

NTSB. "Hard Landing, Gear Collapse, Federal Express Flight 647, Boeing MD-10-10F, Memphis, Tennessee." NTSB/AAR-05/01, May 2005.

Gil Wittlin. "The Effect of Aircraft Size on Cabin Floor Dynamic Pulses." Lockheed. DOT/FAA/CT-88/15, March 1990.

E. Widmayer and O. Brende. "Commercial Jet Transport Crashworthiness." Prepared for NASA Langley, DOT/FAA/CT-82/86, March 1982.

REFERENCES

Anita Grierson and Lisa Jones. "Recommendations for Injury Prevention in Transport Aviation Accidents." SAE 2001-01-2658, 2001.

Federal Aviation Regulations. Part 25, Airworthiness Standards: Transport Category Airplanes, Subpart C Structure, Landing Load Conditions and Assumptions, Section 25.473.

FAA. "Shock Absorption Tests." AC 25.723, May 21, 2001.

H. Jamshidiat, E. Widmayer, and J. McGrew. "Airplane Size Effects on Occupant Crash Loads." SAE Paper 922035, 1992.

3. In-Flight Breakup

David Kocieniewski. "Center Fuel Tank Seen as Jets' Deadly Weak Spot." *New York Times*, August 30, 1996.

Matthew L. Wald. "Fourth Luggage Bin Found in Air Crash As Mystery Deepens." *New York Times*, August 12, 1996.

Adam Bryant. "First the Quick Answers, Then the Truth." *New York Times*, July 21, 1996.

Matthew L. Wald. "Last Seconds of Flight." New York Times, July 27, 1996.

Joe Sexton. "While Cause of Crash is Sought, Parallel Criminal Inquiry Goes On." *New York Times*, July 28, 1996.

David Rohde. "FBI Ends Inquiry of Flight 800, Find No Sign of a Crime." *New York Times*, November 13, 1997.

Lisa Anderson and Joseph A. Kirby. "TWA Jet Explodes near New York, 229 Feared Dead on Flight to Paris." *Chicago Tribune*, July 18, 1996.

Don Van Natta, Jr. and Matthew Purdy. "Salinger the Crash Theorist Raises More Eyebrows Than New Questions." *New York Times*, November 17, 1996.

Los Angeles Times. "Bomb Explodes on TWA Jet; 4 Die, Plane Lands in Athens, No Group Claims It's Responsible." April 2, 1986.

NTSB. "In-Flight Breakup Over the Atlantic Ocean, Trans World Airlines Flight 800, Boeing 747-131, Near East Moriches, New York, July 17, 1996." NTSB/AAR-00/03, August 23, 2000.

Craig R. Whitney. "Jetliner in Crash Blew Apart in Air, Officials Report." *New York Times*, December 23, 1988.

Michael Wines. "Flaw Not Ruled Out, but Sabotage Is Still Suspected." *New York Times*, December 24, 1988.

Sheila Rule. "Faint Unknown Noise Is Detected at End of Crashed Jets Recording." *New York Times*, December 24, 1988.

Sheila Rule. "Signs in Wreckage." *New York Times*, December 29, 1988.

Malcolm W. Browne. "Terror's Signature, Writ in Metal, Is Round Dent." *New York Times*, July 30, 1996.

Air Accidents Investigation Branch. "Boeing 747-121, N739PA, at Lockerbie, Dumfriesshire, Scotland." (Appendix F: The position of the bomb; Appendix G: Mach stem shock wave effects.) Accident Report No. 2/90, 1990.

Macarthur Job. *Air Disaster, Volume 1*. Canberra, Australia: Aerospace Publications, 1995.

William Triplett. "The Reconstruction." *Air & Space Magazine*, Smithsonian Institution, August/September 1997.

Brett D. Steele. "Muskets and Pendulums: Benjamin Robins, Leonhard Euler, and the Ballistics Revolution." *Technology and Culture*, April 1994.

F.R.W. Hunt. "The Reaction of the Air to Artillery Projectiles." In: Drysdale et al., editors, *The Mechanical Properties of Fluids*. London: Blackie and Son, 1925.

NTSB. "Trajectory Study, TWA Flight 800." Docket No. SA-516, Exhibit No. 22A, July 17, 1996.

Matthew Cox and Tom Foster. *Their Darkest Day: The Tragedy of Pan Am 103 and Its Legacy of Hope*. New York, NY: Grove Weidenfeld, 1992.

Robert Banks. "Tests Take to the Wing." *Environmental Engineering*, Summer 2002.

Macarthur Job. *Air Disaster, Volume 2*. Osceola, WI: Motorbooks International, 1996.

Aircraft Accident Investigation Committee. "Lauda Air Luftfahrt Aktiengesellschaft Boeing 767-300ER, Ban Dan Chang District, Suphan Buri Province, Thailand, 26 May B.E. 2534 (A.D. 1991)." July 21, 1993.

National Transportation Safety Committee. "Silkair Flight MI 185, Musi River, Palembang, Indonesia, 19 December 1997." Aircraft Accident Report, Department of Communications, Republic of Indonesia, December 14, 2000.

NTSB. "China Airlines Boeing 747-SP, February 19, 1985." NTSB/AAR-86/03, 1986.

NTSB. "Trans World Airlines Boeing 727-31, April 4, 1979." NTSB/AAR-81/8, 1981.

NTSB. "In-Flight Separation of Vertical Stabilizer, American Airlines Flight 587." NTSB/AAR-04/04, 2004.

Aviation Week & Space Technology. "AA 587: It Should Not Have Been So Shocking." April 22, 2002.

4. Pressure, Explosive Decompression, and Burst Balloons

NTSB. "Aloha Airlines Flight 243, Boeing 737-200, N73711, near Maui, Hawaii, April 28, 1988." NTSB/AAR-89/03, 1989.

Smithsonian National Air and Space Museum. "Lockheed XC-35 Electra." www .nasm.si.edu/research/aero/aircraft/lockheed_xc35.htm.

Air Accidents Investigation Branch. Bulletin No: 5/97, Ref: EW/C Incident Investigation, November 2, 1996.

Wikipedia. "De Havilland Mosquito." en.wikipedia.org/wiki/De_Havilland_Mosquito.

British Ministry of Transport and Civil Aviation. "Civil Aircraft Accident: Report of the Court of Inquiry into the Accident to Comet G-ALYP on 10th January, 1954." London, 1955.

British Ministry of Transport and Civil Aviation. "Civil Aircraft Accident: Report of the Court of Inquiry into the Accident to Comet G-ALYY on 08th April, 1954." London, 1955.

Macarthur Job. *Air Disaster, Volume 1*. Canberra, Australia: Aerospace Publications, 1995.

Macarthur Job. *Air Disaster, Volume 2*. Osceola, WI: Motorbooks International, 1996.

Matthew L. Wald. "A Mysterious Flight to Death." *New York Times*, October 31, 1999.

REFERENCES

NTSB. "Crash of Sunjet Aviation, Learjet Model 35, N47BA, Aberdeen, South Dakota." Report No. AAR-00-01, October 25, 1999.

Anthony Colarossi. "Jury Clears Learjet in '99 Crash Death of Golfer." *Ottawa Citizen*, June 9, 2005.

Eugene Rodgers. *Flying High: The Story of Boeing and the Rise of the Jetliner Industry*. 1st ed. New York, NY: Atlantic Monthly Press, 1996.

NTSB. "National Airlines, Inc., DC-10-10 near Albuquerque, New Mexico, November 3, 1973." NTSB/AAR-75-02, January 1975.

Air Accidents Investigation Branch. "British Airways BAC-111, June 10, 1990." Aircraft Accident Report No. 1/92, 1992.

John J. Swearingen. "An Evaluation of Potential Decompression Hazards in Small Pressurized Aircraft." FAA Office of Aviation Medicine, FAA AM 67-14, June 1967.

FAA. "Damage Tolerance: Assessment Handbook. Volume II: Airframe Damage Tolerance Evaluation." DOT/FAA/CT-93/69.II, October 1993.

C. Seher and C. Smith. "Managing the Aging Aircraft Problem." The AVT Symposium on Aging Mechanisms, Manchester, England, October 8, 2001.

Martin Aubury. "Hidden Hazards, Aloha." *Flight Safety Australia*, September–October 2003.

Aviation Safety Council. "In-Flight Breakup over the Taiwan Strait Northeast of Makung, Penghu Island. China Airlines Flight CI611, Boeing 747-200, May 25, 2002." ASC-AOR-05-02-001, Taiwan, 2005.

Michael Chun-Yung Niu. *Airframe Structural Design*. 2nd ed. Hong Kong: Conmilit Press Ltd., 1999.

M. Wolf. "Under Pressure." *Flight Safety Australia*, November–December 2000.

David Learmount. "Investigation Reveals Greek Helios Crash Was Waiting to Happen." *Flight International*, March 3, 2006.

Richard Witkin. "Altitude Limit Put on Older 737's Because of Accident Over Hawaii." *New York Times*, May 1, 1988.

Richard Witkin. "FAA Widens Inspection for Boeing Cracks." *New York Times*, May 5, 1988.

Richard Witkin. "FAA to Require Fuselage Repairs in Older 737 Jets." *New York Times*, October 28, 1988.

Aviation Week & Space Technology. News Brief, p. 19, March 18, 1996.

Aviation Week & Space Technology. "Crack Detection Guidelines Set for DC-9." April 22, 1996.

NTSB. Accident Report LAX95FA303, August 23, 1995.

NTSB. Accident Report CHI96IA157, May 12, 1996.

NTSB. Accident Report DCA00RA065, June 13, 2000.

NTSB. Incident Report ANC051A016, December 5, 2004.

FAA. "Operations of Aircraft at Altitudes above 25,000 Feet MSL and/or Mach Numbers (Mmo) Greater Than .75." AC 61-107A, January 2, 2003.

U.S. Air Force. *Flight Surgeon's Guide*. School of Aerospace Medicine, undated.

WHO (World Health Organization). "International Travel and Health." 2005.

REFERENCES

Airworthiness Assurance Working Group. "Recommendations for Regulatory Action to Prevent Widespread Fatigue Damage in the Commercial Airplane Fleet." Final Report, March 11, 1999. Revision A, June 29, 1999.

Steve Swift. "Damage Tolerant Repairs." *Flight Safety Australia*, Spring 1996.

Hellenic Republic, Ministry of Transport & Communications. "Aircraft Accident Report, Helios Airways Flight HCY522, Boeing 737-31S at Grammatiko, Hellas on 14 August 2005." November 2006.

5. Jet Propulsion, Burst Engines, and Reliability

James Barron. "Engine Fan May Have Come Apart, Crippling Jet." *New York Times*, July 21, 1989.

James Barron. "United Could Not Tell Crew How to Control Badly Crippled Jet." *New York Times*, July 22, 1989.

John H. Cushman, Jr. "Experts Try to Salvage Lessons from Wreckage of Jet in Iowa." *New York Times*, July 22, 1989.

Captain Al Haynes. "The Crash of United Flight 232." Transcript of presentation to NASA Dryden Flight Research Facility, Edwards, California, May 24, 1991.

Aviation Safety Network. "Cockpit Voice Recorder Transcript of the July 19, 1989 Emergency Landing of a United Airlines DC-10-10 at Sioux Gateway Airport." August 2003.

John H. Cushman, Jr. "Crash Tape Shows a Pessimistic Crew." *New York Times*, September 19, 1989.

David M. North. "Finding Common Ground in Envelope Protection Systems." *Aviation Week & Space Technology*, August 28, 2000.

NTSB. "United Airlines Flight 232, McDonnell Douglas DC-10-10, Sioux City Iowa, July 19,1989." NTSB/AAR-90/06, 1990.

Macarthur Job. *Air Disaster, Volume 2*. Osceola, WI: Motorbooks International, 1996.

K.S. Chan et al. "Constitutive Properties of Hard-Alpha Titanium." *Metallurgical and Materials Transactions* A, Vol. 31A, December 2000.

Clifford E. Shamblen and Andrew P. Woodfield. "Progress in Titanium-Alloy Hearth Melting." *Industrial Heating* 69: 49–52, 2002.

FAA. "Manufacturing Process of Premium Quality Titanium Alloy Rotating Engine Components." AC 33.15-1, September 1998.

FAA. "Contaminated Billet Study." DOT/FAA/AR-05/16, September 2005.

FAA. "Large Engine Uncontained Debris Analysis." DOT/FAA/AR-99/11, May 1999.

FAA. "Design Considerations for Minimizing Hazards Caused by Uncontained Turbine Engine and Auxiliary Power Unit Rotor Failure." AC 20-128A, March 25, 1997.

FAA. "Damage Tolerance for High Energy Turbine Engine Rotors." AC 33.14-1, January 2001.

FAA. "Turbine Rotor Material Design." DOT/FAA/AR-00/64, December 2000.

Barry J. Schiff. *The Boeing 707*. Fallbrook, CA: Aero Publishers, 1982.

Eugene Rodgers. *Flying High: The Story of Boeing and the Rise of the Jetliner Industry*. 1st ed. New York, NY: Atlantic Monthly Press, 1996.

REFERENCES

H. Kinnison. "Heading ETOP." *Flight Safety Australia*, January–February 2002.

FAA. "Turbine Engine Foreign Object Ingestion and Rotor Blade Containment Type Certification Procedures." AC 33-1B, April 22, 1970.

Wall Street Journal. "After Engine Blew, Deciding to Fly on 'As Far As We Can.'" September 23–24, 2006.

NTSB. "Jet Engine Problems." www.ntsb.gov/aviation/aviation.htm.

6. Metal Fatigue

M. Stone. "DC-10 Full-Scale Fatigue Test Program." Paper No. 73-803, American Institute of Aeronautics and Astronautics, 5th Aircraft Design, Flight Test and Operations Meeting, St. Louis, Missouri, August 6–8, 1973.

D. Paul and D. Pratt. "History of Flight Vehicle Structures, 1903–1990." *Journal of Aircraft*, September–October 2004.

Michael Mohaghegh. "Validation and Certification of Aircraft Structures." AIAA 2005-2162, 48th AIAA/ASME Structures, Structural Dynamics and Materials Conference, Austin, Texas, April 18–21, 2005.

Michael Mohaghegh. "Evolution of Structures Design Philosophy and Criteria." AIAA 2004-1785, 45th AIAA/ASME/ASCE/AHS/ASC Structures, Structural Dynamics and Materials Conference, Palm Springs, California, April 19–22, 2004.

C.V. Oster, C.K. Zorn, and J.S. Strong. *Why Airplanes Crash: Aviation Safety in a Changing World.* Oxford University Press, 1992.

Carl Hoffman. "20,000 Hour Tuneup." *Air & Space Magazine*, Smithsonian Institution, October/November 1997.

Paul Beuter et al. "747 Scheduled Maintenance." *Aero Magazine*, Boeing Commercial Airplanes, No. 20, October 2002.

Lufthansa German Airline. "Diary of a D Check: Aircraft Maintenance at Its Peak." Lufthansa Report, Press and Public Relations Department, Deutsche Lufthansa AG, undated.

NTSB. "Panel Summary: Aging Aircraft." Hearing—TWA800, December 11, 1997.

Boeing. "Boeing 777 Fatigue Test Airplane Completes Record Number of Flights." News Release, March 17, 1997.

Aviation Week & Space Technology. "777 Finishes Fatigue Tests." March 24, 1997.

Michael C. Walton et al. "Winds of Change: Domestic Air Transport Since Deregulation." Special Report 230, Transportation Research Board, National Research Council, 1991.

Aydin Akdeniz. "The Impact of Mandated Aging Airplane Programs on Jet Transport Airplane Scheduled Inspection Programs." *Aircraft Engineering and Aerospace Technology*, Vol. 73., No. 1, 2001.

Ralph Stephens et al. *Metal Fatigue in Engineering.* 2nd ed. New York, NY: John Wiley & Sons, Inc., 2001.

James F. Shackelford. *Introduction to Material Science for Engineers.* 5th ed. Upper Saddle River, NJ: Prentice Hall, 2000.

REFERENCES

7. Combustion

Matthew L. Wald. "Focus on Finding Bodies, Not Flight Recorders." *New York Times*, September 4, 1998.

Robert D. McFadden. "No Indication of Foul Play Is Detected." *New York Times*, September 4, 1998.

Transportation Safety Board of Canada. "In-Flight Fire Leading to Collision with Water, Swissair Transport Limited, McDonnell Douglas MD-11 HB-IWF, Peggy's Cove, Nova Scotia, 2 September 1998." Report Number A98H0003, March 27, 2003.

FAA. "In-Flight Fires." AC 120-80, January 8, 2004.

International Civil Aviation Organization. "Saudi Arabian Airlines Flight SV163 Cockpit Voice Recorder Transcript." Circular 178-AN/111, August 19, 1980.

New York Times. "265 Are Feared Dead as Saudi Plane Burns in Landing at Riyadh." August 20, 1980.

Scott Roberts. "Airbus's Landing Point a Factor in Crash." *Toronto Star*, August 8, 2005.

FAA. "Emergency Evacuation Demonstrations." AC 25.803-1, November 13, 1989.

New York Times. "Jet with 153 Safe after Engine Fire." June 29, 1965.

FAA. "Design Considerations Concerning the Use of Titanium in Aircraft Turbine Engines." AC 33-4, July 28, 1983.

R.H. Wood and R.W. Sweginnis. *Aircraft Accident Investigation.* Casper, WY: Endeavor Books, 1995.

Naval Safety Center, Aeromedical Division. *The Naval Flight Surgeon's Pocket Reference to Aircraft Mishap Investigation.* 5th ed., 2001.

Timothy R. Marker. "Full Scale Test Evaluation of Aircraft Fuel Fire Burnthrough Resistance Improvements." DOT/FAA/AR-98/52, January 1999.

Ray Cherry and Kevin Warren. "Fuselage Burnthrough Protection for Increased Postcrash Occupant Survivability: Safety Benefit Analysis Based on Past Accidents." DOT/FAA/AR-99/57, September 1999.

D.L. Greer et al. "Crashworthy Design Principles." Technical Report ADS-24, Contract FA-WA-4583, November 1965.

I. Irvin Pinkel, Solomon Weiss, G. Merritt Preston, and Gerard J. Pesman. "Origin and Prevention of Crash Fires in Turbojet Aircraft." NACA Report 1363, 1958.

NTSB. "In-Flight Breakup Over the Atlantic Ocean, Trans World Airlines Flight 800 Boeing 747-131, N93119, July 17, 1996." AAR-00/03, August 2000.

FAA. "Fuel Tank Flammability Minimization." AC 25.981-2, April 2001.

FAA. "A Review of the Flammability of Hazard of Jet A Fuel Vapor in Civil Transport Aircraft Fuel Tanks." DOT/FAA/AR-98/26, June 1998.

Joseph E. Shepherd et al. "Spark Ignition Energy Measurements in Jet A." Prepared for NTSB, California Institute of Technology, January 24, 2000.

Joseph E. Shepherd. "Learning from a Tragedy: Explosions and Flight 800." *Engineering & Science*, Vol. LXI, No. 2, 1998.

Michael A. Dornheim. "Accidents Drive NTSB Push for New Rules." *Aviation Week & Space Technology*, July 14, 1997.

Baer et al. "Extended Modeling Studies of the TWA 800 Center-Wing Fuel Tank Explosion." Sandia National Labs, March 2000.

M. Birky et al. "TWA Flight 800, Accident DCA-96-MA-070, Summary and Conclusions of Explosion Research Team." NTSB Docket SA-516, August 2000.

M.A. Uman and V.A. Rakov. "The Interaction of Lightning with Airborne Vehicles." *Progress in Aerospace Sciences*, 39: 61–81, 2003.

FAA. "Protection of Aircraft Fuel Systems Against Fuel Vapor Ignition Due to Lightning." AC 20-53A, April 1985.

Leslie Miller. "FAA Proposes New, Safer Fuel Tank Systems." *San Francisco Chronicle*, November 14, 2005.

John B. Bdzil et al. "High Explosives Performance." *Los Alamos Science*, Number 28, 2003.

Michael A. Dornheim. "Explosive Residue Easily Detectable." *Aviation Week & Space Technology*, July 22, 1996.

Carey Goldberg and Michael Cooper. "A Timeless Fishing Village Is Transformed in a Moment." *New York Times*, September 4, 1998.

FAA. "Fuel Tank Ignition Source Prevention Guidelines." AC 25.981-1B, April 18, 2001.

8. Crash Testing

E.L. Fasanella, E. Alfaro-Bou, and R.J. Hayduk. "Impact Data from a Transport Aircraft During a Controlled Impact Demonstration." NASA-TP-2589, L-16125, September 1, 1986.

FAA. "Summary Report: Full-Scale Transport Controlled Impact Demonstration Program." DOT/FAA/CT-87/10, September 1987.

Robert J. Hayduk. "Full-Scale Transport Controlled Impact Demonstration." NASA Conference Publication 2395, Langley Research Center, 1986.

David Nolan. "A Fuel Test Goes Up in Smoke, but Proponents Haven't Given Up—Anti-Misting Kerosene—Special Report: Airline Safety." *Discover*, October 1986.

G. Wittlin and D. Lackey. "Analytical Modeling of Transport Aircraft Crash Scenarios To Obtain Floor Pulses." NASA Contractor Report 166089, April 1983.

R.J. Hayduk and A.K. Noor. "Research in Structures and Dynamics." NASA-CP-2335, October 1, 1984.

S. Soltis et al. "FAA Structural Crash Dynamics Program Update—Transport Category Aircraft." SAE Paper No. 851887, October 1985.

E.L. Fasanella, H.D. Carden, R.L. Boitnott, and R.J. Hayduk. "A Review of the Analytical Simulation of Aircraft Crash Dynamics." NASA-TM-102595, January 1, 1990.

S.A. Williams and R.J. Hayduk. "Vertical Drop Test of a Transport Fuselage Center Section Including the Wheel Wells." NASA TM 85706, October 1983.

REFERENCES

Edwin Fasanella and Emilio Alfaro-Bou. "Vertical Drop Test of a Transport Fuselage Section Located Aft of the Wing." NASA TM 89025, September 1986.

S.A. Williams and R.J. Hayduk. "Vertical Drop Test of a Transport Fuselage Section Located Forward of the Wing." NASA TM 85679, August 1983.

Thomas Logue, Robert McGuire, John Reinhardt, and Tong Vu. "Vertical Drop Test of a Narrow-Body Fuselage Section with Overhead Stowage Bins and Auxiliary Fuel Tank System on Board." DOT/FAA/CT-94/116, April 1995.

Allan Abramowitz. "Vertical Drop Test of a Narrow-Body Transport Fuselage Section with Overhead Stowage Bins." SAE Paper No. 2002-01-2995, 2002.

E.L. Fasanella and K.E. Jackson. "Crash Simulation of a Vertical Drop Test of a B737 Fuselage Section with Auxiliary Fuel Tank." 3rd Triennial International Fire & Cabin Safety Research Conference, Atlantic City, New Jersey, October 22–25, 2001.

K.E. Jackson and E.L. Fasanella. "Crash Simulation of a Vertical Drop Test of a B737 Fuselage Section with Overhead Bins and Luggage." 3rd Triennial International Fire and Cabin Safety Research Conference, Atlantic City, New Jersey, October 22–25, 2001.

S.P. Desjardins et al. "Discussion of Transport Passenger Seat Performance Characteristics." SAE Paper 881378, October 1988.

Federal Aviation Regulations. Dynamic Evaluation of Seat Restraint Systems and Occupant Protection on Transport Airplanes, Section 25.562.

FAA. "Transport Aircraft Crash Dynamics." DOT/FAA/CT-82/69, March 1982.

FAA. "Transport Airplane Cabin Interiors Crashworthiness Handbook." AC 25-17, July 1991.

J. Marcus. "Dummy and Injury Criteria for Aircraft Crashworthiness." DOT/FAA/AM-96/11, April 1996.

Code of Federal Regulations, Title 49, Part 572 Anthropomorphic Test Dummy, Subpart B—50th Percentile Male.

V. Hodgson and L. Thomas. "Comparison of Head Acceleration Injury Indices in Cadaver Skull Fracture." SAE Paper 710854, February 1971.

J.W. Coltman et al., editors. "Aircraft Crash Survival Design Guide, Volume II—Aircraft Design Crash Impact Conditions and Human Tolerance." USAAVS-COM TR 89-D-22B, December 1989.

J.A. Tennant, R.H. Jensen, and R.A. Potter. "GM-ATD 502 Anthropomorphic Dummy-Development and Evaluation." SAE 740590, 3rd International Conference on Occupant Protection, Troy, Michigan, 1974.

Karen E. Jackson and Edwin L. Fasanella. "Development of an LS-DYNA Model of an ATR42-300 Aircraft for Crash Simulation." 8th International LS-DYNA Users Conference, Dearborn, Michigan, May 2–4, 2004.

FAA. "Design and Fabrication of a Head Injury Criteria-Compliant Bulkhead." DOT/FAA/AR-02/98, December 2002.

FAA. "Parametric Study of Crashworthy Bulkhead Designs." DOT/FAA/AR-02/103, December 2002.

REFERENCES

FAA. Seat Experimental Results Full-Scale Transport Aircraft Controlled Impact Demonstration. DOT/FAA/CT-85/25, July 1986.

Andy Pasztor. "Putting Air Bags in the Air." *Wall Street Journal*, November 16, 2006.

9. Human Tolerances to G Loads and Crash Forces

Gregory Freeman. "22 Seconds." *Reader's Digest*, February 2005.

Exponent Failure Analysis Associates. *Investigation of Amusement Park and Roller Coaster Injury Likelihood and Severity.* August 2002.

Andrew Duffy. "Franks' Incredible Flying Suit: The Canadian Invention that Changed the Face of the War." *Ottawa Citizen*, November 11, 2001.

Lydia Dotto. "Canada's Aviation Medicine Pioneers." Canadian Space Agency, www.space.gc.ca/asc/eng/astronauts/osm_aviation.asp.

PBS Nova. "The Mysterious Crash of Flight 201." Flight 201 Accident Investigation of Panamanian B737 Crash, Nova Adventures in Science, 1993.

G.J. Pesman and A.M. Eiband. "Crash Injury." Technical Note 3775, National Advisory Committee for Aeronautics, Washington, November 1956.

Mary Collins. "Ejection Seats." *Air & Space Magazine*, Smithsonian Institution, June/July 2002.

Christopher T. Carey. "Ejection Seats: A Brief History of the Development of Western Aircraft Ejection Seat Systems." webs.lanset.com/aeolusaero/Articles/seat_history.htm.

Martin Herker. "A Physics Primer for Understanding Ejections." showcase.netins.net/web/herker/ejection/physics.html.

NASA. "History of Research in Space Biology and Biodynamics at the U.S. Air Force Missile Development Center Holloman Air Force Base, New Mexico, 1946–1958." Historical Division, Office of Information Services, Holloman Air Force Base, undated.

Nick T. Spark. "The Story of John Paul Stapp: The Fastest Man on Earth." *Wings/Airpower Magazine*, July 2003.

Jon Franklin and John Sutherland. *Guinea-Pig Doctors: The Drama of Medical Research Through Self-Experimentation.* New York, NY: William Morrow & Co., 1984.

J.W. Coltman et al., eds. "Aircraft Crash Survival Design Guide, Volume II: Aircraft Design Crash Impact Conditions and Human Tolerance." USAAVSCOM TR 89-D-22B, December 1989.

Naval Safety Center, Aeromedical Division. *The Naval Flight Surgeon's Pocket Reference to Aircraft Mishap Investigation.* 5th ed., 2001.

Richard Synder. "Human Impact Tolerance." SAE Transactions, 1971.

S.H. Backaitis and H.J. Mertz, editors. *Hybrid III: The First Human-Like Crash Test Dummy.* Warrendale, PA: Society of Automotive Engineers, 1994.

P. Prasad and H. Mertz. "The Position of the United States Delegation to the ISO Working Group 6 on the Use of HIC in the Automotive Environment." Technical Paper 851246, Society of Automotive Engineers, June 1985.

REFERENCES

Federal Aviation Regulations. Part 25 Airworthiness Standards: Transport Category Airplanes, Subpart C Structure, Emergency Landing Conditions, Section 25.561.

E.L. Stech and P.R. Payne. "Dynamic Models of the Human Body." Aerospace Medical Research Laboratory, Wright-Patterson Airforce Base, Ohio, AMRL Technical Report 66-157, November 1969.

Index